Organoboranes
in Organic Synthesis

Studies in Organic Chemistry

Executive Editor
Paul G. Gassman
Department of Chemistry
The Ohio State University
Columbus, Ohio

Volume 1: Organoboranes in Organic Synthesis

OTHER VOLUMES IN PREPARATION

Organoboranes
in Organic Synthesis

GORDON M. L. CRAGG

Department of Chemistry
University of Cape Town
Rondebosch, Cape, South Africa

MARCEL DEKKER, INC., New York   1973

MARCEL DEKKER, INC.

95 Madison Avenue, New York, New York 10016

LIBRARY OF CONGRESS CATALOG CARD NUMBER: 72-90962

ISBN: 0-8247-6018-2

PRINTED IN THE UNITED STATES OF AMERICA

TO

Herbert C. Brown

PREFACE

The hydroboration of alkenes followed by oxidation with alkaline hydrogen peroxide is generally accepted by organic chemists as the best method for achieving anti-Markownikoff hydration of alkenes. In recent years, however, many reports have appeared dealing with new applications of organoboranes to the synthesis of a wide variety of organic compounds. The potential usefulness of these methods prompted me to write a brief review of the recent developments in the field in 1969. Similar reviews have appeared in various journals, but as yet these methods have not found general application in organic synthesis.

In writing this book I have not confined myself to the recent developments but have tried to provide a comprehensive review of the uses of organoboranes in organic synthesis. I have therefore devoted chapters to the hydroboration of alkenes and alkynes and related compounds, as well as to the reduction of functional groups. In order to make the book a useful reference tool I have included numerous equations and diagrams and have cited relevant references. I have tried to cover the literature up to mid-1972. I have included not only those reactions proceeding readily and in high yield, but also less obviously successful reactions which might prove useful under different circumstances.

The syntheses of the various classes of organic compounds developed using organoborane intermediates appear to provide useful, and in some cases, superior alternatives to those using more established methods. In order to emphasize this I have classified the reactions of organoboranes in chapter one according to the synthesis of compound types. I hope that many of these

reactions will come to be included in general organic chemistry
courses at both the undergraduate and graduate levels.

I wish to express my gratitude to Drs. A. S. Howard and
G. McGillivray for reading the manuscript and offering many
useful suggestions. I also wish to thank Drs. H. C. Brown,
J. Klein, and E. Negishi for helpful discussions. Finally, I
am very grateful to Mrs. R. du Toit and Miss J. Moore for typing
the manuscript, and to my wife, Jacqui, for her encouragement.

Gordon Cragg

Department of Chemistry
University of Cape Town

CONTENTS

ABBREVIATIONS

Ac , acetyl

Ac_2O , acetic anhydride

Ar , aryl

9-BBN , 9-borabicyclo[3.3.1]nonane

R-B-9-BBN , B-alkyl-9-borabicyclo[3.3.1]nonane

Bu , normal butyl

i-Bu , isobutyl

s-Bu , secondary butyl

tert-Bu , tertiary butyl

Bz , benzyl

$(C_6H_{11})_2BH$, dicyclohexylborane

DG , diglyme[$(CH_3OCH_2CH_2)_2O$]

DME , 1,2-dimethoxyethane (glyme)

DMSO , dimethylsulfoxide

Et , ethyl

Et_2BH , diethylborane

Et_2O , diethylether

HMPA , hexamethylphosphoramide

HOAc , acetic acid

$(IPC)BH_2$, monoisopinocampheylborane

$(IPC)_2BH$, diisopinocampheylborane

LiBPH , lithium perhydro-9b-boraphenalylhydride

Me , methyl

Ms , methanesulfonyl

[O] , oxidation with alkaline hydrogen peroxide

Ph , phenyl

Py , pyridine

R , alkyl

r.t. , room temperature

Sia$_2$BH , disiamylborane

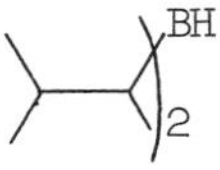

TG , triglyme $\left[(CH_3OCH_2CH_2OCH_2)_2 \right]$

Th , thexyl

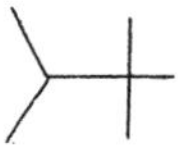

ThBH$_2$, thexylborane
THF , tetrahydrofuran
THP , tetrahydropyranyl
Ts , p-toluenesulfonyl
TsH , p-toluenesulfonic acid

TERMS

Hydroboration refers to reaction with borane unless otherwise
specified.
Oxidation refers to reaction with alkaline hydrogen peroxide
unless otherwise specified.

YIELDS

Yields given in equations and diagrams refer to reaction with
borane unless otherwise specified.
Figures given in parentheses indicate yields determined by vapor
phase chromatography. Otherwise figures refer to isolated yields.

Organoboranes
in Organic Synthesis

Chapter 1

ORGANOBORANES IN ORGANIC SYNTHESIS

Organoboranes were first synthesized in 1859 by the reaction
of dialkylzinc compounds and triethoxyborane ($\underline{1}$).

$$3(C_2H_5)_2Zn + 2(C_2H_5O)_3B \longrightarrow 2(C_2H_5)_3B + 3Zn(OC_2H_5)_2$$

The discovery of Grignard reagents led to the development of a
more versatile synthesis based on the reaction of these reagents
with boron trifluoride etherate or alkoxyboranes ($\underline{2}$). However,
since these methods of synthesis involve the prior formation of
more reactive organometallic reagents, little attention was paid
to the applications of organoboranes to organic synthesis.

Another possible route to organoboranes involves the reaction
of alkenes or alkynes with diborane, but early studies indicated
that the reaction requires the use of elevated temperatures and
prolonged reaction times, thus making it unsuitable as a preparative method ($\underline{3}$). In 1956, however, it was observed that the
addition of aluminum chloride to a solution of sodium borohydride
in diglyme gives a solution which rapidly reduces carboxylic acids,
esters, and nitriles, groups which are normally resistant to
reduction by sodium borohydride alone ($\underline{4}$). Furthermore, the
reduction of the unsaturated ester, ethyl oleate, proceeded with
utilization of 2.4 to 2.5 hydrides per molecule instead of the 2
hydrides per molecule required for reduction of the ester group to
the alcohol stage ($\underline{4}$). This indicated participation by the double
bond in the reaction, and it was soon shown that alkenes readily

1

react with the reagent in diglyme to give organoboranes in high
yield (5).

$$9RCH=CH_2 + 3NaBH_4 + AlCl_3 \xrightarrow[\text{3 hr.}]{\text{DG, r.t.}} 3(RCH_2CH_2)_3B + AlH_3 + 3NaCl$$
$$80-90\%$$

Use of boron trichloride or boron trifluoride in place of aluminum
chloride in the above reaction likewise results in the rapid
formation of organoboranes, while also utilizing the four available
hydrides in sodium borohydride (6).

$$12RCH=CH_2 + 3NaBH_4 + BF_3 \xrightarrow[\text{r.t.}]{\text{DG}} 4(RCH_2CH_2)_3B + 3NaBF_4$$

Finally, it was shown that diborane adds rapidly to alkenes in
ether solvents at room temperature to give high yields of the
corresponding organoboranes (6).

$$6RCH=CH_2 + B_2H_6 \xrightarrow{\text{DG, r.t.}} 2(RCH_2CH_2)_3B$$

The discovery of the marked effect of ethers on the hydroboration reaction provided a convenient and efficient method for
the synthesis of a wide variety of organoboranes from alkenes and
alkynes (Sec. 2.2.1). This ready availability of organoboranes
has led to extensive studies of their chemistry, and to the
development of many reactions which are of great value to synthetic
organic chemistry.

The hydroboration of alkenes, followed by oxidation of the
resultant organoborane with alkaline hydrogen peroxide, provides
a convenient method for the anti-Markownikoff hydration of alkenes.
The general applicability of the method was demonstrated in 1956;
since then it has been extensively used in organic synthesis, and
its application to a wide range of alkenes and alkynes is discussed
in later chapters of this book. More recent studies have shown
that organoboranes are extremely versatile intermediates in the
synthesis of many other types of organic compounds. Thus, in
addition to their conversion to alcohols (Sec. 1.1), organoboranes
can be readily converted to aldehydes and ketones (Sec. 1.2), carboxylic acids and their derivatives (Sec. 1.3), nitriles (Sec. 1.4),
alkyl halides (Sec. 1.5), amines (Sec. 1.5), alkenes and dienes

(Sec. 1.6), and various other functional derivatives (Sec. 1.5). Organoboranes also serve as useful intermediates in the extension of carbon chains (Sec. 1.7) and the construction of carbocyclic compounds (Sec. 1.8).

Borane can be used for the reduction of many functional groups (Chap. 10), but the ease of reduction is often dependent on the reaction conditions used. Thus, by use of controlled reaction conditions, unsaturated substrates containing common functional groups, such as esters, nitriles, and halogen, can be hydroborated without affecting the functional group (Sec. 10.13). Selective hydroboration can also be achieved using a selective hydroborating reagent such as disiamylborane (Sec. 10.13), or a reactive functional group can be protected during the hydroboration reaction (Sec. 10.14). This <u>tolerance of various common functional groups by the hydroboration reaction</u> constitutes a major advantage of this reaction over other commonly used reactions, such as Grignard reactions.

This chapter provides a brief survey of the utility of organoboranes in organic synthesis. The full scope of the various reactions is not discussed, but, in each case, reference is made to the relevant section elsewhere in the book where such a discussion is given. Throughout this book the term oxidation refers to reaction with alkaline hydrogen peroxide (Sec. 4.3) unless otherwise stated.

1.1 SYNTHESIS OF ALCOHOLS

1.1.1 Synthesis of Saturated Monoalcohols

Hydroboration-oxidation of alkenes provides a most useful method for the synthesis of alcohols. The full scope of the hydroboration reaction as applied to alkenes is discussed in Sec. 3.1, while the most commonly used method of oxidation, which involves treatment of the organoborane with alkaline hydrogen peroxide, is discussed in Sec. 4.3. This method of oxidation

proceeds with retention of configuration; an alternative method
involving reaction with an amine N-oxide also proceeds with
retention of configuration (Sec. 4.4). Autoxidation of organoboranes also gives alcohols in high yield but it lacks the
stereospecificity of the other two methods (Sec. 4.1.1). The
hydroboration-oxidation sequence of reactions results in predominant anti-Markownikoff hydration of alkenes. Both the
regioselectivity and stereoselectivity of the hydroboration
reaction is increased by use of selective hydroborating reagents
(Secs. 2.2.1 (c) and 3.1). Use of asymmetric hydroborating
reagents (Secs. 2.2.1 (d) and 3.6) results in the asymmetric
hydroboration of alkenes; oxidation of the resultant organoboranes
gives optically active alcohols (Sec. 3.6).

$$>C=C< \quad \xrightarrow[\text{THF}]{\overset{\displaystyle >B-H}{}} \quad -\overset{|}{\underset{|}{C}}-\overset{|}{\underset{|}{C}}- \quad \xrightarrow{[o]} \quad -\overset{|}{\underset{|}{C}}-\overset{|}{\underset{|}{C}}-$$
$$\qquad\qquad\qquad\qquad\quad \overset{}{\underset{B\;\;\;\;\;H}{}} \qquad\qquad \overset{}{\underset{HO\;\;\;\;H}{}}$$

Dihydroboration-oxidation of 1-alkynes gives the corresponding
primary alcohols (Sec. 7.2.2).

$$RC\equiv CH \quad \xrightarrow[\text{THF}]{2\;>B-H} \quad RCH_2CH(B<)_2 \quad \xrightarrow{[o]} \quad RCH_2CH_2OH$$

   In the methods discussed above the basic carbon structure of
the alcohol formed is determined by the structure of the reactant
alkene or alkyne. However, the development of reactions involving
the transfer of alkyl or aryl groups from boron to carbon permits
the synthesis of many alcohols of widely differing structures.
The carbonylation reaction (Sec. 8.1.1), which involves the
reaction of organoboranes with carbon monoxide, has been adapted
to the synthesis of primary (Sec. 8.3.1), secondary (Sec. 8.3.2),
and tertiary alcohols (Sec. 8.3.3) in high yields.

$$RCH_2OH \quad \xleftarrow[\substack{\text{LiAlH(OCH}_3)_3 \\ 2)\ KOH}]{1)\ THF} \quad R_3B + CO \quad \xrightarrow[2)\ KOH]{1)\ DG,\ H_2O} \quad R_2CHOH$$

$$\Big\downarrow \substack{1)\ THF,\ (CH_2OH)_2 \\ 2)\ [o]}$$

$$R_3COH$$

Less satisfactory methods for the synthesis of primary alcohols
utilize the reaction of organoboranes with various ylids
(Sec. 8.3.1), while secondary alcohols are formed in the reaction
of methoxy- or chlorocarbene with organoboranes (Sec. 8.3.2).

Other convenient methods for the synthesis of tertiary
alcohols have been developed based on the reaction of trialkylboranes with trihalosubstituted methanes, and the treatment of
trialkylcyanoborates with trifluoroacetic anhydride (Sec. 8.3.3).

$$R_3B + HCClF_2 \xrightarrow[\text{2) [o]}]{\text{1) LiOCEt}_3,\ \text{THF}} R_3COH$$

$$R_3B + NaCN \xrightarrow{\text{THF}} \underset{Na^{\oplus}}{\overset{\theta}{R_3B\text{-CN}}} \xrightarrow[\text{2) NaOH} \quad \text{3) [o]}]{\text{1) (CF}_3\text{CO)}_2\text{O}} R_3COH$$

The scope of the above two reactions, as well as that of the
carbonylation reaction, in the synthesis of tertiary alcohols
has been greatly widened by the development of convenient methods
for the synthesis of mixed trialkylboranes (Sec. 2.2.4). A
further method for the synthesis of tertiary alcohols involves
the α-bromination of trialkylboranes, followed by rearrangement
and oxidation (Sec. 8.3.3).

$$(C_2H_5)_3B \xrightarrow[\substack{\text{CH}_2\text{Cl}_2 \\ h\nu}]{\text{Br}_2} \underset{\text{Br}}{\overset{C_2H_5}{CH_3\text{CHB-C}_2H_5}} \xrightarrow{\text{H}_2\text{O}} \underset{\text{OH}}{\overset{C_2H_5}{CH_3\text{CHBC}_2H_5}} \xrightarrow[\text{2) [o]}]{\text{1) Br}_2,\text{H}_2\text{O}} \underset{\text{CH}_3}{(C_2H_5)_2\text{COH}}$$

This method has also been adapted to the synthesis of secondary
alcohols (Sec. 8.3.2).

Application of the carbonylation-oxidation reaction sequence
to cyclic organoboranes gives cyclic alcohols (Sec. 8.3.3).

Dihydroboration of 1-bromo-2-propyne or 1-tosyloxy-3-butyne with
9-BBN, followed by treatment with base and oxidation, gives
cyclopropanol and cyclobutanol, respectively (Sec. 7.2.2).

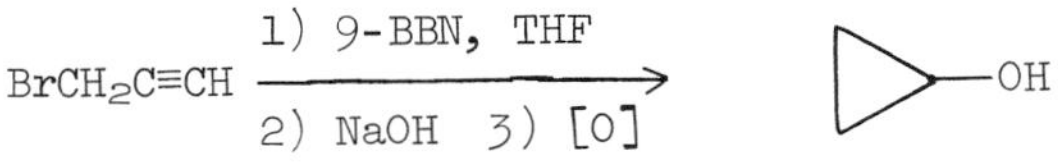

$$BrCH_2C{\equiv}CH \xrightarrow[\text{2) NaOH 3) [O]}]{\text{1) 9-BBN, THF}} \quad \triangleright\text{-OH}$$

1.1.2 Synthesis of Unsaturated Monoalcohols

Monohydroboration of acyclic and cyclic dienes with a
selective hydroborating reagent, such as disiamylborane, followed
by oxidation, gives unsaturated alcohols (Secs. 6.5.1 and 6.5.2).
Symmetrical acyclic conjugated dienes, however, undergo preferential dihydroboration (Sec. 6.5.1).

Acyclic Δ^3- and Δ^4-1-alcohols can be synthesized by hydroboration of 2,3-dihydrofurans and Δ^2-dihydropyrans, respectively,
followed by treatment with boron trifluoride etherate and
hydrolysis (Sec. 5.9.1).

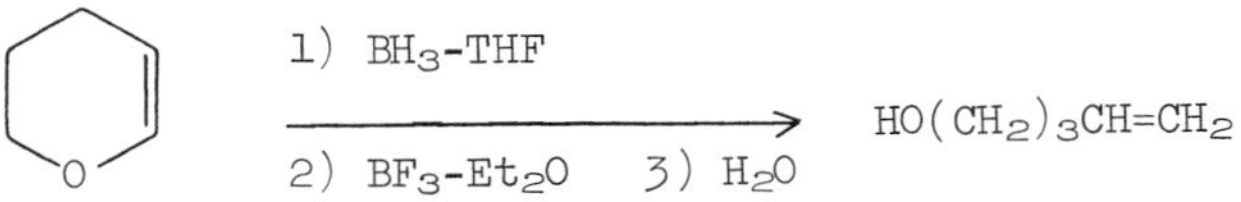

$$\xrightarrow[\text{2) BF}_3\text{-Et}_2\text{O   3) H}_2\text{O}]{\text{1) BH}_3\text{-THF}} \quad HO(CH_2)_3CH{=}CH_2$$

2-Cyclohexenol derivatives can be prepared via hydroboration
of the corresponding cyclohexanone enamines as shown below
(Sec. 5.2.4).

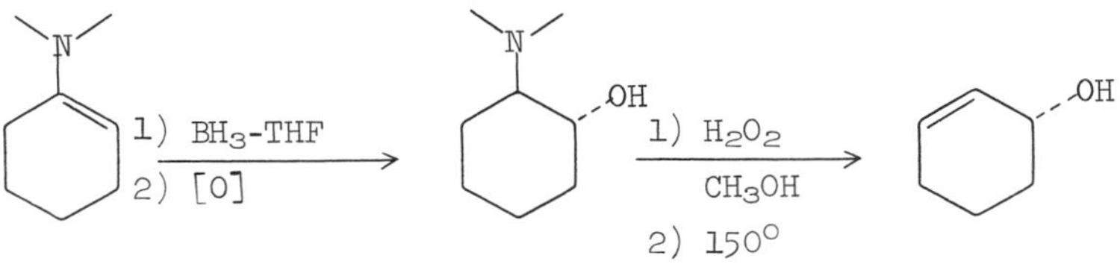

$$\xrightarrow[\text{2) [O]}]{\text{1) BH}_3\text{-THF}} \qquad \xrightarrow[\text{CH}_3\text{OH}]{\text{1) H}_2\text{O}_2} \qquad \text{2) 150}^\circ$$

1.1.3 Synthesis of Vicinal Diols

The hydroboration of acyclic allylic alcohols often proceeds
with the occurrence of a considerable amount of elimination
followed by rehydroboration (Sec. 5.1). Oxidation of the resultant

organoborane mixtures thus gives mixtures of monoalcohols and diols
(Sec. 5.3.2). The elimination reaction can, however, usually be
avoided by conversion of the allylic alcohol to the disiamylborinate ester; hydroboration-oxidation of this ester gives the vicinal
diols (Sec. 5.3.2).

$$RCH=CHCH_2OH \xrightarrow[\text{THF}]{Sia_2BH} RCH=CHCH_2OBSia_2 \xrightarrow[\text{2) [O]}]{\text{1) } BH_3\text{-THF}} RCH_2\underset{\overset{|}{OH}}{C}HCH_2OH$$

Vicinal diols may be prepared in a similar manner by reduction of
α,β-unsaturated aldehydes and ketones with disiamylborane, followed
by hydroboration-oxidation (Sec. 5.4).

Hydroboration of the tetrahydropyranyl- or benzyl-ethers of
allylic alcohols, followed by oxidation and removal of the ether
group by conventional procedures, gives the corresponding vicinal
diols (Sec. 5.3.3).

$$RCH=CHCH_2OTHP \xrightarrow[\text{2) [O] 3) p-TsH}]{\text{1) } BH_3\text{-THF}} RCH_2\underset{\overset{|}{OH}}{C}HCH_2OH$$

Hydroboration-oxidation of 3-hydroxycyclohexenes gives mainly
trans-diequatorial vicinal diols (Sec. 5.3.2). Similar results
are obtained in the hydroboration-oxidation of conjugated cyclohexenones and steroid enones (Sec. 5.4). The formation of
trans-1,2-diols in variable yields has also been reported in the
hydroboration-oxidation of cyclic enol acetates (Sec. 5.2.3).

1.1.4 Synthesis of Vicinally Substituted Alcohols

Hydroboration-oxidation of enol ethers (Sec. 5.2.2) and
allylic ethers (Sec. 5.3.3) gives mainly vicinal hydroxy ethers
with the trans isomers predominating in the case of cyclic ethers.
Hydroboration-oxidation of enamines gives vicinal amino-alcohols
(Sec. 5.2.4). However, the similar reaction of enol acetates
(Sec. 5.2.3) and vinyl or allylic halides (Secs. 5.2.1 and 5.3.1,
respectively) is usually not suitable for the synthesis of the
corresponding vicinally substituted alcohols due to the occurrence
of elimination-rehydroboration reactions.

1.1.5 Synthesis of Nonvicinal Diols

A variety of acyclic and cyclic diols may be prepared by the dihydroboration-oxidation of the appropriate dienes (Secs. 6.3 and 6.4). Oxidation of cyclic organoboranes (Secs. 2.2.3 and 6.2) also gives diols; in this respect, hydroboration of alkenes with thexylborane, followed by pyrolysis and oxidation provides a convenient route to acyclic 1,5-diols (Sec. 2.2.3).

$$(CH_3)_2CHCH_2CH=CH_2 \xrightarrow[\text{THF}]{\text{ThBH}_2} \cdots \xrightarrow[(-H_2)]{200^\circ} \cdots \xrightarrow{[O]}$$

Reaction of α-lithium furan with trialkylboranes, followed by oxidation, gives 4,4-dialkyl-cis-2-buten-1,4-diols (Sec. 8.3.4).

$$\xrightarrow[\text{2) CH}_3\text{CO}_2\text{H}]{\text{1) R}_3\text{B, THF}} $$
3) [O]

Hydroboration-oxidation of the allyllithium derivatives formed from 1-arylpropenes gives 1,3-diols (Sec. 8.3.4).

$$PhCH_2CH=CH_2 \xrightarrow[\text{2) BH}_3\text{-THF } \text{3) [O]}]{\text{1) BuLi}} \underset{\overset{|}{OH}}{PhCHCH_2CH_2OH}$$

1.1.6 Synthesis of Nonvicinally Substituted Alcohols

The reaction of a variety of 1-substituted-3-butenes with disiamylborane, followed by oxidation, gives the corresponding primary alcohols as the major products (Sec. 5.7).

$$CH_2=CHCH_2CH_2X \xrightarrow[\text{2) [O]}]{\text{1) Sia}_2\text{BH, THF}} HOCH_2CH_2CH_2CH_2X$$

$(X=OH, OCH_3, OAc, C\ell, NH_2)$

Similar reaction of ω-unsaturated esters gives the corresponding primary hydroxy esters as the major products (Sec. 5.6), while in the hydroboration-oxidation of various cyclohexene mono- and dicarboxylate derivatives, the trans hydroxy esters are the major products (Sec. 5.6).

Reaction of allyl chloride or its β-substituted derivatives with disiamylborane, followed by oxidation, gives the corresponding γ-chlorohydrins (Sec. 5.3.1).

$$CH_2=C-CH_2Cl \quad \xrightarrow[\text{2) [o]}]{\text{1) Sia}_2\text{BH, THF}} \quad HOCH_2CHCH_2Cl$$

with R below the first carbon and below the product carbon.

(R=H, alkyl, aryl)

Cyclic organoboranes react with 3-buten-2-one to give, after oxidation, ω-hydroxy ketones (Sec. 8.5.4).

$$\xrightarrow[\text{2) 170}^\circ]{\text{1) BH}_3\text{-THF}} \quad B-(CH_2)_4-B \quad \xrightarrow[\text{2) [o]}]{\text{1) CH}_2=CHCOCH_3, \text{ THF, H}_2O}$$

$HO(CH_2)_6COCH_3$

1.2 SYNTHESIS OF ALDEHYDES AND KETONES

1.2.1 Synthesis of Aldehydes

Reaction of 1-alkynes with disiamylborane, followed by oxidation, gives the corresponding aldehydes (Sec. 7.1.2).

$$RC\equiv CH \quad \xrightarrow[\text{2) [o]}]{\text{1) Sia}_2\text{BH, THF}} \quad RCH_2CHO$$

Carbonylation of organoboranes in the presence of lithium trimethoxy- or tributoxyaluminum hydride gives homologated aldehydes. Use of the latter hydride permits the presence of reducible groups such as cyano and ester groups in the alkyl moiety (Sec. 8.4).

$$R\text{-}B\text{-}9\text{-}BBN + CO + LiAlH(tert\text{-}BuO)_3 \xrightarrow[\text{2) [O]}]{\text{1) THF, }<\text{-}20^\circ} RCHO$$

Reaction of trialkylboranes with diazoacetaldehyde gives the corresponding alkylated acetaldehydes, while reaction with propenal or 2-substituted propenals gives 3-alkylated propanals (Sec. 8.4).

$$R_3B + N_2CHCHO \xrightarrow{\text{THF, H}_2\text{O, }25^\circ} RCH_2CHO$$

$$R_3B + CH_2{=}\underset{\underset{R'}{|}}{C}CHO \xrightarrow{\text{THF, H}_2\text{O, }25^\circ} RCH_2\underset{\underset{R'}{|}}{C}HCHO$$

(R'=H, alkyl, Br)

With 3-alkylated propenals the reaction only proceeds readily in the presence of air (Sec. 8.4).

$$R_3B + R'CH{=}CHCHO \xrightarrow[25^\circ]{\text{THF, H}_2\text{O, air}} R\underset{\underset{R'}{|}}{C}HCH_2CHO$$

Reaction of triallylborane with 1-alkynes, followed by oxidation, gives 2-alkyl-4-pentenals (Sec. 8.8.1).

$$RC{\equiv}CH + (CH_2{=}CHCH_2)_3B \xrightarrow[\text{3) [O]}]{\text{1) }20^\circ \quad \text{2) }CH_3OH}$$

1.2.2 Synthesis of Ketones

Oxidation of organoboranes, derived from the hydroboration of alkenes, with 8 N chromic acid gives the corresponding ketones (Sec. 4.6).

Acyclic ketones may be synthesized from disubstituted alkynes by monohydroboration with disiamylborane, followed by oxidation (Sec. 7.1.2).

$$RC \equiv CR \xrightarrow[\text{2) [O]}]{\text{1) Sia}_2\text{BH, THF}} RCH_2COR$$

Since the reaction of unsymmetrically disubstituted alkynes with disiamylborane occurs with a high degree of regioselectivity (Sec. 7.1), a large variety of ketones may be synthesized by the above method.

Reaction of 1-halo-1-alkynes with dicyclohexylborane, followed by treatment with sodium hydroxide and oxidation, gives <u>alkyl-cyclohexyl ketones</u> (Sec. 7.1.2).

(X=Br, I)

n-Alkylcyclohexyl ketones may also be prepared by the carbonylation of n-alkyldicyclohexylboranes (Sec. 8.5.1).

Carbonylation also provides convenient synthetic routes to <u>symmetrical and unsymmetrical dialkyl ketones</u> (Sec. 8.5.1).

$$R_3B + CO \xrightarrow[\text{2) [O]}]{\text{1) DG, H}_2\text{O, 100}^\circ} R_2CO$$

$$\text{Alkene A} \xrightarrow[\text{THF}]{\text{ThBH}_2} R_A\underset{H}{\overset{|}{B}}Th \xrightarrow[\text{THF}]{\text{Alkene B}} R_A\underset{Th}{\overset{|}{B}}R_B \xrightarrow[\text{2) [O]}]{\text{1) CO, 70 atm}} R_A R_B CO$$

Symmetrical and unsymmetrical dialkyl ketones may also be synthesized from trialkylcyanoborates (Sec. 8.5.2).

$$R_3B \xrightarrow[\text{Na}^{\oplus}]{\text{NaCN, DG}} R_3\overset{\ominus}{B}CN \xrightarrow[\text{2) [O]}]{\text{1) (CF}_3\text{CO)}_2\text{O}} R_2CO$$

$$\text{ThBH}_2 \xrightarrow[\text{2) Alkene B}]{\text{1) Alkene A}} R_A \overset{\text{BR}}{\underset{\text{Th}}{|}} B \xrightarrow[\substack{\text{2) (CF}_3\text{CO)}_2\text{O} \\ \text{3) [O]}}]{\text{1) NaCN, \quad DG}} R_A R_B CO$$

Reaction of B-alkyl-9-BBN derivatives with α-bromo ketones in the presence of potassium tert-butoxide gives dialkyl ketones (Sec. 8.5.3).

$$\text{R-B-9-BBN} + \text{R}'\text{COCH}_2\text{Br} \xrightarrow[\text{THF, 0}^\circ]{\text{tert-BuOK}} \text{R}'\text{COCH}_2\text{R}$$

$$(\text{R}' \neq \text{CH}_3)$$

Use of B-aryl-9-BBN derivatives gives the corresponding α- arylated ketones.

The extension of the above reaction to α-bromo acetone, and hence to the synthesis of methyl ketones, requires the use of potassium 2,6-di-tert-butylphenoxide as base in place of potassium tert-butoxide (Sec. 8.5.3).

$$\text{R-B-9-BBN} + \text{CH}_3\text{COCH}_2\text{Br} + \text{[phenoxide]} \xrightarrow[\text{2) EtOH}]{\text{1) THF}} \text{CH}_3\text{COCH}_2\text{R}$$

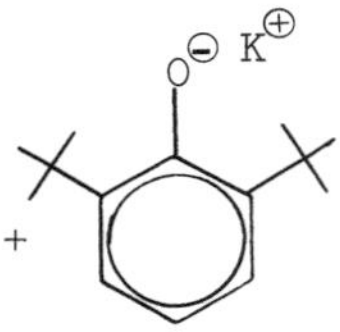

Reaction of trialkylboranes with diazo acetone also gives methyl ketones (Sec. 8.5.3).

$$\text{R}_3\text{B} + \text{CH}_3\text{COCHN}_2 \xrightarrow[\text{2) KOH}]{\text{1) THF}} \text{CH}_3\text{COCH}_2\text{R} + \text{R}_2\text{BOH}$$

A further synthesis of methyl ketones involves the reaction of trialkylboranes or B-alkylboracyclohexane derivatives with 3-buten-2-one (Sec. 8.5.4).

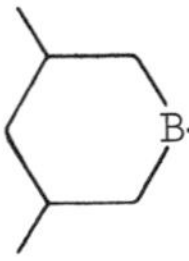

$$\text{B-R} + \text{CH}_2={=}\text{CHCOCH}_3 \xrightarrow[\text{2) [O]}]{\text{1) THF, H}_2\text{O}} \text{RCH}_2\text{CH}_2\text{COCH}_3$$

However the above reaction only proceeds with β-substituted enones or cyclic enones in the presence of air (Sec. 8.5.4).

$$R_3B + CH_3CH=CHCOCH_3 \xrightarrow[\text{air, 25}°]{\text{THF, H}_2\text{O}} \underset{\underset{CH_3}{|}}{R}CHCH_2COCH_3$$

Reaction of acetylethyne with trialkylboranes in the presence of air gives α,β-unsaturated ketones, which may be reacted further with a different trialkylborane to give a wide variety methyl ketones (Sec. 8.5.4).

$$HC≡CCOCH_3 \xrightarrow[\text{H}_2\text{O, air}]{\text{R}_3\text{B, THF}} RCH=CHCOCH_3 \xrightarrow[\text{H}_2\text{O, air}]{\text{R}_3'\text{B, THF}} RR'CHCH_2COCH_3$$

The reaction of trialkylboranes with α,β-unsaturated ketones may be adapted to permit the introduction of a second alkyl group, α to the carbonyl group (Sec. 8.5.4).

$$R_3B + CH_2=CHCOCH_3 \xrightarrow[\text{air}]{\text{THF, H}_2\text{O}} \underset{\underset{CH_3}{|}}{RCH_2CH=CO}BR_2 \xrightarrow[\text{2) BuLi}]{\text{1) Distill}}$$

$$\underset{\underset{CH_3}{|}}{RCH_2CH=CO}Li \xrightarrow{R'I} \underset{\underset{R'}{|}}{RCH_2CHCOCH_3} + RCH_2CH_2COCH_2R'$$
$$\text{(Minor isomer)}$$

Dialkyl ketones may also be synthesized from alkoxydialkyl-boranes via α-bromination followed by rearrangement (Sec. 8.5.5).

$$\underset{\underset{CH_3}{|}}{C_6H_{13}BO}C_4H_9 \xrightarrow[\text{hν}]{\text{Br}_2\text{, CCl}_4} \underset{\underset{CH_3}{|}}{C_5H_{11}\overset{\overset{Br}{|}}{CH}BO}C_4H_9 \xrightarrow[\text{2) [O]}]{\text{1) Br}_2} C_5H_{11}COCH_3$$

Reaction of tris(allylic)boranes with alkoxyalkynes gives 2-alkoxy-1,4-pentadienes, which are converted to allylic methyl ketones on treatment with acid (8.8.1).

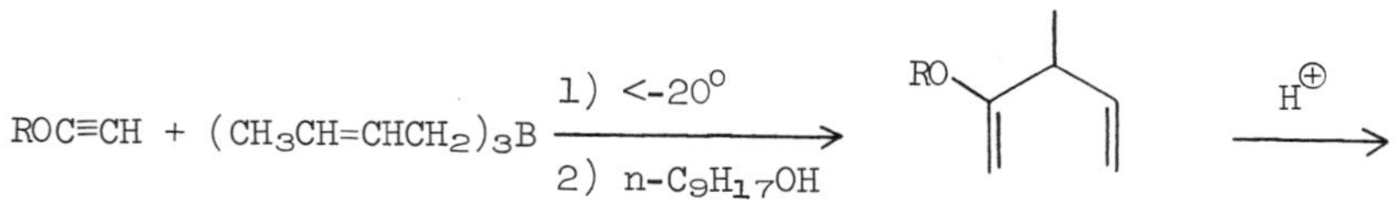

$$ROC≡CH + (CH_3CH=CHCH_2)_3B \xrightarrow[\text{2) n-C}_9\text{H}_{17}\text{OH}]{\text{1) <-20}°} \quad \text{(structure)} \quad \xrightarrow{H^{\oplus}}$$

$$CH_3COCHCH=CH_2$$
$$|$$
$$CH_3$$

Cyclic organoboranes react with 3-buten-2-one to give, after oxidation, ω-hydroxy ketones (Sec. 8.5.4).

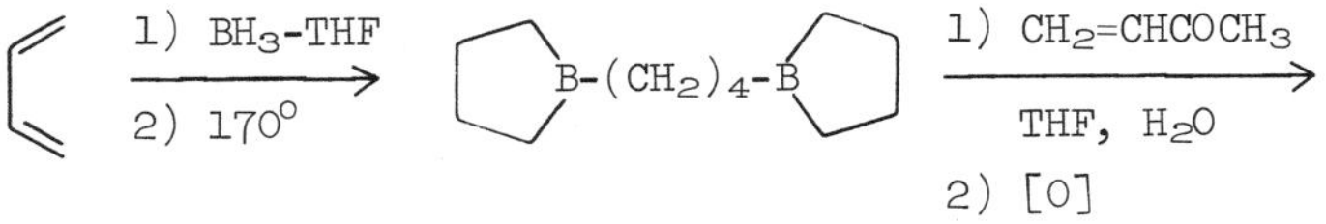

$$HO(CH_2)_6COCH_3$$

Cyclic organoboranes can be converted to <u>cyclic ketones</u> either via carbonylation (Sec. 8.5.1) or via formation of the trialkylcyanoborate (Sec. 8.5.2).

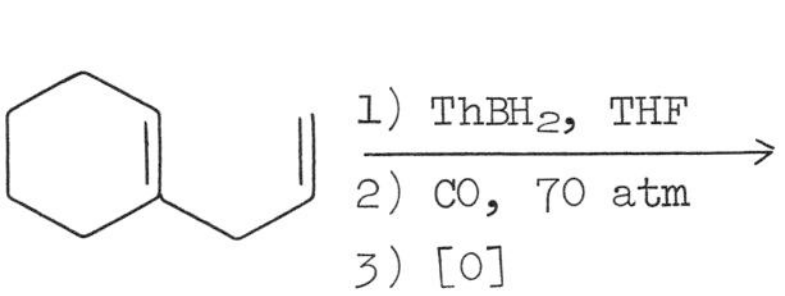

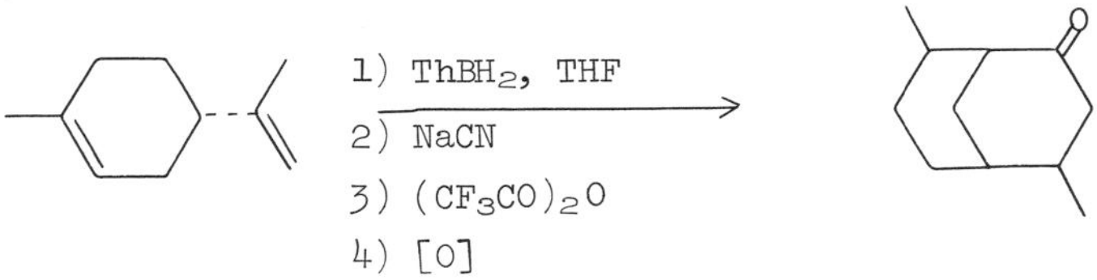

<u>α-Alkylcycloalkanones</u> can be synthesized from the α-bromo- or α-diazo cycloalkanones by reaction with trialkylboranes (Sec. 8.5.3).

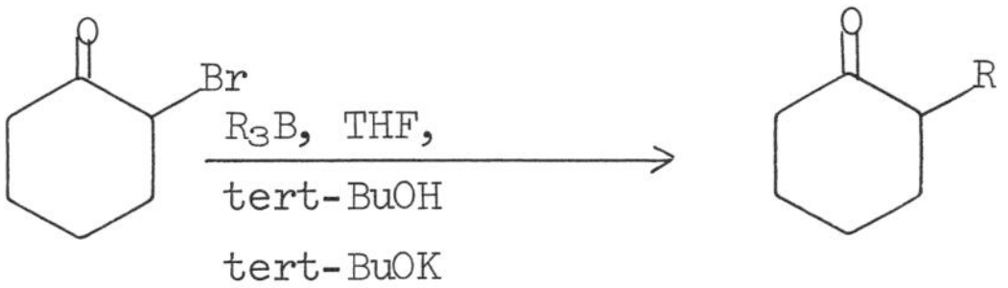

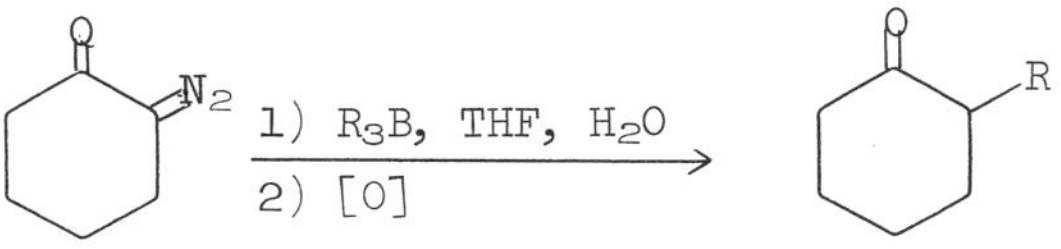

The latter method can be adapted to the synthesis of α,α-di-
alkylated cycloalkanones (Sec. 8.5.3). α-Alkylcycloalkanones
may also be prepared from the appropriate Mannich bases via in
situ formation of α,β-unsaturated ketones, followed by reaction
with an organoborane (Sec. 8.5.4).

The transposition of a keto group to the neighboring position
may be achieved via the hydroboration-oxidation of enamines
(Sec. 5.2.4).

1.3 SYNTHESIS OF CARBOXYLIC ACIDS AND DERIVATIVES
      Dihydroboration of 1-alkynes with dicyclohexylborane,
followed by oxidation with m-chloroperbenzoic acid gives the
corresponding carboxylic acids (Sec. 7.2.2).

$$RC\equiv CH \xrightarrow[\text{THF}]{2(C_6H_{11})_2BH} RCH_2CH\left[B(C_6H_{11})_2\right]_2 \xrightarrow[\text{THF}]{m\text{-}ClC_6H_4CO_3H} RCH_2CO_2H$$

Homologated carboxylic acids may be synthesized by Baeyer-Villager
oxidation of cyclohexyl ketones derived from the carbonylation of
n-alkyldicyclohexylboranes (Sec. 8.5.1).

$$RCH=CH_2 \xrightarrow[\text{2) CO \quad 3) [O]}]{\text{1) }(C_6H_{11})_2BH} RCH_2CH_2COC_6H_{11} \xrightarrow[\text{2) KOH}]{\text{1) }RCO_3H} RCH_2CH_2CO_2H$$

Reaction of 1-boro-1-lithio derivatives with carbon dioxide gives geminal dicarboxylic acids (Sec. 7.2.2).

$$RC\equiv CH \xrightarrow[\text{2) 2 BuLi}]{\text{1) }2(C_6H_{11})_2BH} \underset{\underset{C_4H_9}{|}}{RCH_2\overset{\overset{Li}{|}}{CH}-\overset{\ominus}{B}(C_6H_{11})_2} \xrightarrow[\text{2) H}^{\oplus}]{\text{1) }CO_2} RCH_2CH(CO_2H)_2$$

Esters may be synthesized by reaction of ethyl bromoacetate with organoboranes in the presence of base; ethyl dihaloacetates give α-halocarboxylates which may be reacted further with a different organoborane to give dialkylated derivatives (Sec. 8.6.1).

$$R_3B + Br_2CHCO_2C_2H_5 \xrightarrow[\text{Base}]{\text{THF}} \underset{\underset{Br}{|}}{RCHCO_2C_2H_5} \xrightarrow[\text{Base}]{R'_3B} RR'CHCO_2C_2H_5$$

Similar reaction occurs between trialkylboranes and ethyl diazo acetate (Sec. 8.6.1).

$$R_3B + N_2CHCO_2C_2H_5 \xrightarrow[\text{2) }H_2O]{\text{1) THF}} RCH_2CO_2C_2H_5$$

Ethyl 4-bromo-2-butenoate reacts with trialkylboranes in the presence of base to give trans-β,γ-unsaturated esters (Sec. 8.6.2).

$$R_3B + BrCH_2CH=CHCO_2C_2H_5 \xrightarrow[\text{Base}]{\text{THF}} \text{Trans-}RCH=CHCH_2CO_2C_2H_5$$

1.4 SYNTHESIS OF NITRILES

Chloroacetonitrile reacts with organoboranes in the presence of base to give the corresponding nitriles; use of dichloroaceto-nitrile permits the introduction of two different alkyl groups (Sec. 8.7).

$$R\text{-}B\text{-}9\text{-}BBN + Cl_2CHCN \xrightarrow[\text{Base}]{\text{THF}} \underset{\underset{Cl}{|}}{RCHCN} \xrightarrow[\text{THF, Base}]{R'\text{-}B\text{-}9\text{-}BBN} RR'CHCN$$

Ethyl bromocyanoacetate and bromodicyanomethane react in a similar manner to give the corresponding alkylated derivatives (Sec. 8.7). Diazo acetonitrile likewise gives the corresponding nitriles (Sec. 8.7).

$$R_3B + N_2CHCN \xrightarrow[\text{2) KOH}]{\text{1) THF}} RCH_2CN$$

1.5 SYNTHESIS OF OTHER FUNCTIONAL DERIVATIVES

Autoxidation of trialkylboranes under controlled conditions, followed by oxidation with hydrogen peroxide, gives <u>alkyl hydroperoxides</u>

$$R_3B + O_2 \xrightarrow[\text{3) H}_2O_2]{\text{1) -78}^\circ \quad \text{2) 0}^\circ} 2RO_2H + ROH \qquad \underline{\text{Sec. 4.1.2}}$$

The conversion of organoboranes to <u>alkyl halides</u> is demonstrated by the following equations.

$$RCl \xleftarrow[\text{H}_2O, \text{ NaOH}]{R_2' NCl, \text{ THF}} R_3B \xrightarrow[\text{THF, H}_2O]{\text{CuCl}_2} RCl \quad \underline{\text{Sec. 9.1.1}}$$

$$RBr \xleftarrow[]{CH_2Cl_2} R_3B + Br_2 \xrightarrow[\text{THF}]{CH_3ONa, \ CH_3OH} RBr \quad \underline{\text{Sec. 9.1.2}}$$

(Retention of configuration)   (Inversion of configuration)

$$RI \xleftarrow[\text{CH}_3\text{OH,THF}]{I_2, \text{NaOH}} R_3B \xrightarrow[\text{air, THF}]{CH_2=CHCH_2I} RI \qquad \underline{\text{Sec. 9.1.3}}$$

<u>Primary amines</u> are formed by the reaction of organoboranes with hydroxylamine-O-sulfonic acid or chloramine.

$$RNH_2 \xleftarrow[]{H_2NOSO_3H, \ DG} R_3B \xrightarrow[\text{NaOH, THF}]{H_2NCl,} RNH_2 \qquad \underline{\text{Sec. 9.2.1}}$$

Organic azides react with triethylborane to give secondary ethylamines, while use of dialkylchloroboranes gives a wide variety of <u>secondary amines</u>.

$$RNHC_2H_5 \xleftarrow{\begin{array}{l} 1)\ (C_2H_5)_3B \\ 2)\ Xylene,\ \Delta \\ 3)\ CH_3OH \end{array}} RN_3 \xrightarrow[\Delta]{R'_2\ BCl,\ PhCH_3} RNHR' \quad \underline{Sec.\ 9.2.1}$$

Organoboranes may also be converted into <u>sulfides</u> (Sec. 9.3.1), <u>disulfides</u> (Sec. 9.3.2), and <u>organomercury</u> compounds (Sec. 9.4).

The synthesis of a variety of <u>deuterated or tritiated compounds</u>, either by the use of labeled hydroborating reagents or by the reaction of organoboranes with suitably labeled reagents is discussed in Sec. 9.5.

1.6 SYNTHESIS OF ALKENES, DIENES AND DERIVATIVES

1.6.1 Synthesis of Alkenes

Hydroboration of disubstituted alkynes or cycloalkynes, followed by protonolysis, gives the corresponding cis-alkenes (Secs. 7.1.1 and 7.3, respectively)

$$RC \equiv CR \xrightarrow{\begin{array}{l} 1)\ Sia_2BH,\ DG \\ 2)\ CH_3CO_2H \end{array}} \underset{H}{\overset{R}{\diagdown}} C = C \underset{H}{\overset{R}{\diagup}}$$

Similar reaction of 1-alkynes gives terminal alkenes (Sec. 7.1.1).

Reaction of α,β-unsaturated aldehydes with disiamylborane, followed by hydroboration and treatment with methanesulfonic acid, gives the corresponding terminal alkenes (Sec. 5.4). Similarly, alkenes may be prepared from acyclic and cyclic α,β-unsaturated ketones (Sec. 5.4).

$$RCH = CHCHO \xrightarrow[THF]{1)\ Sia_2BH} RCH = CHCH_2OBSia_2 \xrightarrow{\begin{array}{l} 1)\ BH_3\text{-}THF \\ 2)\ CH_3SO_3H \end{array}} RCH_2CH = CH_2$$

Application of the above reaction sequence to allylic alcohols
also gives alkenes (Sec. 5.3.2). Hydroboration of steroid
4-en-3-ketones, followed by refluxing with acetic anhydride, gives
3-alkenes (Sec. 5.4).

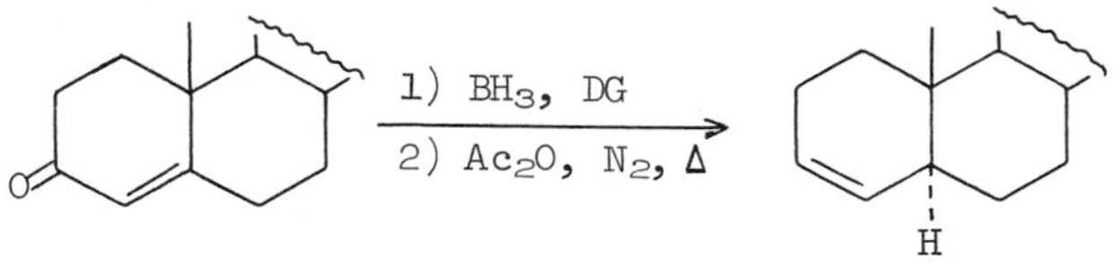

The above reaction sequence has been applied to a variety of
cyclic α,β-unsaturated ketones but the yields of alkenes reported
are generally low (Sec. 5.4).

   Hydroboration of enamines, derived from straight-chain
aldehydes, acyclic, and cyclic ketones, followed by treatment
with propanoic acid, gives the corresponding alkenes (Sec. 5.2.4).

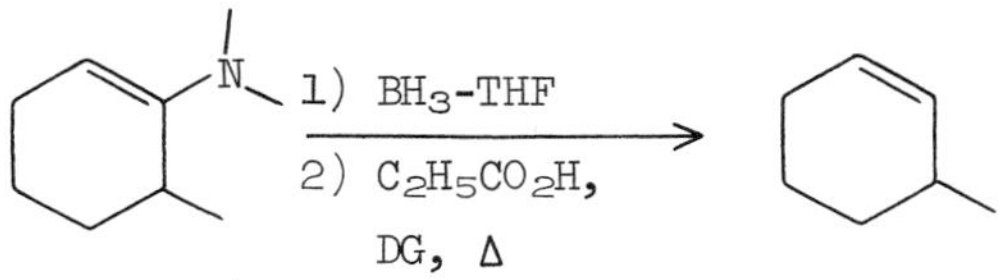

Steroid enamines, however, give low yields of the corresponding
alkenes. The synthesis of various steroid alkenes via the
hydroboration of enol ethers (Sec. 5.2.2) or enol acetates
(Sec. 5.2.3), followed by treatment with sodium hydroxide or
acetic anhydride, has been reported.

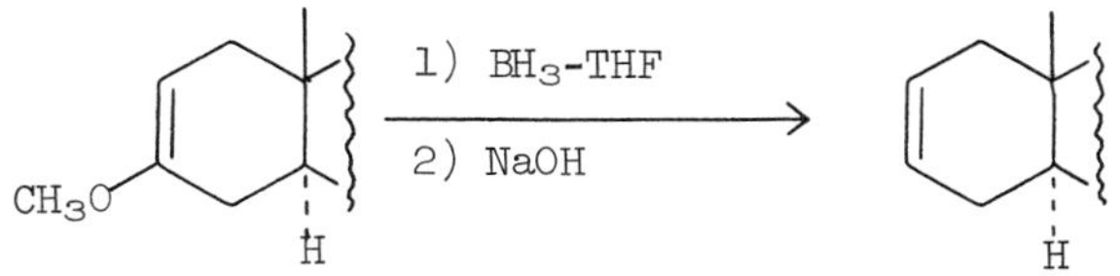

   Cis- and trans-cyclohexyl alkenes may be synthesized from
1-alkynes and 1-halo-1-alkynes, respectively, via hydroboration
with dicyclohexylborane followed by migration of a cyclohexyl
group from boron to carbon (Sec. 7.1.1).

$$RC\equiv CH \xrightarrow[\text{THF}]{(C_6H_{11})_2BH} \underset{H}{\overset{R}{>}}C=C\underset{B(C_6H_{11})_2}{\overset{H}{<}} \xrightarrow[2)\ I_2]{1)\ NaOH} \underset{H}{\overset{R}{>}}C=C\underset{H}{\overset{C_6H_{11}}{<}}$$

$$RC\equiv CX \xrightarrow[\text{THF}]{(C_6H_{11})_2BH} \underset{H}{\overset{R}{>}}C=C\underset{B(C_6H_{11})_2}{\overset{X}{<}} \xrightarrow[2)\ CH_3CO_2H]{1)\ NaOCH_3} \underset{H}{\overset{R}{>}}C=C\underset{C_6H_{11}}{\overset{H}{<}}$$

Use of disubstituted alkynes in the former reaction sequence gives the corresponding trisubstituted alkenes (Sec. 7.1.1). Reaction of 1-alkynes with 2 mole equivalents of dicyclohexylborane followed by 2 mole equivalents of butyllithium gives 1-boro-1-lithio derivatives, which react with aldehydes and ketones to give the corresponding alkenes (Sec. 7.2.2).

$$RC\equiv CH \xrightarrow[\text{2) 2BuLi}]{\text{1) } 2(C_6H_{11})_2BH} RCH_2\overset{\overset{Li}{|}}{C}H\text{-}\overset{\ominus}{B}(C_6H_{11})_2 \xrightarrow{R_2'CO} R_2'C=CHCH_2R$$

Trialkylboranes derived from C_n-terminal alkenes react with phenyl (bromodichloromethyl)mercury to give the corresponding C_{2n+1} internal alkenes (Sec. 8.8.2).

$$(RCH_2CH_2)_3B + PhHgCCl_2Br \xrightarrow[60\text{-}70^\circ]{1)\ PhH} RCH_2CH_2CH=CHCH_2R + PhHgBr$$
$$2)\ H_2O$$

Terminal alkenes may be formed from internal alkenes by the hydroboration-isomerization-displacement sequence of reactions (Sec. 8.8.3).

$$CH_3CH=C(C_2H_5)_2 \xrightarrow[\text{3) 1-Decene, }\Delta]{\text{1) } BH_3,\ DG\ \ 2)\ \Delta} CH_2=CHCH(C_2H_5)_2$$

1.6.2 Synthesis of Vinyl Derivatives

Reaction of 1-alkynes with disiamylborane gives trans-vinylboranes which react with bromine to give the corresponding dibromo adducts. Treatment of the adducts with sodium hydroxide

gives cis-vinyl bromides, while thermal decomposition gives
trans-vinyl bromides (Sec. 7.1.1).

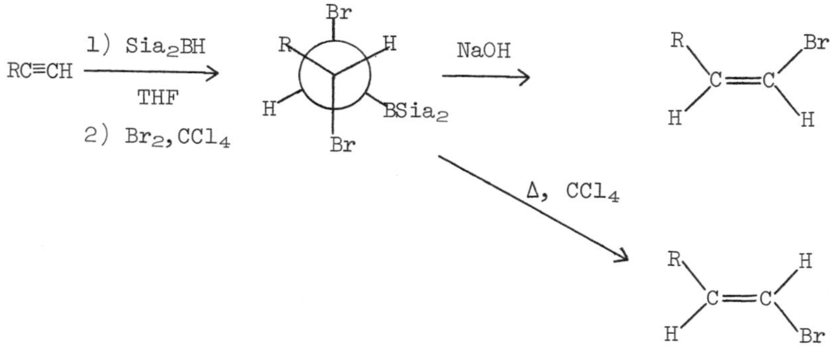

Cis-vinyl halides may also be prepared by reaction of 1-halo-1-
alkynes with dicyclohexylborane, followed by protonolysis
(Sec. 7.1.1).

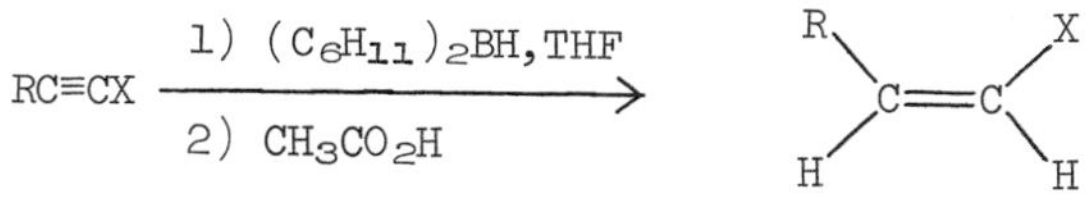

1.6.3 Synthesis of Dienes

1-Alkynes and disubstituted alkynes react with thexylborane
to give the corresponding divinylthexylboranes, which may be
converted to conjugated cis, trans-dienes as shown below
(Sec. 7.1.1).

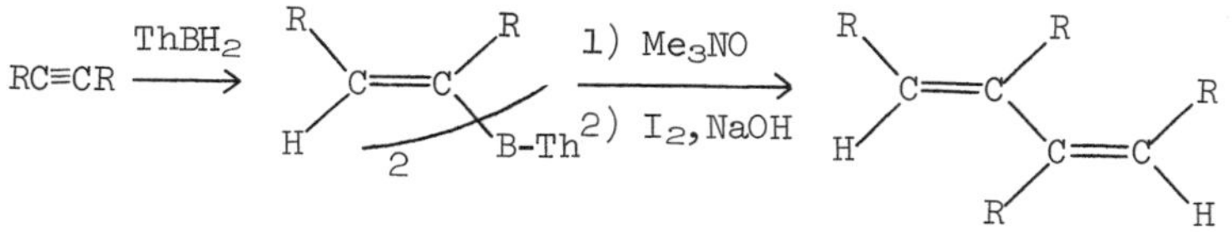

Conjugated cis, cis-dienes may be synthesized by dihydroboration-
protonolysis of conjugated diynes using dicyclohexylborane as the
hydroborating reagent (Sec. 7.4.2).

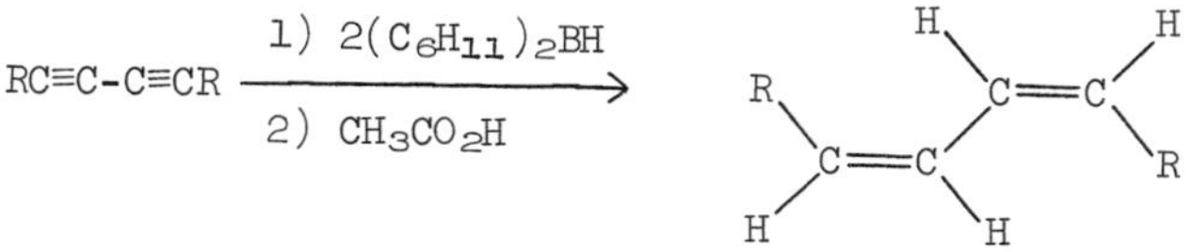

Only monohydroboration occurs on reaction with disiamylborane thus permitting the synthesis of cis-enynes (Secs. 7.4.1 and 7.4.2). However, enynes react with disiamylborane to give, after protonolysis, dienes (Sec. 7.5).

Reaction of triallylborane or tris (allylic) boranes with 1-alkynes or alkoxyalkynes gives substituted 1,4-pentadienes as shown below (Sec. 8.8.1).

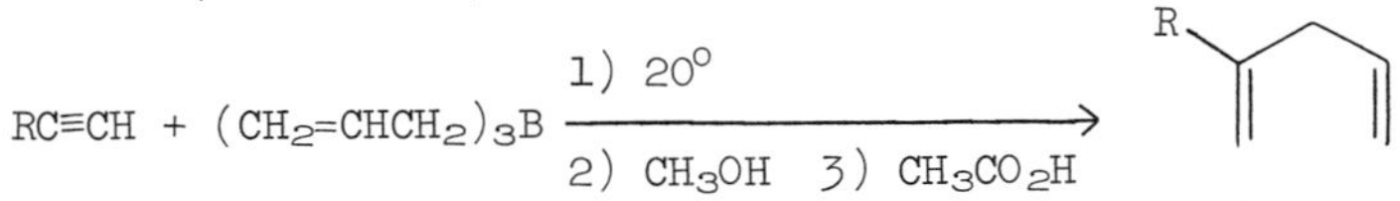

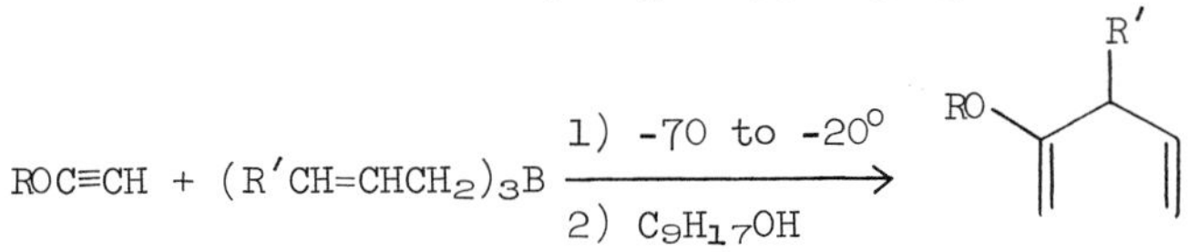

$$RC{\equiv}CH + (CH_2{=}CHCH_2)_3B \xrightarrow[\text{2) } CH_3OH \quad \text{3) } CH_3CO_2H]{\text{1) } 20°}$$

$$ROC{\equiv}CH + (R'CH{=}CHCH_2)_3B \xrightarrow[\text{2) } C_9H_{17}OH]{\text{1) } -70 \text{ to } -20°}$$

A variety of substituted 1,4-pentadienes may also be prepared by reaction of tris (allylic) boranes with vinyl ethers (Sec. 8.8.1).

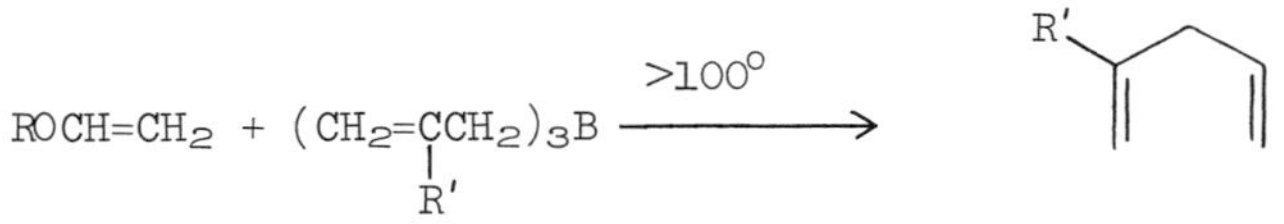

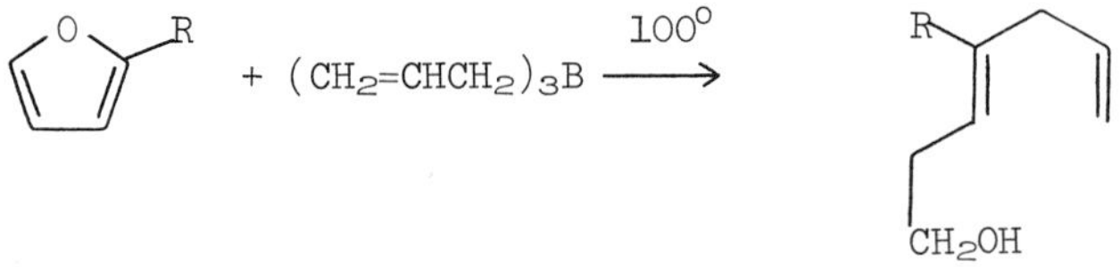

$$ROCH{=}CH_2 + (CH_2{=}\overset{\underset{\displaystyle R'}{|}}{C}CH_2)_3B \xrightarrow{>100°}$$

$$\text{(furan-2-yl with } R) + (CH_2{=}CHCH_2)_3B \xrightarrow{100°}$$

1,5-pentadienes may be prepared by hydroboration of suitably substituted cyclohexenyl methanesulfonates, followed by treatment with sodium hydroxide (Sec. 5.8).

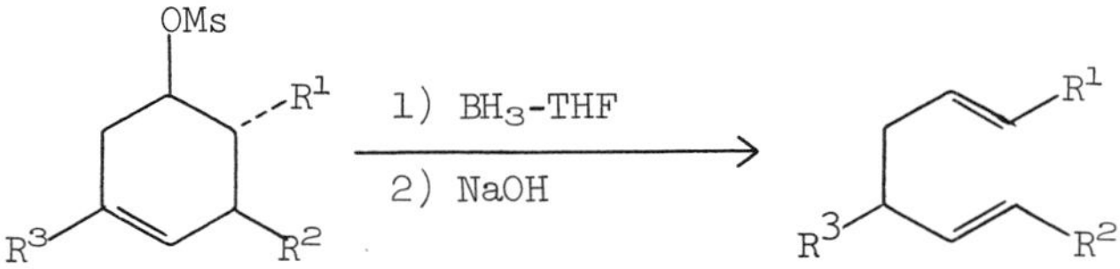

$$\xrightarrow[\text{2) NaOH}]{\text{1) } BH_3\text{-THF}}$$

Application of the reaction sequence to suitably substituted
unsaturated bicyclic methanesulfonates gives cyclic 1,5- and
1,6- dienes (Sec. 5.8).

Terminal allenes may be synthesized from 1-chloro-2-alkynes
by reaction with disiamylborane followed by sodium hydroxide
(Sec. 7.1.1).

$$RC{\equiv}CCH_2Cl \xrightarrow[\text{2) NaOH}]{\text{1) Sia}_2\text{BH, THF}} RCH{=}C{=}CH_2$$

1.6.4   Resolution of Alkene and Diene Mixtures and Racemates

The rates of reaction of alkenes with selective hydroborating
reagents such as disiamylborane are dependent on the structures of
the alkenes. This difference in reactivity thus permits the
separation of reactive from less reactive alkenes by reaction of
the alkene mixtures with a controlled amount of selective
hydroborating reagent (Sec. 3.3).

In a similar manner, alkene racemates may be resolved by
reaction of the racemates with a controlled amount of an asymmetric
hydroborating reagent such as diisopinocampheylborane (Sec. 3.8).
This method has also been applied to the resolution of diene and
allene racemates (Sec. 6.7).

1.7   CHAIN EXTENSION REACTIONS

A number of the reactions discussed in earlier sections of
this chapter provide convenient methods for achieving the
extension of carbon chains by a definite number of atoms. These
methods are summarized in Table 1.1.

TABLE 1.1

Methods for Achieving Carbon Chain Extension

Reactants[a]	Products	Section
One-carbon extension		
1) R_3B, CO, $LiAlH(OCH_3)_3$ 2) KOH	RCH_2OH	8.3.1
1) R_3B, CO, $LiAlH(OCH_3)_3$ 2) [O]	RCHO	8.4
1) $(C_6H_{11})_2BR$, CO 2) [O] 3) $R'CO_3H$	RCO_2H	8.5.1
Two-carbon extension		
R_3B, N_2CHCHO, H_2O	RCH_2CHO	8.4
R_3B, $BrCH_2CO_2C_2H_5$, base	$RCH_2CO_2C_2H_5$	8.6.1
R_3B, $N_2CHCO_2C_2H_5$, H_2O	$RCH_2CO_2C_2H_5$	8.6.1
R_3B, $ClCH_2CN$, base	RCH_2CN	8.7
Three-carbon extension		
R_3B, $CH_2{=}CHCHO$, H_2O	RCH_2CH_2CHO	8.4
1) R_3B, $N_2CHCOCH_3$ 2) KOH	RCH_2COCH_3	8.5.3
R-B-9-BBN, $BrCH_2COCH_3$, base	RCH_2COCH_3	8.5.3
Four-carbon extension		
1) R_3B, $CH_2{=}CHCH\overset{\diagdown}{\underset{O}{\diagup}}CH_2$, air 2) [O]	$RCH_2CH{=}CHCH_2OH$	8.3.1
R_3B, $CH_2{=}CHCOCH_3$, H_2O	$RCH_2CH_2COCH_3$	8.5.4
R_3B, $HC{\equiv}CCOCH_3$, air, H_2O	cis-$RCH{=}CHCOCH_3$	8.5.4
R_3B, $BrCH_2CH{=}CHCO_2C_2H_5$, base	trans-$RCH{=}CHCH_2CO_2C_2H_5$	8.6.2
Four or more carbon extension		
1) R-BH-Th, $CH_2{=}CH(CH_2)_nOAc$ 2) CO, H_2O 3) [O]	$RCO(CH_2)_{n+2}OAc$	8.5.1

[a]Tetrahydrofuran is used as solvent.

Carbon chain construction may also be achieved by the <u>coupling</u> of organoboranes on treatment with alkaline silver nitrate (Sec. 8.9), and the conversion of vinylboranes to dienes on treatment with iodine under alkaline conditions (Sec. 7.1.1).

$$R_3B \xrightarrow{\text{AgNO}_3, \text{ KOH, } H_2O} R\text{-}R$$

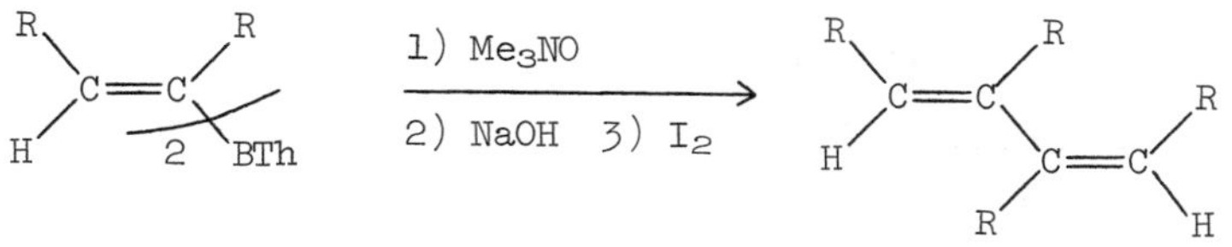

1.8 SYNTHESIS OF CARBOCYCLIC COMPOUNDS

The conversion of organoboranes into a variety of carbocyclic compounds is summarized in Table 1.2.

TABLE 1.2

Synthesis of Carbocycles

Reactants	Products	Section
1) Cl-C-C=CH$_2$, 9-BBN 2) KOH	Cyclopropanes	5.3.1
1) Br-C-C≡CH, 9-BBN 2) NaOH 3) [O]	Cyclopropanols	7.2.2
1) TsO-C-C-C≡CH, 9-BBN 2) CH$_3$Li 3)[O]	Cyclobutanols	7.2.2
1) Cyclic organoboranes, CO 2) [O]	Tertiary cyclic alcohols	8.3.3
1) B-Thexylboracyclanes, CO, H$_2$O 2) [O]	Cyclic ketones	8.5.1
1) B-Thexylboracyclanes, NaCN 2) (CF$_3$CO)$_2$O 3) [O]	Cyclic ketones	8.5.2

1.9 REDUCTION OF FUNCTIONAL GROUPS

The use of borane and selective reagents, such as disiamyl-borane, in the reduction of various functional groups is discussed in Chap. 10.

REFERENCES

1. E. Frankland and B. F. Duppa, Proc. Roy. Soc. (London), 10, 568 (1859).

2. E. Krause and R. Nitsche, Chem. Ber., 54, 2784, (1921).

3. D. T. Hurd, J. Am. Chem. Soc., 70, 2053 (1948).

4. H. C. Brown and B. C. Subba Rao, J. Am. Chem. Soc., 78, 2582 (1956).

5. H. C. Brown and B. C. Subba Rao, J. Am. Chem. Soc., 78, 5694 (1956); H. C. Brown and B. C. Subba Rao, J. Am. Chem. Soc., 81, 6423 (1959).

6. H. C. Brown and B. C. Subba Rao, J. Am. Chem. Soc., 81, 6428 (1959).

Chapter 2

SYNTHESIS OF ORGANOBORANES

This chapter briefly reviews the various methods available for the synthesis of organoboranes, with major emphasis being placed on the procedures used in the hydroboration of unsaturated compounds (Sec. 2.2.1). A number of useful methods have recently been developed for the synthesis of mixed alkylboranes and these are discussed in Sec. 2.2.4, while the methods of synthesis of cyclic organoboranes are discussed in Sec. 2.2.3.

The structure of alkylboranes is discussed in Sec. 2.1, while the use of spectroscopic methods in the structure determination of organoboranes is discussed in Sec. 2.4. Under certain conditions organoboranes undergo disproportionation and isomerization reactions; these reactions are discussed in Sec. 2.3.1 and 2.3.2, respectively.

2.1 STRUCTURE OF DIBORANE AND ALKYLBORANES

Diborane exists as a bridged molecule having two two-electron three-center bonds (1). It undergoes symmetrical cleavage reactions with many ligands such as amines, sulfides, and phosphines, and can also undergo unsymmetrical cleavage in certain cases (2a). A stepwise mechanism involving formation of a singly bridged intermediate has been proposed for both modes of cleavage (2a).

$$\underset{H}{\overset{H}{B}}\underset{H}{\overset{H}{}}\underset{H}{\overset{H}{B}} \quad \rightleftharpoons \xrightarrow{:L} \quad H_2B\text{-}H\text{-}BH_3 \quad \rightleftharpoons \xrightarrow{:L} \quad H_2BL_2^{\oplus} \ BH_4^{\ominus} \quad \text{Unsymmetrical cleavage}$$

$$\underset{L}{\overset{|}{}} \quad \updownarrow : L$$

$$2BH_3:L \quad \text{Symmetrical cleavage}$$

Ethers react with diborane to produce symmetrical cleavage products, though, except in the case of tetrahydrofuran, these products have been shown to be very unstable (2b). However, it has been firmly established that diborane in tetrahydrofuran exists predominantly as the borane-tetrahydrofuran adduct, $BH_3:THF$. Thus, the active species in hydroboration reactions with diborane in tetrahydrofuran is borane, and this is often assumed to be the case when using other ether solvents as well (Sec. 3.4). The unsymmetrical cleavage of diborane in tetrahydrofuran has also been invoked to explain certain reactions (Secs. 5.1, 10.1, and 10.5.3). While unsymmetrical cleavage as shown above has been postulated (3a), kinetic studies of the reduction of ketones have led to the following equilibrium being proposed (Sec. 10.1) (3b).

$$3\,BH_3:THF \ \rightleftharpoons \ H_2B(THF)_2^{\oplus} \ B_2H_7^{\ominus} + THF$$

The $B_2H_7^{\ominus}$ ion has been shown to exist as the singly hydrogen bridged species, $H_3B\text{-}H\text{-}BH_3^{\ominus}$ (3c).

Mono- and dialkylboranes usually exist as dimers even in tetrahydrofuran solution (4a). The failure of these compounds to dissociate in tetrahydrofuran has been attributed to the polar and steric influences of the alkyl substituents which reduce the acceptor ability of the boron atom (4a). Dialkylboranes containing two bulky alkyl groups, such as dithexylborane, have, however, been shown to exist in the monomeric form which is not associated with the ether solvent (4b).

$$\nu_{max.} = 2470 \ cm.^{-1}$$

The existence of such monomeric dialkylboranes is due to the large
steric requirements of the two alkyl groups (4b). Mono- and
dialkylboranes are often used as selective hydroborating agents
(Sec. 2.2.1.b), and it is usual to represent these compounds as
monomers; however, the fact that they often exist as dimers in
ether solutions is of importance to the interpretation of their
reactions (Secs. 3.4, 3.7, and 10.1).

Trialkylboranes exist only as monomers (5a).

2.2   SYNTHESIS OF ORGANOBORANES

Before discussing the various methods available for the
synthesis of organoboranes it must be stressed that boranes have
been reported to produce toxic effects in humans (5b), and hence
the use of adequate ventilation is recommended.

2.2.1   Synthesis of Organoboranes by the Hydroboration of Unsaturated Compounds

Hydroboration constitutes one of the most important methods
of synthesis of organoboranes and has been applied to a wide
variety of unsaturated compounds (Chaps. 3, 5, 6, and 7). The
various procedures available for carrying out hydroboration
reactions are discussed in this section.

The most common hydroborating reagent is diborane itself, but
mono- or dialkylboranes are often used in cases where a greater
degree of selectivity of reaction is required. Hydroboration
reactions are usually carried out in anhydrous, peroxide-free
ether solvents, such as tetrahydrofuran (THF), diglyme (DG),
triglyme (TG), or diethyl ether (Et_2O). Due to the sensitivity
of many organoboranes to oxygen (Sec. 4.1), the reactions are
carried out in an inert atmosphere. Temperatures of room
temperature or lower are normally used since the use of elevated
temperatures can lead to isomerization of the products (Sec. 2.3.2).

2.2.1.a   Use of Diborane as Hydroborating Reagent

Many methods have been developed for the generation of
diborane. These methods generally involve the treatment of a

metal hydride or complex metal hydride with an acidic reagent in
a suitable ether solvent, and a number of them have been reviewed
from the point of view of the efficiency and convenience of the
reaction (6,7). By far the most widely used method involves the
reaction of sodium borohydride and boron trifluoride etherate in
tetrahydrofuran or diglyme.

$$3NaBH_4 + 4BF_3:O(C_2H_5)_2 \xrightarrow[DG]{THF\ or} 3NaBF_4 + 2B_2H_6 + 4(C_2H_5)_2O$$

If diethyl ether is used as solvent, 10 mole % of zinc chloride
is added in order to catalyze the reaction (6). In certain
reactions the use of boron trifluoride leads to undesirable side
effects (Sec. 10.3.4), and in such cases mercurous chloride (8) or
particularly iodine (8,9) can be used as the acidic reagent.

$$2NaBH_4 + I_2 \xrightarrow{DG} 2NaI + B_2H_6 + H_2$$

Other acids which have been used include aluminum chloride,
boron trichloride, hydrogen chloride, and sulfuric acid (6,7).

Another convenient method for the generation of diborane
involves the reaction of lithium aluminum hydride and boron
trifluoride etherate in diethyl ether (10), though it has been
reported that subsequent isolation of the organoborane is
complicated by formation of a precipitate of aluminum hydroxide (6).

$$3LiAlH_4 + 4BF_3:O(C_2H_5)_2 \xrightarrow{Et_2O} 3LiAlF_4 + 2B_2H_6$$

Among the other hydride sources which have been investigated are
lithium or potassium borohydride, and sodium or lithium hydride
(6,7).

Three hydroboration procedures are normally used (7). The
first procedure involves the reaction of the unsaturated compound
with diborane generated in situ, and is useful for large scale
preparations in which the presence of the reagents and inorganic
reaction products (e.g., NaBF_4) does not produce undesired side

reactions. In this procedure a 10-20% excess of the acidic
reagent, usually in a suitable ether solvent, is added slowly to
a stirred mixture of an equivalent amount of the hydride source
and the unsaturated compound in an ether solvent, the reaction
being performed in an inert atmosphere at room temperature. After
completion of the addition the mixture is stirred at room
temperature for a further period (1-4 hr) after which the excess
hydride is carefully decomposed by addition of water or
1,2-ethanediol. The product organoborane is then reacted
further as desired.

The second procedure involves the external generation of
diborane in a suitable generator (7). When using sodium
borohydride as the hydride source a solution of the hydride in
diglyme is added to the acid in diglyme. This mode of addition
is necessitated by the fact that sodium borohydride in diglyme
absorbs a half molar equivalent of the diborane generated to form
sodium diborohydride. The generated diborane is swept over into
a solution of the unsaturated compound in tetrahydrofuran using a
stream of nitrogen. If boron trifluoride etherate is used as the
acid in the generator, traces of the reagent can be swept into the
reaction mixture; this results in undesirable effects (Sec. 10.3.4).
This difficulty may be avoided by passing the diborane through a
dilute solution of sodium borohydride in a suitable ether solvent,
or by use of an alternative acid such as iodine (9). The
remainder of the procedure is similar to that given for the in
situ method discussed above. This method has the advantage of
avoiding the possibility of unwanted side reactions due to the
presence of the hydride and acid reagents, and also results in
the organoborane being the sole product in the reaction mixture.

The third procedure involves the addition of a solution
of diborane in tetrahydrofuran to a solution of the unsaturated
compound in tetrahydrofuran. An approximately 1 M solution
of diborane in tetrahydrofuran may be prepared by passing
diborane, generated from 0.95 moles of sodium borohydride and 1.9

moles of boron trifluoride etherate (50% excess), into 500 ml
tetrahydrofuran (7). The concentration of the solution may be
determined, either by measuring the volume of hydrogen evolved on
treating an aliquot with dilute acid, or by titration as boric
acid. In the latter method an aliquot of the solution is pipeted
into excess acetone to form diisopropoxyborane which, on hydrolysis
with water, gives boric acid. Mannitol is added and the acid
titrated with standard sodium hydroxide (11). Provided pre -
cautions are taken to exclude traces of boron trifluoride,
solutions may be kept under nitrogen at 0° for periods of several
weeks (7). Studies at various temperatures have shown that
solutions are also stable at room temperature, but concentrations
decrease rapidly with increasing temperature (12a). Reports on
the instability of solutions of diborane in tetrahydrofuran have
appeared (9), and it should be noted that diborane reacts slowly
with ethers, particularly at elevated temperatures (Sec. 10.6.3).
Solutions of diborane in tetrahydrofuran are commercially
available (12b).

Hydroboration of alkenes and alkynes using diborane provides
a method of synthesis of uniform trialkyl- and trivinylboranes
(Chaps. 3 and 7, respectively), while reaction with dienes gives
cyclic organoboranes (Secs. 2.2.3 and 6.2).

2.2.1.b Use of Borane Adducts as Hydroborating Agents

The use of borane-ether adducts has been discussed in
Sec. 2.2.1.a. Amine-boranes have been used in the hydroboration
of unsaturated compounds, but the method usually requires the use
of temperatures exceeding 100°, and hence can lead to isomerization
of the product organoboranes (Sec. 2.3.2) (6). Amine-boranes may
be prepared by reduction of amine hydrochlorides with sodium
borohydride (13), or by the reduction of trialkoxyboranes (borate
esters) with lithium aluminum hydride in the presence of amines
(14a).

Dimethyl sulfide-borane is a stable liquid which, when stored
under nitrogen at room temperature, retains its activity for

several months (14b). Its reactivity in hydroboration reactions
is reported to parallel that of borane in ether solvents.

2.2.1c Use of Selective Hydroborating Reagents

Certain highly substituted alkenes react rapidly with diborane
to give mono- or dialkylboranes, with further reaction being
relatively slow (Sec. 3.2). The greater steric bulk of these
alkylboranes compared to borane itself results in these compounds
being useful selective hydroborating reagents. Although these
compounds exist as dimers (Sec. 2.1), they are usually represented
as monomers. Some useful selective reagents are listed in Table
2.1.

TABLE 2.1

Some Selective Hydroborating Reagents

Alkene	Reagent Structure	Reagent Name	Reference
a	$(C_2H_5)_2BH$	Diethylborane	15
$(CH_3)_2C=CHCH_3$	$\left[(CH_3)_2CH-\underset{CH_3}{CH-}\right]_2 BH$	Disiamylborane(Sia_2BH) [Bis-(3-methyl-2-butyl) borane]	7, 16
		Dicyclohexylborane $[(C_6H_{11})_2BH]$	17
		Bis(3,5-dimethyl) boracyclohexane	18
		9-Borabicyclo[3.3.1] nonane(9-BBN)	19
$(CH_3)_2C=C(CH_3)_2$	$(CH_3)_2CHC(CH_3)_2$ BH_2	Thexylborane($ThBH_2$)	20

[a]Reagent prepared by hydrogenation of triethylborane.

These reagents are usually prepared in situ by the addition
of the calculated amount of boron trifluoride etherate to a
stirred solution of equivalent quantities of the alkene and
sodium borohydride in diglyme at 0° (7). Alternatively, a
solution of the reagent in tetrahydrofuran may be prepared by
addition of the calculated amount of borane-tetrahydrofuran to
the alkene in tetrahydrofuran at 0° (7). The reaction mixture
from either of the above two procedures is stirred for a period
of 1-4 hr at 0° to ensure complete formation of the reagent, after
which a solution of the unsaturated compound in diglyme or
tetrahydrofuran is added. After completion of the reaction,
excess hydride is destroyed by addition of water, and the product
organoborane reacted further as desired.

In general, these reagents are sensitive to oxygen, but it
is reported that disiamylborane can be stored at 0-5° under
nitrogen for several days without marked change in hydride
content (21). 9-Borabicyclo[3.3.1]nonane, however, is a crystalline
solid which is both thermally stable and stable to atmospheric
oxygen. It can thus be easily stored and weighed and is hence a
very convenient reagent for hydroboration reactions (19).

2.2.1.d Use of Asymmetric Hydroborating Reagents

The hydroboration of α-pinene at 0° proceeds via cis addition
of borane to the less hindered side of the double bond (remote
from the gem-dimethyl group) to give the tetraalkyldiborane,
sym-tetraisopinocampheyldiborane (Sec. 3.1.6.b) (22a).

The product formed from (+)-(1R:5R)-α-pinene exhibits a rotation
in tetrahydrofuran of $[\alpha]_D^{20}$-37.1°, and has been assigned the
absolute configuration (1R:2S:3R:5R) for each isopinocampheyl
group (22a).

$$\left[\ \left[\ \text{H} \overset{}{\underset{}{}} \text{-BH}\ \right]_2\ \right]_2 \quad \equiv \quad (-)-(\text{IPC})_2\text{BH}$$

Despite the existence of these reagents in the dimeric form, it is usual to discuss them in terms of the monomeric form. Thus, hydroboration of (+)-α-pinene gives (-)-diisopinocampheylborane , while (-)-α-pinene gives (+)-diisopinocampheylborane.

In the absence of excess α-pinene the tetraisopinocampheyl-diboranes undergo significant dissociation to triisopinocampheyl-diborane and α-pinene.

$$\begin{array}{c} \text{IPC}\quad\text{H}\quad\text{IPC} \\ \diagdown\ \diagup\ \diagdown \\ \text{B}\qquad\text{B} \\ \diagup\ \diagdown\ \diagup\ \diagdown \\ \text{IPC}\quad\text{H}\quad\text{IPC} \end{array} \underset{\longleftarrow}{\overset{\longrightarrow}{}} \begin{array}{c} \text{IPC}\quad\text{H}\quad\text{H} \\ \diagdown\ \diagup\ \diagdown \\ \text{B}\qquad\text{B} \\ \diagup\ \diagdown\ \diagup\ \diagdown \\ \text{IPC}\quad\text{H}\quad\text{IPC} \end{array} + \ \alpha\text{-pinene}$$

Thus, a 0.5 M solution in tetrahydrofuran shows the presence of 10% of the original amount of α-pinene used to synthesize the derivative (22a). The product is less soluble in diglyme and precipitates from solution when prepared in similar concentration. In this medium the amount of residual α-pinene is only about 4%, possibly the result of the smaller amount of the reagent in solution. It is thus preferable to synthesize and utilize the reagent in diglyme solution (7,22b).

While diisopinocampheylborane is the most widely used asymmetric hydroborating reagent, triisopinocampheyldiborane (23a) and monoisopinocampheylborane (23b) have also been used, particularly in reactions involving hindered substrates (Sec. 3.6.2).

The preparation and use of the above asymmetric hydroborating reagents are similar to the procedures used in the case of selective hydroborating reagents (Sec. 2.2.1.c). α-Pinene of high optical purity is available from natural sources. In addition, (-)-α-pinene of high optical purity can be prepared on a large scale

by the isomerization of β-pinene using either iron pentacarbonyl
(24) or benzoic acid (25).

2.2.2 Synthesis of Organoboranes Using Organometallic Reagents

The reaction of an organometallic compound with a suitable
boron compound results in the transfer of the organic group from
the metal to boron. Commonly used organometallic reagents are
Grignard reagents and derivatives of aluminum, zinc, and mercury,
while boron trifluoride etherate, boron trichloride, and
trialkoxyboranes are common boron substrates. The use of these
methods in the synthesis of organoboranes has been comprehensively
reviewed (26) and is not discussed in this section. These
methods are useful for the synthesis of arylboranes (27),
vinylboranes (28), allylboranes (29), and alkynylboranes (30a).
In addition, the synthesis of organoboranes by reaction of
borane-tetrahydrofuran with organomercury compounds (30b) or
Grignard reagents (30c) has recently been reported.

The reaction of tert-alkylmagnesium halides with boron
trifluoride etherate occurs with partial isomerization to give
organoboranes containing secondary alkyl groups. However, with
primary and secondary alkylmagnesium halides no isomerization of
the alkyl groups is observed (30d).

2.2.3 Synthesis of Cyclic Organoboranes (Boracyclanes)

In this section only those cyclic organoboranes containing
a boron atom as the sole hetero atom are discussed. The synthesis
of cyclic organoboranes has been reviewed (31,32), and the purpose
of the following section is to briefly discuss the scope of the
various methods employed.

The pyrolysis of alkyl-or arylboranes at temperatures
exceeding 200° gives a variety of boracycloalkanes, the nature
of the product depending on the type and chain length of the
groups bonded to boron (5a,31,32,34). Trialkylboranes containing
two and three carbon atoms in the main alkyl chain give products
resulting from isomerization (Sec. 2.3.2), while tri-n-butylborane
gives B-butylboracyclopentane (5a). Trialkylboranes containing

between four and eight carbon atoms in the main alkyl chain give
complex products resulting from isomerization, while those
containing eight or more carbon atoms in the main alkyl chain
give bicyclic products (5a). The pyrolysis proceeds via initial
elimination of a mole of alkene to form an intermediate dialkylborane. In the case of the formation of a monocyclic product, the
dialkylborane eliminates a mole of hydrogen; when bicyclic
products are formed, additional moles of alkene and hydrogen are
eliminated (5a).

$$(n\text{-}C_4H_9)_3B \xrightarrow[-C_4H_8]{300^\circ} H_9C_4\text{-}B\overset{H\quad CH_3}{\diagdown} \xrightarrow{-H_2} H_9C_4\text{-}B\bigcirc \quad \sim 90\%$$

$$(n\text{-}C_9H_{19})_3B \xrightarrow[\substack{-C_9H_{18} \\ -H_2}]{250\text{-}300^\circ} \underset{\substack{H_{19}C_9 \\ C_4H_9}}{B} \xrightarrow[-H_2]{-C_9H_{18}} \quad \sim 80\%$$

Where possible, products containing six-membered rings are formed
due to their greater thermodynamic stability (Sec. 2.3.2). The
cyclization proceeds most readily when a primary hydrogen atom is
present at the 5-position, with the efficiency of the reaction
decreasing in the case of secondary and tertiary hydrogens (35).

As is evident from the above equations the maximum possible
conversion of alkenes to boracyclanes via uniform trialkylboranes
is 33%. However, hydroboration of alkenes with thexylborane gives
the corresponding monoalkylthexylboranes which permit 100%
conversion of alkene to boracycloalkanes (35). As before,
cyclization only proceeds satisfactorily for compounds having a
primary hydrogen atom in position 5 relative to boron.

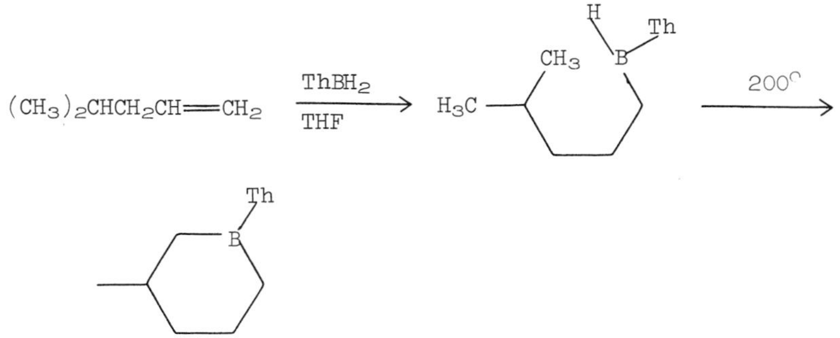

$$(CH_3)_2CHCH_2CH{=\!\!=}CH_2 \xrightarrow[\text{THF}]{\text{ThBH}_2} \quad \xrightarrow{200^\circ}$$

Pyrolysis has also been applied to the synthesis of benzoboracycloalkanes as shown below (5a,31).

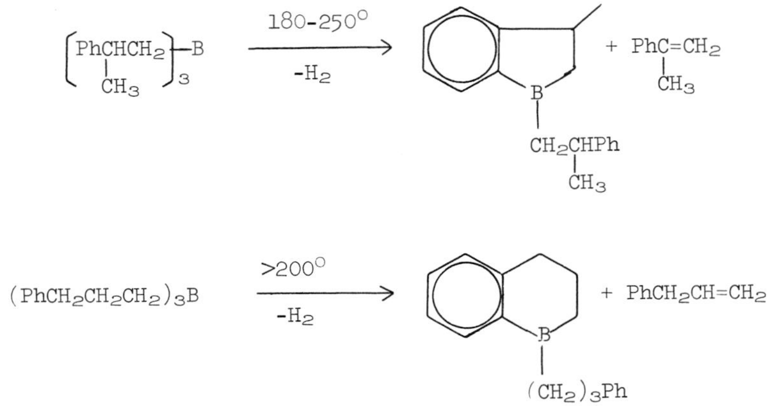

$$\left[\begin{array}{c} \text{PhCHCH}_2 \\ | \\ \text{CH}_3 \end{array} \right]_3 {-}\text{B} \xrightarrow[\text{-H}_2]{180\text{-}250^\circ} \qquad + \ \text{PhC}{=}\text{CH}_2$$

$$(\text{PhCH}_2\text{CH}_2\text{CH}_2)_3\text{B} \xrightarrow[\text{-H}_2]{>200^\circ} \qquad + \ \text{PhCH}_2\text{CH}{=}\text{CH}_2$$

The B-aralkyl derivatives are readily converted to B-alkyl derivatives by treatment with a dialkylborane, such as diethylborane (5a,31).

Similar products can also be formed by the pyrolysis of mixtures of trialkyl- and triaralkylboranes in a molar ratio of 2 to 1, or by pyrolysis of equimolar amounts of trialkyl- and triaralkylboranes in the presence of an amine-borane (5a,31).

$$2(i\text{-}C_4H_9)_3B + \left[\underset{\underset{CH_3}{|}}{PhCHCH_2}\right]_3\!B \xrightarrow[-H_2]{230^\circ} 3 \quad \text{(structure)} + 3C_4H_8$$

$$\sim 90\%$$

$$\left[\underset{\underset{CH_3}{|}}{PhCHCH_2}\right]_3\!B + R_3B \xrightarrow{Et_3N:BH_3} 3 \quad \text{(structure)}$$

$$\xrightarrow{180^\circ} \quad \text{(structure)}$$

These reactions proceed via redistribution reactions (Sec. 2.3.1). Replacement of trialkylboranes by trialkoxyboranes gives the corresponding B-alkoxy derivatives in high yields (5a,31).

The synthesis of a cyclic organoborane by means of photocyclization of a suitable acyclic organoborane has also been reported (36). Thus, reaction of dicyclohexylborane with 2-methylbut-1-en-3-yne gives an intermediate dienylborane (1) which, on irradiation, gives the boracyclopent-3-ene derivative (2) in 60% yield.

$$HC\!\equiv\!C\text{-}\underset{\underset{CH_3}{|}}{C}\!=\!CH_2 \xrightarrow[THF,\ 0^\circ]{(C_6H_{11})_2BH} (C_6H_{11})_2B\text{-}\underset{(1)}{C} \quad \xrightarrow[THF]{h\nu} \underset{(2)}{H_{11}C_6B} \quad \text{(structure)}$$

Further methods for the synthesis of cyclic organoboranes are discussed in other sections of this book. B-Alkylboracyclopentanes may be prepared via the hydroboration of 1,3-butadiene (Sec. 6.1),

while a variety of other boracycloalkanes may be synthesized via
the hydroboration of suitable dienes with borane or thexylborane
(Sec. 6.2). Hydroboration of cyclic dienes gives, in certain
cases, bicyclic organoboranes (Sec. 6.4); these compounds may
also be prepared by the hydroboration of suitable acyclic trienes,
while certain cyclic trienes may be converted to tricyclic
organoboranes (Sec. 6.8). The reaction of bifunctional organometallic reagents with suitable boron compounds provides a further
route to various cyclic organoboranes (Sec. 2.2.2) (_31,33_).

2.2.4 Synthesis of Mixed Organoboranes

The methods of synthesis of organoboranes discussed thus far
usually give uniform products. However, the application of many
of the synthetic procedures involving use of organoboranes also
requires convenient methods of synthesis of mixed alkyl and
arylboranes.

Mixed organoboranes may be prepared by the reaction of
suitable organometallic reagents with mono- or dihaloorganoboranes
(Sec. 2.2.2); the synthesis of haloorganoboranes is discussed in
Sec. 2.2.5. Another possible method of synthesis involves
$>$B-H catalyzed or trialkylaluminum-catalyzed ligand exchange
between pure trialkyl- and triarylboranes (Sec. 2.3.1), but
separation of the exchange products is difficult due to
disproportionation and isomerization of the products during
distillation (Secs. 2.3.1 and 2.3.2).

Dialkylboranes, with the exception of relatively hindered
compounds such as disiamylborane and other selective hydroborating
reagents (Sec. 2.2.1.c), are prone to disproportionate, and are
hence of limited use in the synthesis of mixed trialkylboranes
via reaction with alkenes. This problem has been overcome by the
in situ generation of the dialkylborane in the presence of an
alkene, which results in the rapid formation of the more stable
trialkylborane in situ. Thus, treatment of B-methoxy (_37a_) or
B-aryloxydialkylboranes (_37b_) (Secs. 2.2.5 and 2.3.1) with lithium

aluminum hydride in tetrahydrofuran or hexane in the presence of
an alkene gives the corresponding trialkylboranes in 60-80% yields.

$$R_2BOCH_3 + alkene \xrightarrow[\text{2) } H_2SO_4]{\text{1) } LiAlH_4, THF, r.t.} R_2BR' \qquad Ref.\ \underline{37a}$$

$$60-80\%$$

$$3R_2BOAr \xrightarrow{LiAlH_4, Hexane} \left[3R_2BH\right] \xrightarrow{Alkene} 3R_2BR' \qquad Ref.\ \underline{37b}$$

$(Ar=o-CH_3C_6H_4-)$

These methods are only applicable to derivatives not containing
reducible functional groups. A more versatile method involves
in situ generation of stable addition products comprising
dialkylboranes and aluminum methoxide by treatment of B-methoxydialkylboranes with aluminum hydride in tetrahydrofuran ($\underline{38a}$).

$$3R_2BOCH_3 \xrightarrow[\text{THF, r.t.}]{AlH_3} \left[(CH_3O)_3Al\cdot3R_2BH\right] \xrightarrow[\text{2) } H_2SO_4]{\text{1) Alkene}} R_2BR'$$

$$(78-95\%)$$

Pyridine ↓

$$Al(OCH_3)_3 + 3R_2BH:py$$
$$(stable)$$

$$\xrightarrow{\text{Alkene}} BF_3\cdot Et_2O\ (-BF_3\cdot py)$$

The addition complex can either be reacted directly with the
alkene in situ, or, if the initial reaction is carried out in
the presence of pyridine, the stable pyridine-dialkylborane
complex may be isolated ($\underline{38a}$). These latter complexes provide a
useful method for storing dialkylboranes. Generation of the
dialkylborane by treatment of the pyridine complex with the
theoretical quantity of boron trifluoride etherate in the presence
of an alkene provides an alternative method for the synthesis
of the mixed organoborane ($\underline{38a}$). These methods are particularly
useful in that they tolerate the presence of functional groups
such as acetoxy, chloro, and cyano groups.

Similar methods have been developed for the synthesis of monoalkylboranes which react with alkenes to give the corresponding mixed trialkylboranes in yields exceeding 70% (38b). The most useful method involves the hydroboration of alkenes with benzo-1,3-dioxa-2-borole (3) to give the corresponding B-alkyl derivatives (47a) which, on reduction with aluminum hydride, give the corresponding monoalkylboranes. Addition of an alkene gives the mixed trialkylborane; alternatively, treatment of the monoalkylborane with pyridine gives a stable complex from which the monoalkylborane may be regenerated upon treatment with boron trifluoride etherate (38b).

$$3 \quad \underset{(3)}{\text{(benzodioxaborole–BH)}} \quad \xrightarrow[100^\circ]{\text{Alkene 1}} \quad 3 \quad \text{(benzodioxaborole–BR)}$$

$$\xrightarrow[\substack{C_5H_{12} \\ 0^\circ}]{2\text{AlH}_3} \quad 3\text{RBH}_2 \quad \xrightarrow{\text{Alkene 2}} \quad 3\text{RBR}_2' \quad >70\%$$

$$3\text{RBH}_2 \xrightarrow{\text{Py}} 3\text{RBH}_2{:}\text{Py} \xrightarrow[\text{BF}_3{:}\text{Et}_2\text{O}]{\text{Alkene 2}} >70\%$$

As before, this method tolerates the presence of a variety of functional groups in the reactants. In addition, two different alkyl groups may be introduced by carrying out the hydroboration in two successive stages.

$$3 \quad \text{(benzodioxaborole–BR)} \quad \xrightarrow[\text{Pentane}]{2\text{AlH}_3} \quad 3\text{RBH}_2 \quad \xrightarrow[\text{2) Alkene 2}]{\text{1) Alkene 1}} \quad 3\text{RBR}^1\text{R}^2$$

Organoboranes containing tertiary alkyl groups may be synthesized by the reaction of B-methoxydialkylboranes with tertiary alkyllithium reagents (38c).

$$R_2BOCH_3 + Me_3CLi \longrightarrow R_2BCMe_3$$

Alternatively, B-tert-alkylbenzo-1,3-dioxa-2-boroles, prepared by reaction of tert-alkyldimethoxyboranes with an equimolar quantity of o-dihydroxybenzene ($\underline{38d}$), may be utilized as discussed above.

Monocycloalkylthexylboranes, formed by reaction of thexylborane (Sec. 2.2.1.c) with cycloalkenes, react with hindered alkenes with the displacement of 2,3-dimethyl-2-butene to give dialkylcycloalkylboranes ($\underline{39a}$).

The extent of displacement of 2,3-dimethyl-2-butene, and hence the efficiency of the synthesis of the dialkylcycloalkylboranes, depends on the degree of hindrance of the cycloalkene and the terminal alkene ($\underline{39a}$).

Reaction of trimethylamine tert-butylborane with terminal alkenes gives the corresponding tert-butyldialkylboranes in 35-90% yields ($\underline{39b}$), while B-alkylboracyclohexanes are prepared by the reaction of bisboracyclohexane with alkenes (Sec. 6.2). The synthesis of various B-alkylboracycloalkanes has been discussed in Sec. 2.2.3.

2.2.5 Synthesis of Halo- and Hydroxyboranes and Their Derivatives

The use of halo- and hydroxyborane derivatives in the synthesis of mixed organoboranes has been discussed in Sec. 2.2.4.

A detailed discussion of the synthesis of these derivatives is beyond the scope of this book, but relevant references to their synthesis are given in this section.

The synthesis of a variety of haloorganoborane derivatives has been reviewed (40). More recent methods of synthesis are illustrated by the following equations.

$$R_3B + C_2H_5SH \xrightarrow[(-RH)]{\Delta} R_2BSC_2H_5 \xrightarrow{NH_3} R_2BNH_2 \xrightarrow{PCl_5} R_2BCl \qquad \text{Ref. } \underline{41a}$$
$$\phantom{R_3B + C_2H_5SH \xrightarrow[(-RH)]{\Delta} R_2BSC_2H_5} \sim 90\% \qquad \sim 100\%$$

$$(C_4H_9)_3B + (CH_3)_2NCl \xrightarrow{\text{Galvinoxyl}} (C_4H_9)_2BCl + C_4H_9N(CH_3)_2 \text{ Ref. } \underline{41b}$$

$$LiBH_4 + BCl_3 \xrightarrow{Et_2O, 0^\circ} 2BH_2Cl:OEt_2 \xrightarrow[0^\circ,\ 1\ hr]{\text{Alkene}} R_2BCl \qquad \text{Ref. } \underline{41c}$$
$$\phantom{LiBH_4 + BCl_3 \xrightarrow{Et_2O, 0^\circ} 2BH_2Cl:OEt_2} >80\%$$

$$(C_2H_5)_3B \xrightarrow[CS_2,\ 25^\circ]{Br_2, AlBr_3} \underset{Br\ \searrow_{AlBr_3}}{CH_3\text{-}CH\text{-}\overset{\overset{C_2H_5}{|}}{B}\text{-}C_2H_5} \longrightarrow \underset{Br}{CH_3\text{-}CH\text{-}\overset{\overset{C_2H_5}{|}}{B}\text{-}C_2H_5} \text{ Ref. } \underline{41d}$$
$$\sim 90\%$$

$$2R_3B + BX_3 \xrightarrow{Et_2BH,\ r.t.} 3R_2BX \quad (X=F, Cl, Br) \qquad\qquad \text{Ref. } \underline{42a}$$
$$50\text{-}100\%$$

$$R_3B + 2BX_3 \xrightarrow{Et_2BH,\ r.t.} 3RBX_2 \quad (X=F, Cl, Br) \qquad\qquad \text{Ref. } \underline{42a}$$
$$30\text{-}100\%$$

$$R_4Sn + 2BX_3 \longrightarrow 2R_2BX + SnX_4 \quad (X=Br, I) \qquad\qquad \text{Ref. } \underline{42b}$$
$$60\text{-}90\%$$

$$R_4Sn + 3BX_3 \longrightarrow 3RBX_2 + RSnX_3 \quad (X=Br, I) \qquad\qquad \text{Ref. } \underline{42b}$$
$$>60\%$$

$$ArI + BI_3 \xrightarrow{\Delta} \underset{>80\%}{ArBI_2 + I_2} \qquad\qquad \text{Ref. } \underline{43}$$

The synthesis and reactions of mono- and dihydroxyboranes (borinic and boronic acids, respectively) and their derivatives have been reviewed ($\underline{44}$). More recent syntheses of mono- and dihydroxyboranes involve the redistribution of trialkylboranes with arylborates ($\underline{45}$) and trimethyleneborate (46), respectively (Sec. 2.3.1).

$$2R_3B + (ArO)_3B \xrightarrow[100^\circ]{BH_3-THF} 3R_2BOAr \xrightarrow{H_2O} 3R_2BOH \qquad \text{Ref. } \underline{45}$$

$$R_3B + \underset{}{\text{(cyclic borate)}}\; B\text{-}OCH_2)_2CH_2 \xrightarrow[120^\circ]{BH_3-THF} \underset{80\text{-}90\%}{3R\text{-}B} \xrightarrow[\Delta]{H_2O} \underset{90\%}{3RB(OH)_2}\; \text{Ref. } \underline{46}$$

Dihydroxyboranes are also readily prepared via the hydroboration of alkenes using benzo-1,3-dioxa-2-borole (3), which is prepared by reaction of borane and o-dihydroxybenzene at 0° ($\underline{47a}$).

$$\underset{(3)}{\text{(benzodioxaborole BH)}} \xrightarrow[100^\circ]{\text{Alkene}} \text{(benzodioxaborole BR)} \xrightarrow{H_2O} \underset{80\text{-}95\%}{RB(OH)_2}$$

Highly substituted tertiary alkyldihydroxyboranes may be prepared from dialkylhydroxyboranes containing at least one alkyl group with a tertiary hydrogen atom α to the boron atom by photochemical bromination in the presence of water ($\underline{47b}$).

$$R^1R^2CHBR^3 \underset{\substack{CH_2Cl_2 \\ H_2O}}{\overset{Br_2,\ h\nu}{\xrightarrow{\hspace{1.5cm}}}} \left[R^1R^2\overset{R^3}{\underset{\underset{Br}{\overset{|}{\nwarrow}}}{C\text{-}B\text{-}OH}} \;\; :OH_2 \right] \xrightarrow{\hspace{1.5cm}} R^1R^2R^3CB(OH)_2$$

Dihydroxyboranes and their derivatives containing alkyl, vinyl, alkynyl, and aromatic substituents may be prepared by the reaction of Grignard reagents with trimethoxyborane (48). Mono- and dihydroxyboranes and their derivatives may also be synthesized from the corresponding chloroboranes by treatment with water or alcohols (41c).

2.3 SOME PROPERTIES OF ORGANOBORANES

Only those properties which influence the synthesis of organoboranes are discussed in the following sections.

2.3.1 Redistribution Reactions of Organoboranes

Redistribution reactions involving boron compounds have been reviewed (49), and in this section only those redistribution reactions involving the making and breaking of boron-carbon bonds are discussed.

Trialkyl- and triarylboranes are generally stable and do not undergo exchange reactions at room temperature. However, in the presence of catalytic amounts of $>$B-H bonds (e.g., diborane and mono- or dialkylboranes) redistribution occurs to give alkyldiboranes (50).

$$R_3B + B_2H_6 \longrightarrow RHBH_2BH_2 + R_2BH_2BH_2 + R_2BH_2BHR + R_2BH_2BR_2$$

$$(R=CH_3, C_2H_5, C_3H_7)$$

B-Alkylboracyclopentanes undergo ring cleavage with borane to give the corresponding 1-alkyl-1,2-tetramethylenediborane (Sec. 6.1), but treatment of B-alkylboracyclohexanes with borane results in migration of the alkyl group to the borane moiety rather than ring opening (51).

Compounds containing B-H bonds also catalyze mutual ligand exchange between different boron compounds, such as trialkyl- and triarylboranes. The redistribution reaction proceeds via an unsymmetrical bridged species which breaks down into different alkyl- or arylboranes ($\underline{5a}, \underline{49a}$).

$$R_3B + H-B \rightleftharpoons \overset{R}{\underset{R'}{\diagup}} H \overset{R}{\underset{H}{\diagdown}} B \rightleftharpoons B-R + R_2BH \overset{R_3'B}{\rightleftharpoons} \underset{R \ R' \ R'}{\overset{R \ H \ R'}{B \ B}}$$

$$\rightleftharpoons R_2BR' + R_2'BH \quad \text{etc.}$$

The composition of the trialkylboranes formed depends on the rate of ligand exchange. Alkyl groups in which a secondary carbon is directly attached to boron (e.g., isopropyl) are transferred at a much slower rate than primary alkyl groups, probably due to steric hinderance($\underline{5a}, \underline{49a}$).

Trialkylaluminums undergo alkyl exchange with trialkylboranes, probably via dimeric intermediates involving aluminum-boron alkyl bridges ($\underline{49a}$).

$$R_3B + R_3'Al \rightleftharpoons R_2B \overset{R}{\underset{R'}{\diamondsuit}} AlR_2' \rightleftharpoons R_2BR' + R_2'AlR, \text{ etc.}$$

Trialkylaluminums are thus extremely effective catalysts for promoting alkyl exchange between different trialkylboranes

$$R_3B + R_3'B \overset{R_3''Al}{\rightleftharpoons} R_2BR' + R_2'BR, \text{ etc.}$$

Trialkylaluminums are useful catalysts for promoting the disproportionation of mixed dialkylalkenylboranes which are formed from the partial hydroboration of dienes using diethylborane (Sec. 6.5). The products are triethylborane and the trialkenyl-boranes ($\underline{15}, \underline{49a}$).

Trialkylboranes and boron trihalides react rapidly at temperatures exceeding 100° to give dialkylhaloboranes ($\underline{40}$). In the presence of $\rangle$B-H bonds, however, the exchange reaction proceeds at room temperature to give mixtures of dialkylhalo- and alkyldihaloboranes ($\underline{42a}$).

$$R_3B + BX_3 \xrightarrow[\text{r.t.}]{\text{Et}_2\text{BH}} RBX_2 + R_2BX$$

Trialkylboranes react with trimethoxyborane in the presence of borane above 100° to give B-methoxydialkylboranes ($\underline{52}$). This reaction only proceeds satisfactorily for tri-n-alkylboranes ($\underline{45}$); use of triaryloxyboranes, such as tri-o-tolyloxyborane, results in much more rapid redistribution and gives high yields of the corresponding B-aryloxydialkylboranes, which are readily converted into B-methoxy derivatives ($\underline{45}$).

$$2R_3B + B(OC_6H_4\text{-oCH}_3)_3 \xrightarrow[100^\circ]{\text{BH}_3\text{-THF}} 3R_2BOC_6H_4\text{-oCH}_3$$

2.3.2 Isomerization Reactions of Organoboranes

While trialkylboranes are usually stable at room temperature they isomerize at temperatures exceeding 100° to give mixtures of trialkylboranes containing primary and secondary alkyl groups attached to boron ($\underline{5a}$). Boron-primary carbon bonds are strongly preferred to boron-secondary carbon bonds, while boron-tertiary carbon bonds are usually not detected as illustrated below ($\underline{5a,53}$).

$$(\text{i-C}_3H_7)_3B \xrightarrow[24 \text{ hr}]{160^\circ} (\text{n-C}_3H_7)_2B \text{ i-C}_3H_7 + (\text{n-C}_3H_7)_3B + \text{n-C}_3H_7B(\text{i-C}_3H_7)_2$$

$$\qquad\qquad\qquad\quad (16\%) \qquad\qquad\qquad (83\%) \qquad\qquad\quad (1\%)$$

$$(\text{s-C}_4H_9)_3B \xrightarrow[7 \text{ hr}]{190^\circ} (\text{n-C}_4H_9)_3B + (\text{n-C}_4H_9)_2B \text{ s-C}_4H_9$$

$$\qquad\qquad\qquad\qquad (82\%) \qquad\qquad (18\%)$$

The above isomerizations proceed at a slow rate, but the reaction is markedly catalyzed by a small amount of borane or

other compounds containing B-H bonds (54). Thus, hydroboration of
2-hexene with a 20% excess of hydride in diglyme, followed by
heating at 150° for one hour and oxidation, gives mainly l-hexanol.

$$C_3H_7CH=CHCH_3 \xrightarrow[\substack{2) \ 150°, \ 1 \ hr \\ 3) \ [O]}]{1) \ BH_3, \ DG} C_6H_{13}OH + C_4H_9\underset{\underset{OH}{|}}{C}HCH_3 + C_3H_7\underset{\underset{OH}{|}}{C}HC_2H_5$$

88%   7%   5%

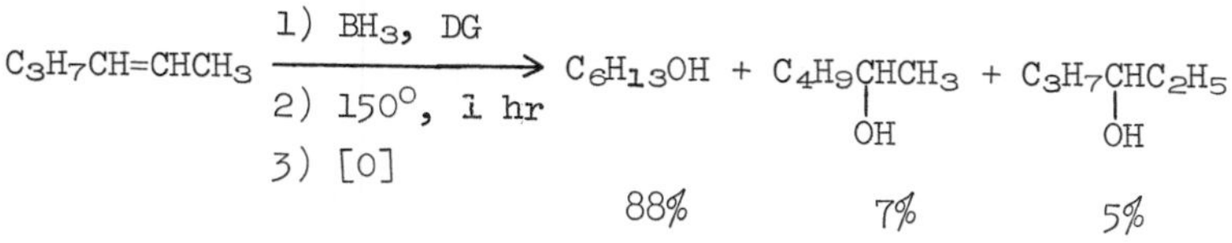

The catalytic effect of the excess borane is illustrated by the
fact that use of a slight excess of 2-hexene in the above reaction
sequence gives 1-, 2-, and 3-hexanol in yields of 13%, 64%, and
23%, respectively (54).

   When the reaction sequence using a slight excess of hydride
is carried out with l-hexene, 2% of 3-hexanol is obtained in
addition to l- and 2-hexanol (94 and 4%, respectively) (54).
This indicates that isomerization results in an equilibrium
distribution of boron at all positions in the chain, with a
strong preference for the terminal position. In the case of
branched-chain alkenes, the boron atom readily isomerizes past
a single branch, but no isomerization occurs past a double branch
(54).

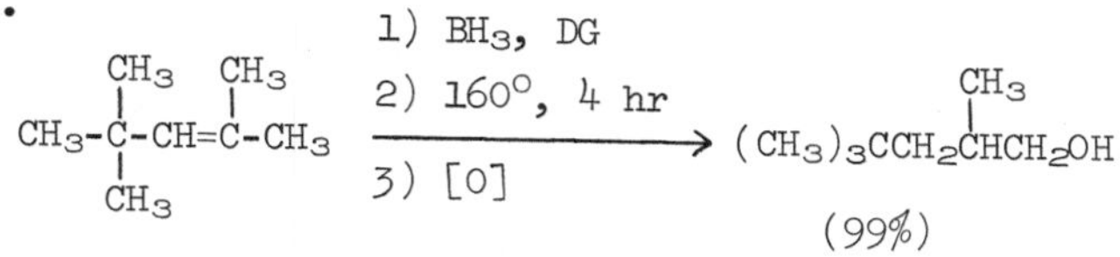

$$CH_3\text{-}\underset{\underset{CH_3}{|}}{\overset{\overset{CH_3}{|}}{C}}\text{-}CH=\overset{\overset{CH_3}{|}}{C}\text{-}CH_3 \xrightarrow[\substack{2) \ 160°, \ 4 \ hr \\ 3) \ [O]}]{1) \ BH_3, \ DG} (CH_3)_3CCH_2\overset{\overset{CH_3}{|}}{C}HCH_2OH$$

(99%)

When two possible primary positions are available in the chain,
preference is shown for the formation of the least hindered
primary alcohol (54).

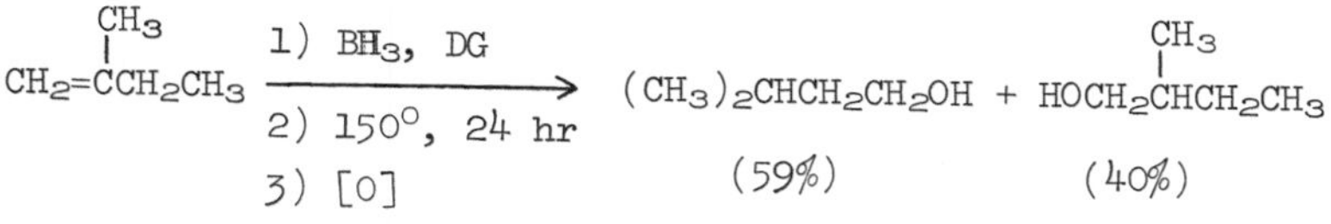

$$CH_2=\overset{\overset{CH_3}{|}}{C}CH_2CH_3 \xrightarrow[\substack{2) \ 150°, \ 24 \ hr \\ 3) \ [O]}]{1) \ BH_3, \ DG} (CH_3)_2CHCH_2CH_2OH + HOCH_2\overset{\overset{CH_3}{|}}{C}HCH_2CH_3$$

(59%)   (40%)

   Isomerization thus achieves a thermodynamic equilibrium
between all the alkylboron moieties with the major constituent

being that containing the boron atom in the sterically least
hindered position. The isomerization proceeds by a mechanism
involving rapid cis-eliminations followed by re-additions (54).
The presence of excess hydride results in the formation of a
small quantity of dialkylborane (or its dimer) which facilitates
the elimination of a dialkylborane moeity from the trialkylborane
to give an alkene and a tetraalkyldiborane (Sec. 3.4).

$$R'CH\text{-}CH_2CH_3 \rightleftarrows R'\text{-}\overset{H}{\underset{B}{C}}\cdots\overset{H}{\underset{B}{C}}\text{-}CH_3 \rightleftarrows R'CH=CHCH_3 + R_2BH_2BR_2$$

$$\rightleftarrows R'CH_2\overset{B}{C}HCH_3 \rightleftarrows R'CH_2CH=CH_2 \rightleftarrows R'CH_2CH_2CH_2BR_2 + R_2BH$$
$$+ R_2BH_2BR_2$$

The above mechanism is consistent with the inhibiting effect of
excess alkene due to its reaction with any catalytically active
dialkylborane species. On the other hand, in the case of alkenes
which are only hydroborated to the mono- or dialkylborane stage,
the presence of excess alkene fails to inhibit isomerization (54).

Application of the hydroboration-isomerization reaction
sequence to cyclic alkenes containing alkyl side chains results
in migration of the boron atom to the primary position of the
alkyl substituent (55). The extent of migration depends on the
ring size and the nature of the side chain.

$$\underset{(CH_2)_n}{\bigcirc}\hspace{-1em} \xrightarrow[\substack{2)\ 160°,\ 2\ hr \\ 3)\ [O]}]{1)\ BH_3,\ DG} \underset{(CH_2)_n}{\bigcirc}\hspace{-1em}^{OH} + \text{ secondary isomers}$$

n = 1	4%	18% (+ ring
n = 2	25%	60% opened
n = 3	45%	38% products)
n = 4	45%	36%

Similar yields of the primary derivatives are obtained
irrespective of the position of the endocyclic double bond (55).
Similar results have been obtained with terpenes such as
(+)-thujene (56) and (+)-3-carene (57) [(6) and (8) respectively;
Sect. 3.1.6.a], while the steroid boranes derived from
5α-cholest-2-ene and cholest-5-ene isomerize to give mainly the
equatorial 3β- and 2α-derivatives (58).

Hydroboration of exocyclic methylene compounds usually
proceeds from the less hindered direction to give the less stable
axial or endo derivatives (Sec. 3.1.2). Isomerization converts
these derivatives to the more stable equatorial or exo derivatives
(55).

Optically active organoboranes having a tertiary carbon atom
β with respect to the boron atom undergo racemization on heating.
Thus, tris [(R)-2-methylbutyl]borane slowly racemizes at 100° with
partial isomerization of the alkyl group (59). However, when the
asymmetric carbon atom is further removed from the boron atom no
racemization occurs (59).

In the case of cyclic organoboranes the stability of the
compounds depends on the ring size as well as the types of boron-

carbon bonds involved (5a,15). In general, six-membered
boracyclanes containing two boron-primary carbon bonds or one
boron-primary carbon bond and one boron-secondary carbon bond
are thermodynamically most stable as illustrated below (5a,15).

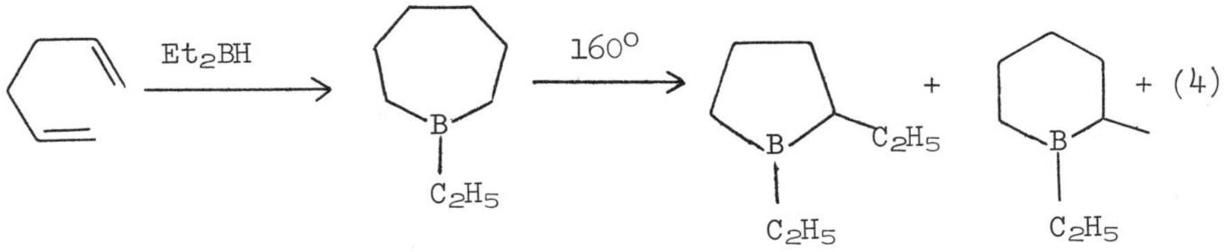

$$\text{(4) (97\%)} \qquad \text{(4\%)} \qquad \text{(80\%) (13\%)}$$

However, six-membered rings are not favored if a boron-tertiary
carbon bond is present (5a,15).

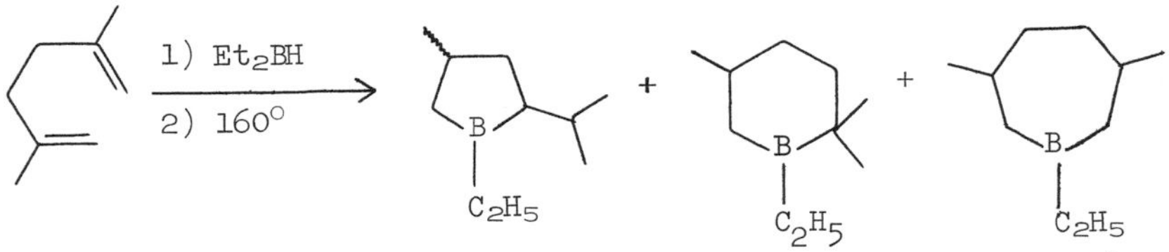

$$\text{(66\%)} \qquad \text{(23\%)} \qquad \text{(11\%)}$$

The same considerations apply to bis-B-boracyclanes derived from
the hydroboration of dienes with diborane (Sec. 6.2), and to
borabicyclic systems (Sec. 6.4). As with the isomerization of
acyclic organoboranes discussed earlier, isomerization of
boracyclanes also proceeds via a process of elimination-
rehydroboration (5a).

Finally, it must be noted that, while isomerizations usually
occur at temperatures exceeding 100°, isomerization of alkylboranes
has been observed in the hydroboration of alkenes at room
temperature. Examples of such isomerizations have been reported
in the hydroboration of steroid cis-propylidene derivatives
[(3; R=CH$_3$); Sec. 3.1.4], α-gurjunene [(17); Sec. 3.1.6.a], and
tricyclic tetrasubstituted alkenes [(31); Sec. 3.1.7].

The hydroboration-isomerization reaction sequence is useful in the synthesis of various functional derivatives such as <u>primary alcohols</u> (via oxidation; Secs. 4.3 and 4.4), <u>primary amines</u> (via treatment with hydroxylamine-o-sulfonic acid; Sec. 9.2), and in the <u>contrathermodynamic isomerization of alkenes</u> (via displacement; Sec. 2.3.3).

2.3.3 Displacement Reactions of Organoboranes

The displacement reactions of organoboranes with alkenes have been reviewed (<u>60</u>). Organoboranes react with less volatile alkenes to liberate more volatile alkenes and give the organoboranes of the less volatile alkenes. Thus, the displacement reaction has been used in the synthesis of organoboranes by the reaction of alkenes with lower trialkylboranes, such as triisobutylborane (<u>61</u>).

$$3C_8H_{17}CH=CH_2 \xrightarrow[170°]{(i-C_4H_9)_3B} (C_8H_{17}CH_2CH_2)_3B + 3(CH_3)_2C=CH_2$$

Triisobutylborane has also been used in the synthesis of
trivinylboranes from disubstituted alkynes (Sec. 7.1.1).

The scope of the displacement reaction, using refluxing
1-decene as the displacing alkene, has been studied with a variety
of trialkylboranes (62,63a). The displacement occurs with no
isomerization of the alkenes or organoboranes, thereby supporting
the mechanism proposed for isomerization outlined in Sec. 2.3.2,
and the rate of displacement parallels the thermodynamic stability
of the alkene liberated. Thus, more stable alkenes are more
readily displaced. Conversely, less stable alkenes such as
terminal alkenes (e.g., 1-decene) are the most efficient
displacing alkenes (62).

The mechanism of the above displacement reaction involves
the dissociation of the trialkylborane at high temperature to
give a small equilibrium concentration of alkene and dialkylborane.
The dialkylborane reacts with the large excess of displacing
alkene, while distillation of the more volatile displaced alkene
ensures completion of the reaction (63a).

Organoboranes also undergo displacement reactions on treatment
with 2-methyl-2-nitrosopropane (63b).

$$R_2B-\overset{|}{\underset{|}{C}}-\overset{|}{\underset{|}{C}}- + (CH_3)_3CNO \xrightarrow[25°]{PhH}$$

The reaction is rapid and quantitative and involves stereospecific
cis-elimination of the alkene. However, only one alkyl group is
readily eliminated, probably due to the decreased Lewis acidity
of the boron atom in the O-dialkylborylhydroxylamine formed in
the reaction (63b).

2.4 SPECTROSCOPIC PROPERTIES OF ORGANOBORANES

In the following section only the use of infrared,
ultraviolet, and nuclear magnetic resonance spectroscopy as
applied mainly to alkylboranes is discussed.

2.4.1 Infrared Spectra of Alkyldiboranes

The infrared spectra of various methyl- and ethyldiboranes have been extensively studied (64). The most characteristic feature of the spectra of all diboranes is a very intense band in the region 1610-1500 cm^{-1} due to asymmetric in-phase stretching of the BH_2B bridge. This band is located in the region 1610-1580 cm^{-1} for all alkyldiboranes except 1,1-dialkyldiboranes, where it occurs in the region 1555-1540 cm^{-1}.

Further characteristic bands occur in the region 2600-2500 cm^{-1} due to terminal B-H stretching. These bands are of medium to strong intensity and are useful in distinguishing between different alkyldiboranes. Compounds containing a BH_2 group show a double absorption due to in-phase and out-of-phase stretching. For diborane itself these bands occur at 2612 and 2525 cm^{-1} and are lowered by 10-15 cm^{-1} for each alkyl substituent on the remote boron atom. Substitution of an alkyl group on the same boron atom as a hydrogen lowers the frequency of the B-H stretching band by about 25 cm^{-1}.

Some characteristic bands are listed in Table 2.2.

TABLE 2.2

Infrared Absorption Bands of Alkyldiboranes

Group	Mode	Position (cm^{-1})	Appearance
BH_2B	B-H bridge stretching	1600-1500	Strong
BH_2	B-H stretching	2600-2500	Doublet; medium-strong
BHR	B-H stretching	2600-2500	Singlet; medium-strong
BH	B-H stretching	2470	Singlet
BH_2	BH_2 scissoring	1170-1140	Medium
BHR	B-H in-plane bending	1180-1110	Medium
BH_2	BH_2 out-of-plane wagging	±930	Medium

2.4.2 Ultraviolet Spectra of Organoboranes

Trialkylboranes do not exhibit ultraviolet absorption above
225 nm, while triarylboranes show broad intense absorption maxima
between 280-360 nm ($\epsilon > 10^4$) (65). Vinylboranes exhibit absorption
maxima in the region 185-234 nm (65).

2.4.3 Nuclear Magnetic Resonance Spectra of Organoboranes

The application of NMR spectroscopy to boron compounds has
been reviewed (66), and data on a large number of boron hydrides
and related compounds has been recorded (67).

Naturally occurring boron contains approximately 80% ^{11}B of
spin 3/2 and 20% ^{10}B of spin 3. Thus, the ^{1}H NMR spectrum of
^{11}B-H shows a quartet of peaks of equal intensity, while ^{10}B-H
shows a seven line pattern. However, the intensity of the latter
lines is about 0.14 that of the former, and hence the ^{1}H spectrum
of B-H has a broad septet buried beneath a more intense quartet
(66). The ^{11}B spectrum of B-H exhibits two peaks of equal
intensity.

In practice the ^{11}B and ^{1}H resonances of boron compounds in
liquids have a band width of 30-60 Hz, and are hence poorly
resolved compared to the ^{1}H spectra of carbon compounds. The
spectra are usually recorded as neat liquids or as solutions in
solvents, such as ethers (67). In ^{11}B spectra, boron trifluoride
etherate is used as an external standard, while tetramethylsilane
is used in ^{1}H spectra.

The ^{11}B chemical shifts of boron hydrides vary over a range
of approximately +50 to -50 ppm relative to boron trifluoride
etherate, while the ^{1}H chemical shifts range over a few ppm about
tetramethylsilane. The coupling constants between ^{11}B and
terminal hydrogens (^{11}B-H$_t$) are 120-150 Hz, while for bridge
hydrogens (^{11}B-H$_b$) coupling constants are usually 40-50 Hz. Thus,
the ^{11}B spectrum of diborane shows a triplet of triplets centered
at -17 to -18 ppm with J_{B-H_t}=135-137 Hz and J_{B-H_b}=46-48 Hz.

The ^{11}B and ^{1}H spectra of various alkyldiboranes have been recorded, and the chemical shift values and coupling constants of the terminal and bridge hydrogens have been shown to be dependent on the nature of the alkyl substitution ($\underline{68}$). In general, monosubstitution and 1,2-disubstitution of terminal hydrogens causes a downfield shift of the resonance of the remaining geminal terminal hydrogen, while, in the case of monosubstitution, the remaining BH_2 group is shifted to higher field; 1,1-disubstitution causes an even greater upfield shift of the remaining BH_2 group. The resonances of the bridge hydrogens and of ^{11}B are generally shifted to lower field.

The ^{11}B chemical shifts of a large number of organoboranes not containing B-H bonds have been recorded ($\underline{69}$).

REFERENCES

1. M. F. Hawthorne, The Chemistry of Boron and its Compounds (E. L. Muetterties, ed.), Wiley, New York, 1967, Chap. 5.

2. (a) D. E. Young and S. G. Shore, J. Am. Chem. Soc., 91, 3497 (1969); and references cited therein; (b) B. Rice, J. A. Livasy, and G. W. Schaeffer, ibid., 77, 2750 (1955).

3. (a) O. P. Shitov, S. L. Ioffe, V. A. Tartakovskii, and S. S. Novikov, Russ. Chem. Revs., 1970, 905; (b) J. Klein, personal communication; (c) D. F. Gaines, Inorg. Chem., 2, 523 (1963).

4. (a) H. C. Brown and G. J. Klender, Inorg. Chem., 1, 204 (1962); (b) E. Negishi, J. Katz, and H. C. Brown, J. Am. Chem. Soc., 94, 4025 (1972).

5. (a) R. Köster, Angew. Chem. Intern. Ed., 3, 174 (1964); (b) R. L. Hughes, I. C. Smith, and E. W. Lawless, Production of Boranes and Related Research, (R. T. Holzmann, ed.), Academic, New York, 1967, pp. 329-331.

6. H. C. Brown, K. J. Murray, L. J. Murray, J. A. Snover, and G. Zweifel, J. Am. Chem. Soc., 82, 4233 (1960).

7. G. Zweifel and H. C. Brown, Org. Reactions, 13, 1 (1963).

8. G. F. Freeguard and L. H. Long, Chem. & Ind. (London), 1965, 471.

9. K. M. Biswas and A. H. Jackson, J. Chem. Soc., C1970, 1667.

10. M. Nussim, Y. Mazur, and F. Sondheimer, J. Org. Chem., 29, 1120 (1964).

11. P. Kohn, R. H. Samaritano, and L. M. Lerner, J. Am. Chem. Soc., 87, 5475 (1965).

12. (a) H. C. Brown, P. Heim, and N. M. Yoon, J. Am. Chem. Soc., 92, 1637 (1970); (b) L. F. Fieser and M. Fieser, Reagents for Organic Synthesis, Wiley, New York, 1967.

13. M. D. Taylor, L. R. Grant, and C. A. Sands. J. Am. Chem. Soc., 77, 1506 (1955); F. M. Tayler, Brit. Patent 909, 390, Oct. 31 1962; Chem. Abstr., 58, 6846 (1962).

14. (a) E. C. Ashby, J. Organometal. Chem., 3, 371 (1965); U. S. Patent 3,257,455, June 21, 1966; Chem. Abstr., 65, 12240(g), 1966; (b) L. M. Braun, R. A. Braun, H. R. Crissman, M. Opperman, and R. M. Adams, J. Org. Chem., 36, 2388 (1971).

15. R. Köster, G. Griasnow, W. Larbig, and P. Binger, Annalen, 672, 1 (1964).

16. H. C. Brown and G. Zweifel, J. Am. Chem. Soc., 82, 3222 (1960).

17. G. Zweifel, N. R. Ayyangar, and H. C. Brown, J. Am. Chem. Soc., 85, 2072 (1963).

18. H. C. Brown and E. Negishi, J. Organometal. Chem., 28, C1 (1971).

19. E. F. Knights and H. C. Brown, J. Am. Chem. Soc., 90, 5280 (1968).

20. G. Zweifel and H. C. Brown, J. Am. Chem. Soc., 85, 2066 (1963).

21. H. C. Brown, D. B. Bigley, S. K. Arora, and N. M. Yoon, J. Am. Chem. Soc.. 92, 7161 (1970).

22. (a) G. Zweifel and H. C. Brown, J. Am. Chem. Soc., 86, 393 (1964); (b) H. C. Brown, N. R. Ayyangar, and G. Zweifel, ibid., 86, 397 (1964).

23. (a) H. C. Brown, N. R. Ayyangar, and G. Zweifel, J. Am. Chem. Soc., 86, 1071 (1964); (b) J. Katsuhara, H. Watanabe, K. Hashimoto, and M. Kobayashi, Bull. Chem. Soc. Japan, 39, 617 (1966).

24. P. A. Spanninger and J. L. von Rosenberg, J. Org. Chem., 34, 3658 (1969).

25. R. L. Settine, J. Org. Chem., 35, 4266 (1970).

26. M. F. Lappert, The Chemistry of Boron and Its Compounds, (E. L. Muetterties, ed.), Wiley, New York, 1967 Chap. 8, pp. 529-535.

27. H. C. Brown and V. H. Dodson, J. Am. Chem. Soc., 79, 2302 (1957).

28. C. D. Good and D. M. Ritter, J. Am. Chem. Soc., 84, 1162 (1962).

29. A. V. Topchiev, A. A. Prokhorova, Ya. M. Paushkin, and M. V. Kurashev, Izvest. Akad. Nauk. SSSR, Otdel. Khim. Nauk., 1958, 370; Chem. Abstr., 52, 12752(f) (1958).

30. (a) W. E. Davidsohn and M. C. Henry, Chem. Rev., 67, 73 (1967); J. Soulie and P. Cadiot, Bull Soc. Chim. France, 1966, 1981; (b) S. W. Breuer, M. J. Leatham, and F. G. Thorpe, Chem. Commun., 1971, 1475; (c) S. W. Breuer and F. A. Broster, J. Organometal. Chem., 35, C5 (1972); (d) A. G. Davies, B. P. Roberts, and R. Tudor, J. Organometal. Chem., 31, 137 (1971); and references cited therein.

31. R. Köster, Advan. Organometal. Chem., 2, 257 (1964).

32. P. M. Maitlis, Chem. Rev., 62, 223 (1962).

33. P. Jutzi, J. Organometal. Chem., 19, P1 (1969).

34. R. Köster, W. Larbig, and G. Rotermund, Annalen, 682, 21 (1965).

35. H. C. Brown, K. J. Murray, H. Muller, and G. Zweifel, J. Am. Chem. Soc., 88, 1443 (1966).

36. G. M. Clark, K. G. Hancock, and G. Zweifel, J. Am. Chem.
Soc., 93, 1308 (1971).

37. (a) H. C. Brown, E. Negishi and S. K. Gupta, J. Am. Chem.
Soc., 92, 6648 (1970); (b) H. C. Brown and S. K. Gupta, J.
Organometal. Chem., 32, C1 (1971)

38. (a) H. C. Brown and S. K. Gupta, J. Am. Chem. Soc., 93,
1818 (1971); (b) H. C. Brown and S. K. Gupta, ibid., 93, 4062
(1971); (c) H. C. Brown and E. Negishi, ibid., 93, 3777 (1971);
(d) E. Negishi and H. C. Brown, Synthesis, 1972, 197.

39. (a) C. F. Lane and H. C. Brown, J. Organometal. Chem., 34,
C29 (1972); (b) M. F. Hawthorne, J. Am. Chem. Soc., 83, 2541
(1961).

40. K. Niedenzu, Organometal. Chem. Rev., 1, 305 (1966).

41. (a) L. F. Hohnstedt, J. P. Brennan, and K. A. Reynard,
J. Chem. Soc., A1970, 2455; (b) A. G. Davies, S. C. W. Hook,
and B. P. Roberts, J. Organometal. Chem., 23, C11 (1970); (c)
H. C. Brown and N. Ravindran, J. Am. Chem. Soc., 94, 2112 (1972);
(d) H. C. Brown and Y. Yamamoto, Chem. Comm., 1972, 71.

42. (a) R. Köster and M. Grassberger, Annalen., 719, 169 (1968);
(b) H. Noth and H. Vahrenkamp, J. Organometal. Chem., 11, 399
(1968).

43. W. Siebert, F. R. Rittig and M. Schmidt, J. Organometal.
Chem., 25, 305 (1970); W. Siebert, Chem. Ber., 103, 2308 (1970).

44. K. Torssel, Progress in Boron Chemistry, (H. Steinberg and
A. L. McCloskey, eds.), Vol. 1, Pergamon, Oxford, 1964, Chap. 9.

45. H. C. Brown and S. K. Gupta, J. Am. Chem. Soc., 93, 2802
(1971).

46. H. C. Brown and S. K. Gupta, J. Am. Chem. Soc., 92, 6983
(1970).

47. (a) H. C. Brown and S. K. Gupta, J. Am. Chem. Soc., 93,
1816 (1971); (b) H. C. Brown and S. K. Gupta, ibid., 94, 4370
(1972).

48. M. J. Blais, J. Soullie, and M. P. Cadiot, Compte Rend. Acad.
Sci. Paris, Ser. C, 271, 589 (1970); E. Favre and M. P. Gaudemar,

Bull. Soc. Chim. France, 1968, 3724; D. J. Pasto, J. Chow, and
S. K. Arora, Tetrahedron, 25, 1557 (1969); I. G. C. Coutts, H. R.
Goldschmid, and O. C. Musgrave, J. Chem. Soc., C1970, 488;
D. S. Matteson, Organometal. Chem. Rev., 1, 1 (1966).

49. (a) R. Köster, Ann. N. Y. Acad. Sci., 159, 73 (1969); (b)
Reference 26, pp. 556-560.

50. H. I. Schlesinger, L. Horvitz, and A. P. Burg, J. Am. Chem.
Soc., 58, 407 (1936).

51. H. C. Brown, E. Negishi, and S. K. Gupta, J. Am. Chem. Soc.,
92, 6649 (1970).

52. B. M. Mikhailov and L. S. Vasilev, Zh. Obshch. Khim., 35,
925 (1965); Chem. Abstr., 63, 7028(e) (1965).

53. P. A. McCusker, F. M. Rossi, J. H. Bright, and G. F. Hennion,
J. Org. Chem., 28, 2889 (1963).

54. H. C. Brown and G. Zweifel, J. Am. Chem. Soc., 88, 1433
(1966).

55. H. C. Brown and G. Zweifel, J. Am. Chem. Soc., 89, 561 (1967).

56. S. P. Acharya, H. C. Brown, A. Suzuki, S. Nozawa, and M. Itoh,
J. Org. Chem., 34, 3015 (1969).

57. W. Cocker, P. V. R. Shannon, and P. A. Staniland, J. Chem.
Soc., C1967, 915.

58. J. E. Herz and L. A. Marquez, J. Chem. Soc., C1969, 2243.

59. L. Lardicci, L. Lucarini, P. Palagi, and P. Pino, J.
Organometal. Chem., 4, 341 (1965).

60. Reference 26; pp. 535-536.

61. R. Köster, Annalen, 618, 31 (1958); U. S. Patent 2,886,599,
May 12, 1959; Chem. Abstr., 53, 19880(b) (1959); H. C. Brown, Ger.
Patent 1, 125,923, Mar. 22, 1962; Chem. Abstr., 57, 9880(g)
(1962); U.S. Patent 3,078,308, Feb. 19, 1963; Chem. Abstr., 58,
13991(a) (1963).

62. H. C. Brown and M. V. Bhatt, J. Am. Chem. Soc., 88, 1440
(1966).

63. (a) H. C. Brown, M. V. Bhatt, T. Munekata, and G. Zweifel,
J. Am. Chem. Soc., 89, 567 (1967); (b) K. G. Foot and B. P.
Roberts, J. Chem. Soc., A1971, 3475.

64. W. J. Lehmann and I. Shapiro, Spectrochim. Acta., 17, 396 (1961).

65. B. G. Ramsey, Electronic Transitions in Organometalloids, Academic, New York, 1969.

66. G. R. Eaton, J. Chem. Educ., 46, 547 (1969); R. Schaeffer, Chap. 10, Ref. 44.

67. G. R. Eaton and W. N. Lipscomb, NMR Studies of Boron Hydrides and Related Compounds, Benjamin, New York, 1969.

68. H. H. Lindner and T. Onak, J. Am. Chem. Soc., 88, 1890 (1966).

69. H. Nöth and H. Vahrenkamp, Chem. Ber., 99, 1049 (1966).

Chapter 3

HYDROBORATION OF ALKENES

Hydroboration constitutes the most important method for the
preparation of organoboranes and has been applied to a wide
variety of functionalized and nonfunctionalized alkenes. The
presence of a functional group in close proximity to the double
bond can have a marked effect on the course of the reaction; the
hydroboration of these alkenes is considered in Chapter 5.

In this chapter the hydroboration of nonfunctionalized
alkenes is discussed, together with those alkenes in which the
functional group is too remote from the double bond to have any
marked influence on the reaction.

3.1 SCOPE OF THE REACTION

By far the most common and useful application of hydroboration in organic synthesis to date has been the anti-Markownikoff
hydration of alkenes, which involves the initial hydroboration of
the alkene followed by oxidation of the organoborane with alkaline
hydrogen peroxide. Since this oxidation procedure gives
essentially quantitative yields of the alcohol and proceeds with
complete retention of configuration (Sec. 4.3), analysis of
the alcohols formed serves to locate the precise position of
the boron atom in the organoboranes, and also establishes the
stereochemistry of the hydroboration reactions. In the following
sections the hydration of the various classes of alkenes using

the above procedure is discussed; such a discussion thus
provides a survey of the scope of the hydroboration reaction.

In general, the hydroboration of alkenes proceeds via the
cis-addition of the boron-hydrogen moiety to the double bond
from the less hindered side, to place the boron atom predominantly
at the less substituted carbon atom. Both the regioselectivity
and stereoselectivity of the reaction are enhanced by use of
selective hydroborating agents (Sec. 2.2.1.c); the application
of these reagents is also included in the following discussion.

Only those reactions which proceed readily at temperatures
below 25°, and in the normal ether solvents such as tetrahydrofuran or diglyme, are considered below.

3.1.1 Terminal Alkenes

Reaction of monosubstituted terminal alkenes ($RCH=CH_2$)
gives mainly the primary alcohol, the regioselectivity being
93-94% (1); the direction of addition is not markedly affected
by branching of the alkyl substituent, R. Use of disiamylborane
(Sia_2BH) (2) or 9-borabicyclo[3.3.1]nonane(9-BBN) (3), however,
increases the regioselectivity to 99%. With vinylcyclopropane
the primary alcohol is formed to the extent of 97% using borane
(4).

$$RCH=CH_2$$
$$(6-7) \quad (88-89)$$
$$(1) \quad (90) \quad (Sia_2BH, \ 9\text{-}BBN)$$

$$-CH=CH_2$$
$$(3) \ (97)$$

In the case of disubstituted terminal alkenes ($RR'C=CH_2$)
almost complete addition of boron to the terminal position
occurs (1).

$$CH_3CH_2-\underset{\underset{CH_3}{|}}{C}=CH_2$$
(94)

$$(CH_3)_3C-\underset{\underset{(CH_3)_3C}{|}}{C}=CH_2$$
60-70
(Ref. 5)

$$C=CH_2$$
CH_3 66
(Ref. 6)

3.1.2 Exocyclic Methylene Compounds

With these compounds attack by boron occurs exclusively at
the terminal position ($\underline{7}$). In the case of substituted methylene
cyclohexanes, the less stable axial hydroxymethyl derivatives
are preferentially formed, because attack from the equatorial
direction is favored ($\underline{7}$).

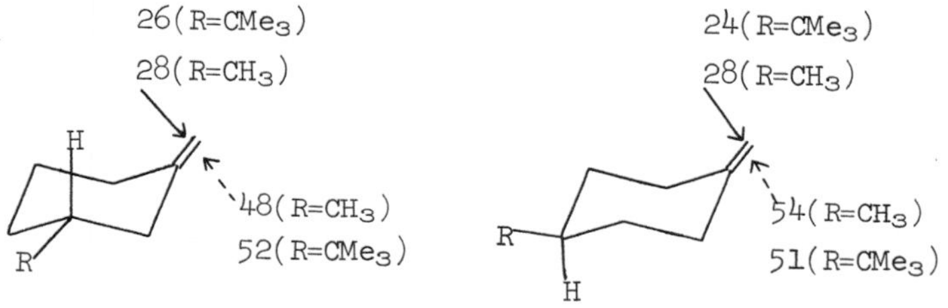

26(R=CMe$_3$)   24(R=CMe$_3$)
28(R=CH$_3$)   28(R=CH$_3$)

48(R=CH$_3$)   54(R=CH$_3$)
52(R=CMe$_3$)   51(R=CMe$_3$)

With 2-methylenenorbornene ($\underline{8}$) and its 3,3-dimethyl
derivative (camphene) ($\underline{9}$), attack occurs preferentially from
the exo direction to give the endo-2-hydroxymethyl derivatives;
on the other hand, reaction of 2-methylene-7,7-dimethylnorbornene
(α-fenchene) occurs preferentially from the endo-direction to
give the exo-2-hydroxymethyl derivative ($\underline{8}$).

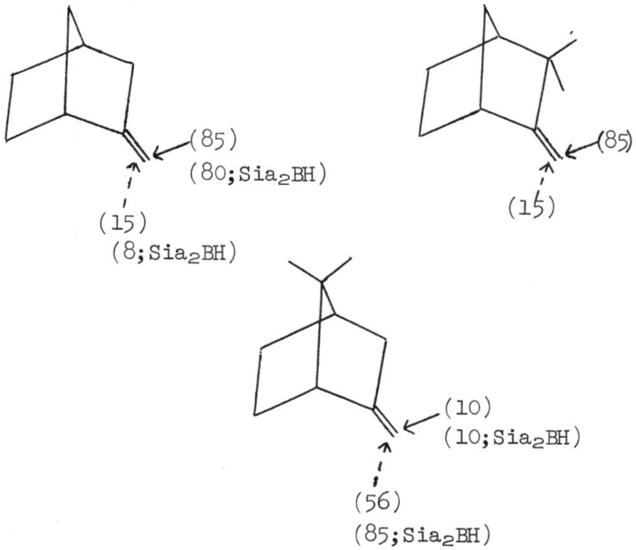

(85)
(80;Sia$_2$BH)

(85)

(15)
(8;Sia$_2$BH)

(15)

(10)
(10;Sia$_2$BH)

(56)
(85;Sia$_2$BH)

Further examples of preferential formation of the less
stable products by attack from the less hindered side of the
double bond are provided by the reaction of β-pinene (1) (10)
and (+)-sabinene (2) (11).

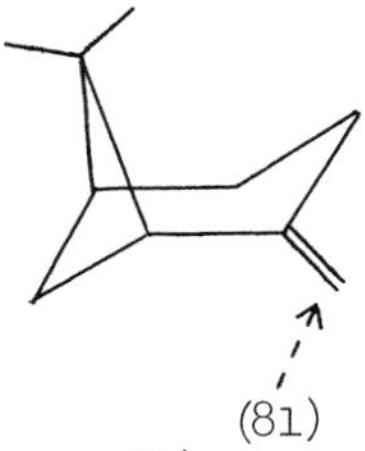

(28; Sia$_2$BH)
(33)

(81)
(1)

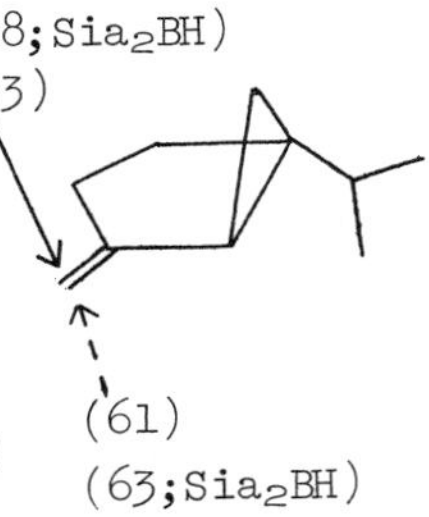

(2)   (61)
        (63; Sia$_2$BH)

3.1.3 Acyclic Internal Alkenes

Reaction of 1,2-di-n-alkyl substituted alkenes with borane
results in approximately equal attack at both possible positions
(1); with selective hydroborating agents some preference is shown
for formation of the less hindered product (12).

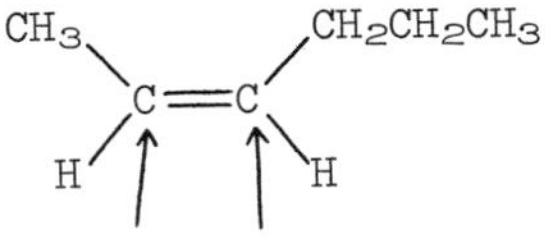

(45) (45)
(57) (33) (Sia$_2$BH)
(59) (31) [(C$_6$H$_{11}$)$_2$BH]

Only small variations in the isomer distribution are
observed in the reaction of cis- and trans-isomers with borane (1),
although the reaction of cis-alkenes with disiamylborane is
considerably faster than that of trans-alkenes (2,13).

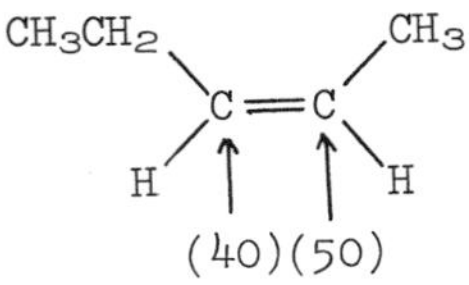

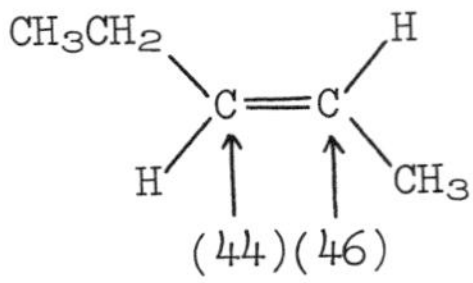

(40)(50)

(44)(46)

α-Branching of one of the two alkyl substituents results in slightly more of the less hindered alcohol being formed when using borane as hydroborating agent ($\underline{1}$); use of disiamylborane ($\underline{2}$) or 9-borabicyclo[3.3.1]nonane ($\underline{3}$), however, results in a marked preference for attack by the boron at the less hindered position.

$$CH_3\diagdown_{}\diagup CH(CH_3)_2$$

$$H \quad \quad H$$
$$(56) \ (43)$$
$$(87) \ (\ 3) \quad (Sia_2BH)$$
$$(90) \ (\ 3) \quad (9\text{-}BBN)$$

$$CH_3\diagdown_{}\diagup H$$

$$H \quad \quad CH(CH_3)_2$$
$$(51) \ (39)$$
$$(86) \ (\ 5) \quad (Sia_2BH)$$

In the case of trisubstituted alkenes the less hindered alcohol is almost exclusively formed ($\underline{1}$); with the exception of 9-borabicyclo[3.3.1]nonane ($\underline{3}$), the reaction with selective hydroborating reagents is very slow.

$$(CH_3)_2C{=}CHCH_3$$
$$(2)(88)$$
$$(1)(94) \ (9\text{-}BBN)$$

$$(CH_3)_2C{=}CHC(CH_3)_3$$
$$(2)(88)$$

C=CHCH₃

$$(4.5)(95.5)$$

(Ref. $\underline{6}$)

Tetrasubstituted alkenes, such as 2,3-dimethyl-2-butene, react readily with borane ($\underline{14,15}$), but, with the exception of 9-borabicyclo[3.3.1]nonane ($\underline{3}$), very slowly with selective reagents ($\underline{2}$).

3.1.4 Exocyclic Substituted Alkenes

The reaction of cis-17-ethylidene steroids ($\underline{3}$; R=H) gives the corresponding 20α-alcohols ($\underline{4}$; R=H) in variable yields (50-95%) ($\underline{16}$).

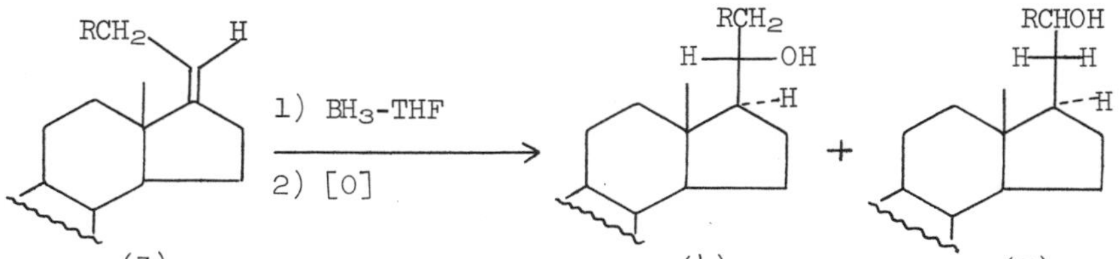

$$(3) \xrightarrow[\text{2) [0]}]{\text{1) BH}_3\text{-THF}} (4) \quad + \quad (5)$$

However, hydroboration of the cis-propylidene derivative (3; R=CH$_3$) at room temperature, followed by oxidation, gives a mixture consisting of 40% of the 20α-alcohol (4; R=CH$_3$) and 20% of the 21-epimeric alcohols (5; R=CH$_3$) (17). Reaction at 0° gives mainly the 20α-alcohol, while at 65° mainly the 21-alcohols are obtained. Isomerization of organoboranes at such low temperatures is unusual (Sec. 2.3.2). When hydroboration was carried out at 160° no terminal 22-alcohols were obtained; reaction of the ethylidene derivative (3; R=H) at 65°, however, gives the terminal alcohol (5; R=H) (17).

Disubstituted exocyclic alkenes appear to be relatively unreactive as illustrated by the selective reaction of dienes shown below (18).

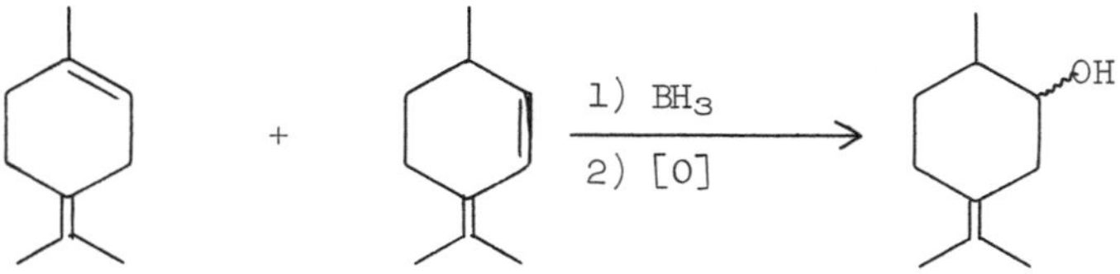

3.1.5 Monocyclic Alkenes

Disubstituted cycloalkenes such as cyclopentene and cyclohexene react readily to give alcohols in high yields (~90%) (14). The presence of an 3-alkyl substituent exerts little directive influence on the reaction with either borane or disiamylborane (10), though a preference for attack trans to the alkyl group is observed (19).

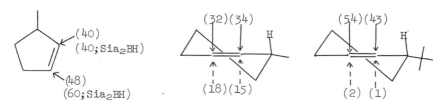

Reaction of 3,3-dimethylcyclohexene gives equal amounts of both alcohols, while disiamylborane shows a small preference for attack at position 1 (10).

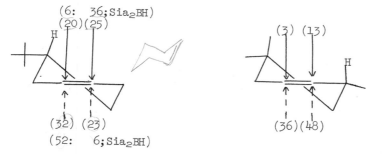

The 3,3-spirocyclopropyl group, however, directs the attack of the boron mainly to the 1-position (4). Reaction of 4-methylcyclohexene gives approximately equal amounts of all four possible alcohols (20), while with 4-tert-butylcyclohexene a preference for cis-attack is shown (19). In the case of 3,5,5-trimethylcyclohexene, predominant attack occurs trans to the axial C-5 methyl group (19,20).

Reaction of 1-alkylcycloalkenes usually gives the corresponding trans-2-alkylcycloalkanols of high purity in yields of 80% or more (10). However, with 1-methylcyclopropene some attack occurs at the methyl-substituted carbon atom (21).

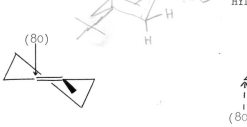

(80)

(80)   (7)

As with disubstituted cycloalkenes, further substituents in the ring exert a steric effect on the reaction.

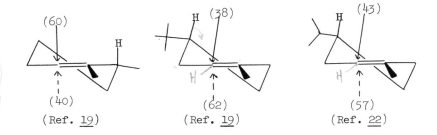

(60)

(38)

(43)

(40)   (62)   (57)
(Ref. 19)   (Ref. 19)   (Ref. 22)

The preference for attack cis to the 4-tert-butyl and isopropyl groups in the compounds shown above is associated with the steric interaction between these bulky groups and the C-4 methine hydrogen leading to distortion of the bond angles about C-4; this leads to added shielding of the double bond due to the C-4 hydrogen being forced closer to the center of the ring (19). Similar results have been obtained with the terpineol derivative shown below (18).

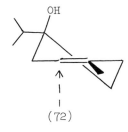

OH

(72)

In the case of 1,4-di-tert-butylcyclohexene, however, predominant attack occurs trans to the 4-tert-butyl group to

give trans,trans-2,5-di-tert-butylcyclohexanol which adopts a
twist boat conformation (19,23).

The preference for trans-attack in this case is ascribed to
the preferred conformation of the 1-tert-butyl group which
results in shielding of the side of the ring cis to the
4-tert-butyl group.

    Reaction of 1,2-dialkylcycloalkenes gives cis-1,2-dialkyl-
cycloalkanols (10).

3.1.6 Bicyclic Systems

    The discussion of the hydroboration-oxidation of these
systems has been divided into bicyclo[n.n'.0]- and [n.n'.1]-
compounds. Very few examples of the reaction of unsubstituted
systems of this nature have been reported; in most cases the
course of reaction is determined by the nature of the
substituents present. Thus, the results presented below cannot
be considered to be typical of the bicyclic system cited.

3.1.6.a Bicyclo[n.n'.0] Systems

    Reaction of fused cyclopropyl systems generally occurs on
the side remote from the cyclopropyl ring as illustrated by the
reactions of (+)-α-thujene (6) (11) and car-2-, -3-, and -4-enes
[(7)-(9) respectively] (24-26).

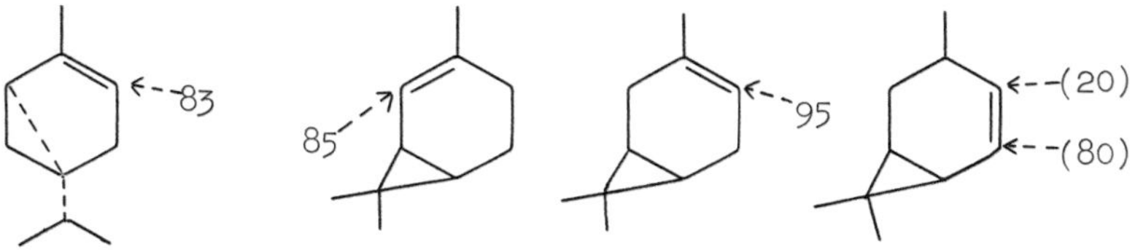

(6)   (7)   (8)   (9)

With cis-bicyclo[3.3.0]-oct-2-ene (10), both alcohols are formed
(27).

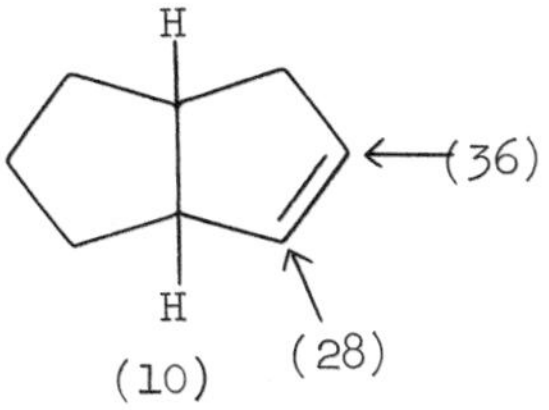

(10)   (28)

1(9)-Octalin (11; R=H) (28) and the corresponding 10-methyl
derivative (11; R=CH₃) (29) give mixtures of the corresponding
cis- and trans-decalols.

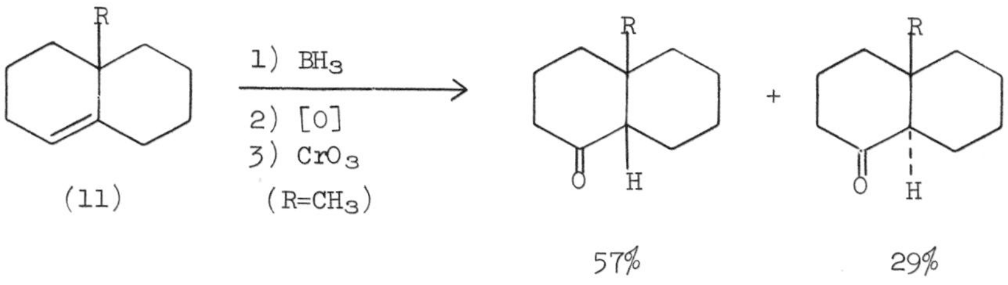

(11)   (R=CH₃)

                57%   29%

Similar results are obtained when equatorial substituents
are present in the system as in compound (12) (30); the presence
of an axial substituent at C-7 as in (13) hinders underside
attack, and results in the exclusive formation of the cis-decalol
in high yields (31).

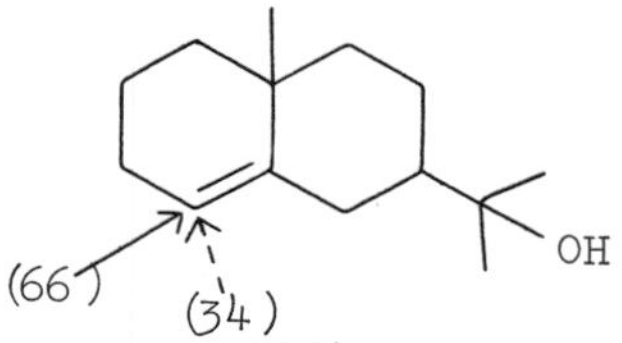

(66) (34)
      (12)

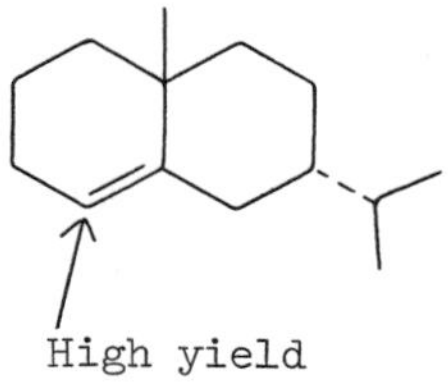

High yield
(13)

Hydroboration-oxidation of the 2,2-cycloethylenedioxy
derivative (14a) gives some of the tertiary alcohol arising from
anti-Markownikoff hydroboration (32); in addition, 40% ketal
cleavage products are also formed. Similar results have been
obtained in the reaction of compound (14b) with excess borane (33);
use of equimolar amounts of the compound and borane-tetrahydrofuran, however, gives only the secondary alcohols (33).

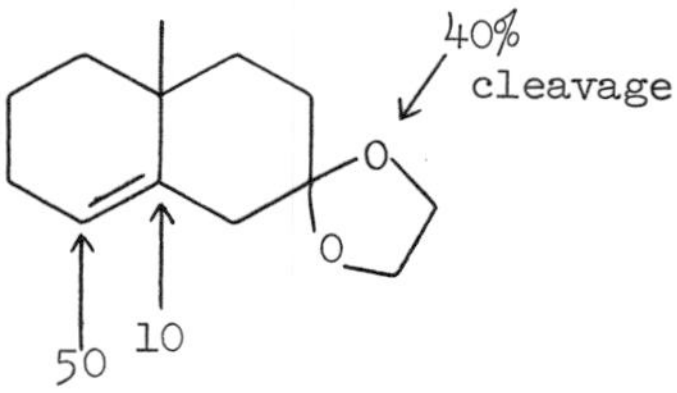

40%
cleavage

50 10

(14a)

PhCH₂O

64

32

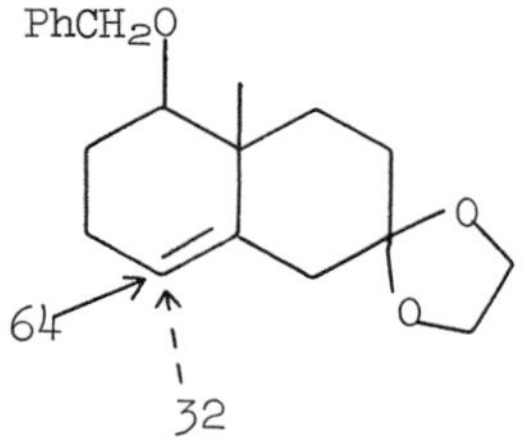

(14b)

Reaction of the 8,8,10-trimethyl derivative (15) is reported
to proceed readily to give the trans-fused alcohol (34); in
contrast, the 2-oxo-1,1-dimethyl-10-phenyl derivative (16) fails
to react (35).

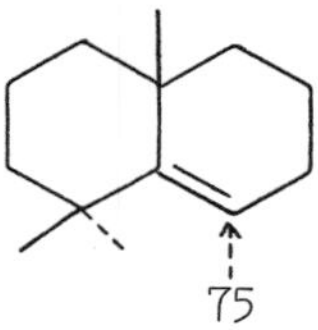

75
(15)

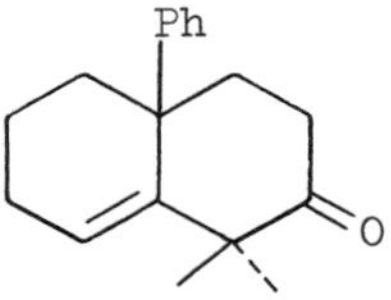

Ph

(16) No reaction

Application of the reaction to a [5.3.0] system is illustrated by the reaction of α-gurjunene (17) (36); in addition to the expected product (18), two other products (19) and (20) are formed. Product (19) is formed by isomerization of the initially formed organoborane, while cleavage of the cyclopropane ring (Sec. 10.12) (37), followed by rehydroboration, accounts for the formation of product (20).

3.1.6.b Bicyclo[n.n'.1]Systems

[2.2.1] Systems. In general, hydroboration of norbornyl systems proceeds via predominant exo attack.

(Ref. 39)   (Ref. 40)   (Ref. 41)

In the case of 7,7-dimethyl derivatives, however, endo-attack is favored, while exo-attack is again predominant with the 7,7-dimethoxy derivative.

(Ref. 8)   (Ref. 42a)   (Ref. 42b)

(Ref. 43)   (Ref. 44)

Attempted reactions of the 7,7-dimethyl derivatives with disiamylborane failed.

[3.1.1] Systems. These systems are illustrated by pinene derivatives where reaction usually occurs on the side remote from the gem-dimethyl group.

(Refs. 10 and 45)   (Ref. 46)

However, hydroboration of compound (21), which is related to β-pinene, followed by oxidation with chromic acid, gives the ketone (22) formed by attack cis to the gem-dimethyl group (47); this result is in contrast to that observed in the reaction of β-pinene (Sec. 3.1.2).

(21)   (22)

[3.2.1] Systems. Again attack occurs on the side remote from the gem-dimethyl group as illustrated by the reactions of α-cedrene (23)(48) and α-patchoulene (24) (49).

(23)   (24)

[3.3.1] Systems. In the reaction of bicyclo[3.3.1]
non-2-ene (25; R=H) both possible exo alcohols are formed (50),
while with the 1,5-dimethyl derivative (25; R=CH₃) the
3-exo-alcohol is the major product (51). Reaction of the
bicyclo[3.3.1]non-1-ene (26), however, gives, in addition to
the expected exo and endo secondary alcohols, a substantial
amount of the tertiary alcohol resulting from anti-Markownikoff
hydroboration (52). The formation of this alcohol is associated
with the high strain of the alkene decreasing the activation
energy of the hydroboration reaction, and thereby lowering the
sensitivity of the reaction to steric and polar effects.

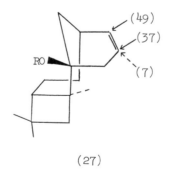

(25)   (26)

[4.3.1] Systems. Reaction of the trimethylsilyl ether of
caryol-9-en-1-ol (27; R=(CH₃)₃Si) proceeds via predominant
exo-attack (53); attempted reaction with disiamylborane failed.

(27)

3.1.7 Tricyclic Systems

As in the case of bicyclic systems, the nature of the
substituents present in the system exerts a strong influence in

determining the course of hydroboration-oxidation of these
systems.

In contrast to the reaction of the carene derivatives
(Sec. 3.1.6.a), the reaction of thujopsene (28a) takes place
via exclusive attack cis to the cyclopropyl ring (54a); however,
with systems (28b) and (28c), attack is directed predominantly
trans to the cyclopropyl ring (54b).

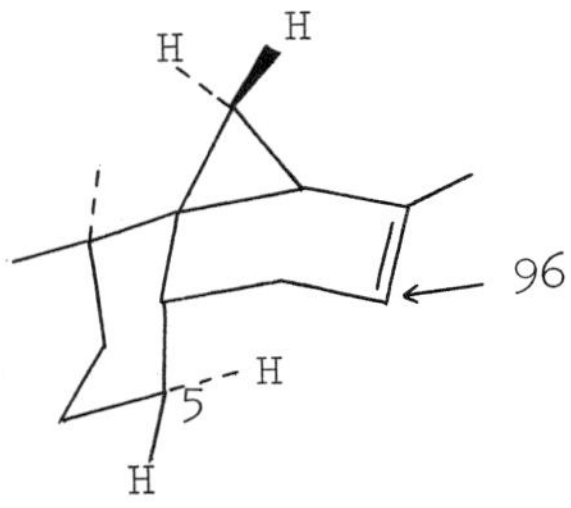

(28a) 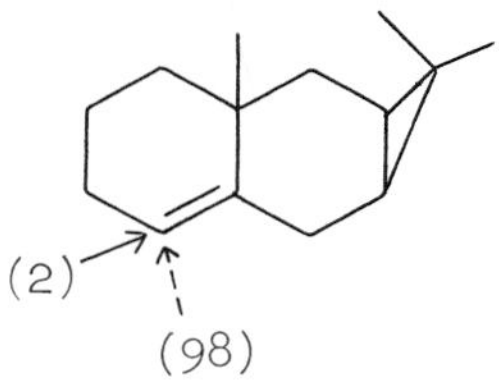   (28b) 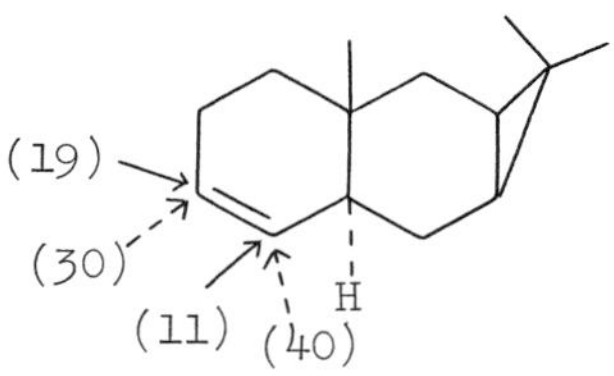   (28c)

In (28a) underside attack is hindered by the C-5 methylene
group.

The application of hydroboration-oxidation to tricyclic
terpenes related to abietane is illustrated below.

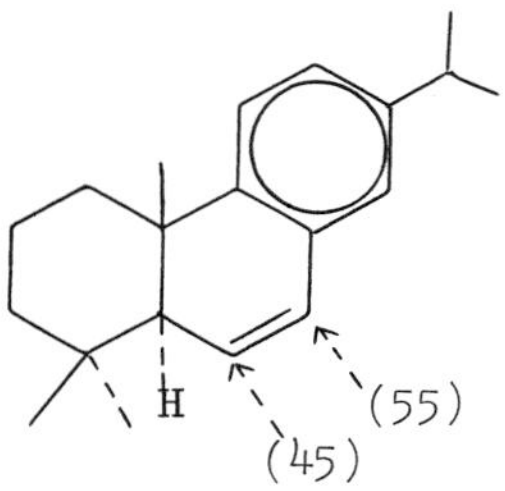

(Ref. 55) 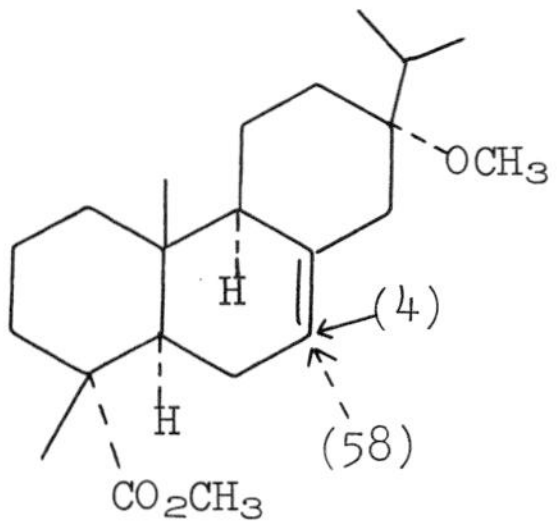   (Ref. 56) 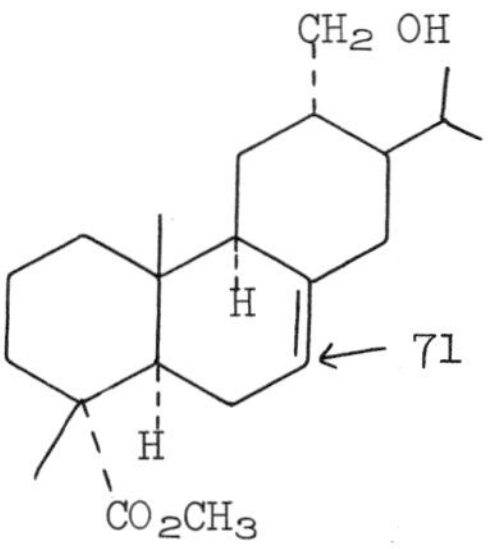   (29) (Ref. 57)

Attack generally occurs from the α-side of the molecule; the unexpected formation of the 7β-alcohol in the case of (29) is possibly due to ring C adopting a conformation in which the hydroxymethyl and isopropyl groups are equatorial, thus leaving the β-face open to attack (57). The β-face attack of the 9(11)-derivative (30) has been explained in a similar manner (58).

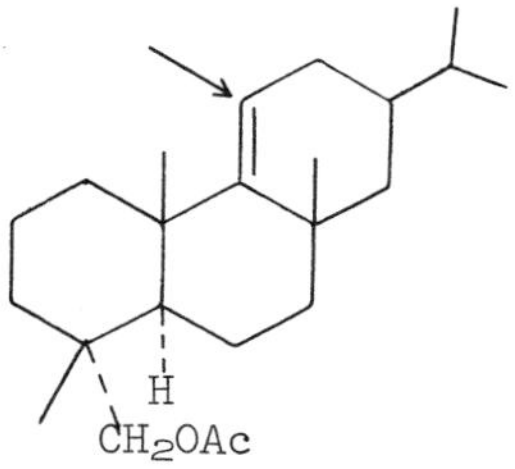

CH₂OAc

(30)

Reaction of the 8(9)-derivative (31) proceeds with isomerization to give the 7α-alcohol (32) (58), while 8(14)-derivatives such as (33) give mainly the corresponding 14α-alcohols (59).

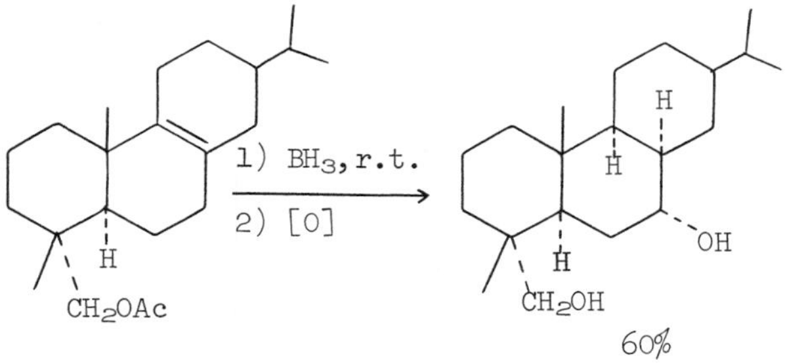

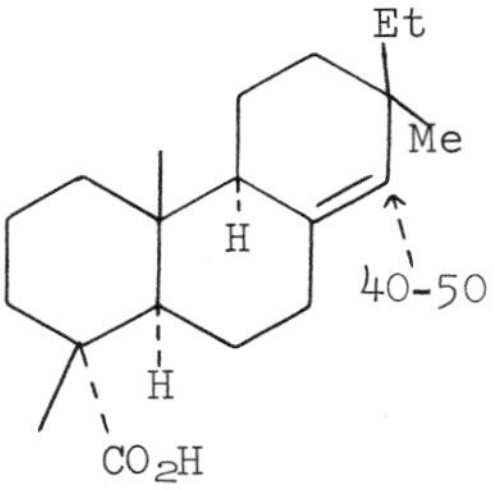

1) BH₃, r.t.

2) [O]

CH₂OAc

(31)

CH₂OH

60%

(32)

Et

Me

40-50

CO₂H

(33)

The 10α,9β-tricyclic derivatives (34a and 34b) react via β-face attack to give the corresponding 6β-alcohols (60).

(34a)   (34b)

3.1.8 Steroid and Polycyclic Alkenes

Hydroboration-oxidation has been extensively applied to steroid alkenes. The majority of compounds studied belong to the 5α-series and in these cases addition occurs predominantly from the less hindered α side of the molecule.

3.1.8a Disubstituted Alkenes

The results of the reactions of a number of steroid alkenes are summarized in Table 3.1.

Reaction of the tetracyclic diterpene, (+)-hibaene (35) gives a 3:2 mixture of alcohols (66).

(35)   60-70   (36)

3.1.8.b Trisubstituted Alkenes

In general, reaction of steroid 4-enes (36; R=H or OH) gives the corresponding 4α-alcohols (61,67).

TABLE 3.1

Hydroboration-oxidation of Disubstituted Steroid Alkenes

Alkene	Reagent	Alcohols formed (%yields)	Reference
5α-Cholest-1-ene	BH$_3$	1α-ol(35); 2α-01(40)	61
	Sia$_2$BH	2α-ol(75)	61
19-Nor-5α-cholest-1-ene	Sia$_2$BH	1α-ol(1); 1β-ol(2) 2α-ol(40); 2β-ol(20)	62
5α-Cholest-2-ene	BH$_3$[a]	2α-ol(35); 3α-ol(45)	61
	Sia$_2$BH	2α-ol(35); 3α-ol(45)	61
5α-Cholest-3-ene	BH$_3$[b]	3α-ol(40); 4α-ol(45)	61
	Sia$_2$BH	3α-ol(45); 4α-ol(35)	61
5β-Cholest-3-ene	BH$_3$	3β-ol(72)	64
5α-Cholest-6-en-3β-ol	BH$_3$	3β,6α-diol and 3β,7α-diol (55 total yield)	61
Methyl 3α,12α-di-acetoxy-6-cholenate (5β series)	BH$_3$	3α,7α,12α-triol (major product)	65
5α,25D-Spirost-11-en-3β-ol acetate	BH$_3$	3β,11α-diol(40) 3β,12α-diol(40)	61

[a]The formation of significant quantities of the 3β-alcohol has been reported (63).

[b]The formation of the 3β-alcohol has been reported (64).

In the reaction of steroid 5-enes the products depend on the nature of the substituent at C-3. With 3β-substituents the 6α-alcohol predominates, while with 3α-substituents, the 6β-alcohol is the major product.

R=H (Ref. 61)   75%
R=β-OH (Ref. 61)   70%   20%
R=β-OTHP (Ref. 61)   45%   35%
R=β-OAc (Ref. 68)   32% (+50% starting material)
R=β-OCH₃ (Ref. 69)   Main product
R=β-Cl (Ref. 68)   80%
R=β-(CH₃)₂N (Ref. 70)   Main product
R=α-OAc (Ref. 65)   8%   92%

The presence of a cycloethylenedioxy group at C-3 also affects the stereochemistry of the 6-alcohol formed. The presence of the gem-4,4-dimethyl group, and reaction in the 19-nor series, also influence the stereochemistry of attack.

R=H;   R'=CH₃ (Ref. 61)   30%   60%
R=CH₃; R'=H   (Ref. 71)   80%
R=R'=CH₃   (Ref. 71)   Both 6α- and 6β-alcohols formed
                                 plus some 7α-alcohol (5α-H)

Steroid 7-enes in the 5α-series (37) fail to react (61), but methyl 3α,12α-dihydroxy-7-cholenate (38) gives a mixture of 7α- and 7β-alcohols (65).

(37) (No reaction)   (38)

In the case of steroid 9(11)-enes (39) members of the
5α-series react readily, while those of the 5β-series fail to
react (61).

$$1) \; BH_3 \quad 90\% \quad 11\alpha\text{-alcohol (5}\alpha\text{-series)}$$
$$2) \; [o] \quad\quad\quad No \; reaction \; (5\beta\text{-series})$$

(39)

Steroid 14-enes (40) generally react to give the 15α-alcohols
(61); surprisingly 5α,6β-dichloro-androst-14-ene-3β,17β-diol (41)
only gives a low yield of a mixture of 15α- and 15β-alcohols (72).

(40) (R=C$_8$H$_7$ or OH)   (41)

Reaction of 17-substituted-16-enes (42) gives the corresponding
16α-alcohols (61).

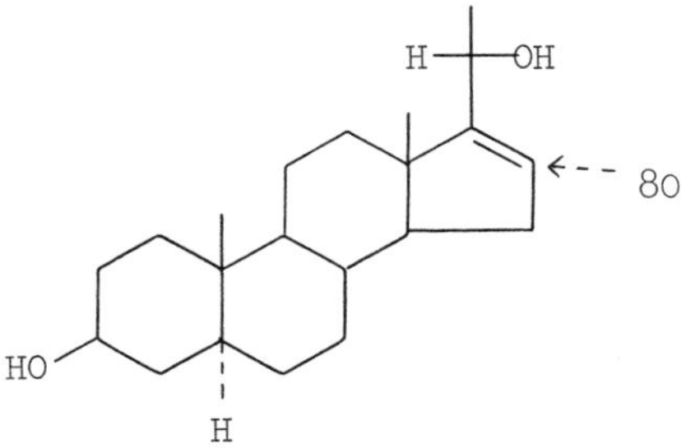

(42) 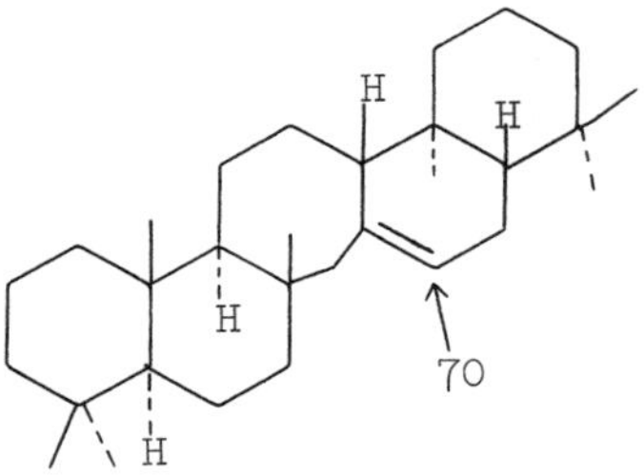   (43)

The pentacyclic triterpene, serratene (43) is reported to give 70% of the β-alcohol (73).

Tetrasubstituted steroid alkenes such as 5α-cholest-8(14)-en-3β-ol have been found to be unreactive (61).

3.2 STOICHIOMETRY OF THE REACTION

Although the hydroboration reaction has been applied to a great number of alkenes, the alkylboranes have not been isolated in many cases but have been directly oxidized to the corresponding alcohols. Thus, in many cases, the stoichiometry of the reaction has not been determined.

The extent of hydroboration depends on the steric requirements of both the alkylboranes and the alkenes (74). In the case of highly hindered alkenes, the reaction of the initially formed monoalkylborane, RBH_2, with a second alkene molecule to form the dialkylborane, R_2BH, is relatively slow, and the monoalkylborane dimerizes to the sym-dialkyldiborane. With less hindered alkenes, however, the reaction of the alkene with the monoalkylborane appears to be faster than that with borane. Consequently, with these alkenes, the monoalkylboranes are converted to dialkylboranes in spite of the presence of excess borane. With relatively unhindered alkenes further reaction occurs between the dialkylborane and alkene to form the trialkylborane. These various possibilities are illustrated below.

Alkene + H$_3$B:OR$_2'$ $\xrightarrow{-R_2'O}$ RBH$_2$ $\xrightarrow{\text{Alkene}}$ R$_2$BH $\xrightarrow{\text{Alkene}}$ R$_3$B

$\downarrow$ H$_3$B:OR$_2'$

(diborane structures) $\xrightarrow{\text{Alkene}}$

The stoichiometry of the reaction of a number of acyclic
and cyclic alkenes is summarized in Table 3.2. While mono- and
dialkylboranes usually exist as dimers (Sec. 2.1), even in ether
solvents, it is usual to discuss them in terms of the monomeric
form, except in cases where the diborane structure is of
significance in the chemistry and properties of the product.

The limited reaction of cyclohexene, as compared with that
of other cycloalkenes such as cyclopentene, has been ascribed to
the low solubility of dicyclohexylborane in the hydroboration
solvent (usually diglyme) which causes it to precipitate before
further reaction occurs. Another contributing factor is the
greater strain to which the double bonds in molecules such as
cyclopentene are subjected, which results in the enhanced
reactivity of such cycloalkenes.

The rates and stoichiometry of the hydroboration of a
number of hindered alkenes have been studied in detail, and have
resulted in the development of a number of useful selective
hydroborating and reducing reagents (74) (Sec. 2.2.1.c).

3.3 RELATIVE REACTIVITIES OF ALKENES WITH SELECTIVE
     HYDROBORATING REAGENTS

The rates of reaction of borane with alkenes are relatively
insensitive to the structure of the alkenes (13). The rates of
reaction with selective reagents such as disiamylborane are,

TABLE 3.2

Stoichiometry of Hydroboration of Some Representative Alkenes

Acyclic alkenes	Reference	Cyclic alkenes	Reference
1. Formation of trialkyl- boranes (R_3B)			
RCH=CH$_2$ (R=n-or branched alkyl)	14,15	Cyclopentene	14,15
		Cyclohexene (slow at	14,15
RR'C=CH$_2$ (R=n-or branched alkyl)	14,15	25°)	
		1-Methylcyclopropene	21
Cis- and trans RCH=CHR' (R=n-alkyl)	1,14	Norbornene	10
		β-Pinene	10
2. Formation of dialkyl- boranes (R_2BH)			
		Cyclohexene (rapid at 0°)15	
(CH$_3$)$_2$C=CHCH$_3$	15	1-Methylcyclopentene	10,15
(CH$_3$)$_2$C=CHC(CH$_3$)$_3$ (slow at 25°)	15	1-Methylcyclohexene	10,15
		α-Pinene	15
		Cholesterol	75
3. Formation of Monoalkyl- boranes (RBH_2)			

(CH$_3$)$_2$C=CHC(CH$_3$)$_3$ (rapid
  at 0°)
	15	1,2-Dimethylcyclo-
		  pentene
			10

(CH$_3$)$_2$C=C(CH$_3$)$_2$	15	1,3-Dimethylcyclo- hexene	10
Trans-(CH$_3$)$_3$CCH=CHC(CH$_3$)$_3$	76		

however, more sensitive to the structure of the alkenes. The
relative rates of reaction of a variety of alkenes with
disiamylborane are given in Table 3.3.

In addition, studies of the rates of reaction of various
2-alkylpropenes with disiamylborane have shown that the rate of

TABLE 3.3

Relative Rates of Reaction of Some Alkenes with Disiamylborane (2, 13)

Alkene	Relative Rate	Alkene	Relative rate
1-Hexene	100	2,4,4-Trimethyl-1-pentene	0.8
3-Methyl-1-butene	57	Cis-4-methyl-2-pentene	0.5
Cyclööctene	26	Trans-2-butene	0.4
Cycloheptene	26	Trans-2-pentene	0.3
2-Methyl-1-pentene	4.9	Trans-3-hexene	0.2
3,3-Dimethyl-1-butene	4.7	Trans-4-methyl-2-pentene	0.1
2,3-Dimethyl-1-butene	3	Cis-4,4-dimethyl-2-pentene	0.08
Cis-2-butene	2.3	Trans-4,4-dimethyl-2-pentene	0.01
Cis-3-hexene	2.0	Cyclohexene	0.01
Cyclopentene	1.4	Tri- and tetrasubstituted alkenes	<0.01

reaction is markedly decreased with increasing steric requirements
of the alkyl group (77). The considerably greater reactivity of
cis-alkenes compared to trans-alkenes has been attributed to the
greater steric strain present in the former isomers (13,78).

Other typical dialkylboranes such as dicyclohexylborane
exhibit similar reactivities to disiamylborane (12).
9-Borabicyclo[3.3.1]nonane, however, reacts rapidly with a
variety of alkenes of different structural types (3); the greater
reactivity of this reagent is associated with its smaller steric
bulk and the greater exposure of the B-H bond. Thexylborane
(tert-$C_6H_{13}BH_2$), while following the same trends as disiamylborane,
is not as selective (79).

The major differences in the reactivities of various alkenes
with disiamylborane permits a wide variety of selective
hydroboration reactions. The reaction of alkene mixtures with
controlled amounts of the reagent results in the separation of
reactive from less reactive alkenes. Thus, 1-alkenes may be
separated from internal alkenes, and simple terminal alkenes,
such as 4-methyl-1-pentene, from 2-alkyl-1-enes, such as
2-methyl-1-pentene (2).

The marked difference in the relative reactivities of cis-
and trans-alkenes enables the selective removal of cis-isomers
from isomeric mixtures. Since the cis-isomer can be regenerated
from the organoborane by reaction with a less volatile alkene
(Sec. 2.3.3), pure cis- and trans-isomers may be obtained in
this manner. In this respect, diisopinocampheylborane has been
found to be a particularly effective reagent (12), while
disiamylborane (2) and thexylborane (79) have also been
successfully used.

The sensitivity of disiamylborane to the steric environment
of the double bond is further illustrated by the separation of
the following alkene mixtures.

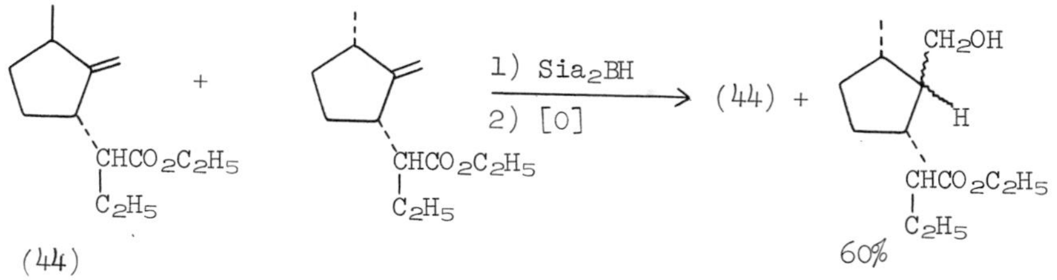

(Ref. 80)

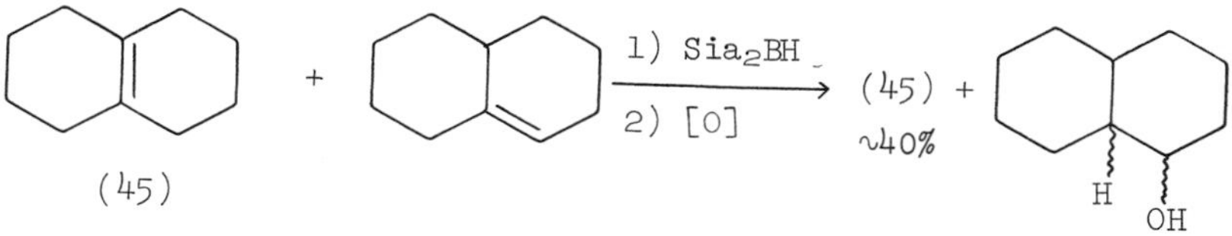

(Ref. 81)

The selective hydroboration of dienes by selective reagents is dealt with in Sec. 6.5.

3.4 MECHANISM OF HYDROBORATION

The importance of steric effects in influencing the hydroboration of alkenes has been illustrated in the foregoing sections.

The directive effects observed in the hydroboration of terminal and trialkylethenes (Secs. 3.1.1 and 3.1.3), as well as the increased regioselectivity exhibited by bulky hydroborating agents such as disiamylborane, could well be ascribed solely to the operation of steric effects. It is significant, however, that the degree of branching of alkyl substituents in monoalkylethenes ($RCH=CH_2$) and dialkylated internal alkenes does not appreciably affect the direction of addition. Steric effects, therefore, cannot fully account for the results obtained.

The inadequacy of the purely steric explanation is further demonstrated by the hydroboration of some substituted styrene

derivatives (82); the results are summarized in Table 3.4.
These results clearly indicate that electronic effects also play
an important role in controlling the direction of addition.

The formation of ethylbenzenes results from the hydrolysis
of α-organoboranes under the oxidation conditions, and occurs
most readily with those organoboranes capable of yielding
relatively stable carbanions (Sec. 10.11) (82).

$$XC_6H_4CH-B \overset{OH^{\ominus}}{\longrightarrow} XC_6H_4CH-B-OH \longrightarrow XC_6H_4\overset{\ominus}{C}HCH_3 \overset{H_2O}{\longrightarrow} XC_6H_4CH_2CH_3$$
$$\qquad CH_3 \qquad\qquad CH_3$$

Ortho- and para-electron-releasing substituents decrease
the amount of α-addition compared to styrene, while electron-
withdrawing substituents increase the amount of α-addition.
The effect of substituents in influencing the direction of
hydroboration parallels qualitatively their σ^+ values, indicating
that the reaction involves an electrophilic attack by borane (and
subsequent intermediates) on the styrene system (82).

Kinetic studies on the hydroboration of methoxy- and
chlorosubstituted styrenes show that the various substituents
produce little change in the overall rate of hydroboration (83)
The Hammett relationship is not followed by the overall reaction,
but is partially obeyed for the reaction at the α- and β-positions
considered separately. Electron-withdrawing substituents
accelerate the reaction at the α-position and diminish the rate
at the β-position, but the low values of ρ reported for these
reactions exclude the formation of an intermediate with a fully
developed charge; the slight polar influences observed, however,
favor the formation of a partially charged transition state in
the reaction.

Studies of the hydroboration-oxidation of a considerable
number of cyclic alkenes (Secs. 3.1.5 to 3.1.8) have provided
convincing evidence that hydroboration proceeds by the cis-
addition of the boron-hydrogen bond to the double bond. In

TABLE 3.4

Directive Effects in the Hydroboration of Substituted Styrene Derivatives (82)

$$XC_6H_4CH=CH_2 \xrightarrow[\quad 2)\ [O]\quad]{1)\ BH_3, DG, 20^{\circ}} XC_6H_4CH_2CH_2OH + XC_6H_4CH(OH)CH_3 + XC_6H_4CH_2CH_2CH_3$$

I (β-addition)

II (α-addition) — III

X	%I %β-addition[a]	%II	%III	% α-addition[a]
H	81	18	1	19
o-CH₃O	86	12	1.6	13.6
m-CH₃O	81	18	1	19
p-CH₃O	93.3	5.4	1.2	6.6
o-Cl	74	23	3	26
m-Cl	70	28	2	30
p-Cl	73.4	25	1.6	26.6
o-CF₃	61.8	12.7	25.5	38.2
m-CF₃	67.7	30	2.3	32.3
p-CF₃	66	12	22	34
m-NO₂	63	14.6	22.4	37

[a]Determined by vapor phase chromatography.

addition, the formation of thermodynamically less stable cis-
dialkyl isomers in the reaction of cyclic alkenes such as
1,2-dimethylcyclopentene (Sec. 3.1.5) (10) illustrates that
product stability does not control the stereochemical course of
the reaction. These results, together with the electronic effects
discussed above, indicate that the hydroboration of alkenes
proceeds via a four-center transition state with partial charges
on the participating atoms.

The fact that the reaction in ether solutions is extremely
fast (13) indicates that the transition state of the reaction
lies close in structure and energy to the reagents (20,84); the
reaction may thus be considered to proceed by attack of the
borane on the alkene to give a transition state possessing a
structure similar to the starting alkene, with the borane situated
perpindicular to the plane of the double bond, thereby ensuring
overlap of the π electrons with the unoccupied orbital of the
boron atom.

Such a transition state readily accounts for the directive
effects observed in the hydroboration of monoalkylethenes, since
the alkyl group, R, exerts a stabilizing effect due to its positive
inductive effect. This explanation can also be extended to
account for the enhanced directive effects observed in the
hydroboration of 1,1-disubstituted and trisubstituted ethenes.
The greater directive effect observed in the reaction of
vinylcyclopropane (Sec. 3.1.1) is associated with the higher
electron-releasing character of the cyclopropyl group compared
to an alkyl group, resulting in greater stabilization of the
transition state. Maximum stabilization is achieved with the

cyclopropyl ring adopting a bisected conformation as shown below
(85).

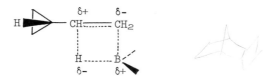

The decreased selectivity observed in the hydroboration of
1,1-dicyclopropyl-1-propene compared to trialkylated ethenes
(Sec. 3.1.3) can possibly be explained by steric hindrance to
the attainment of the optimum bisected conformation of the
cyclopropyl rings. If such a conformation is precluded, the
cyclopropyl rings could exert a destabilizing effect on the
transition state due to their electron-withdrawing inductive
effect (86).

In considering the hydroboration of styrene and its
derivatives, two transition states must be taken into account.

In the case of styrene the phenyl substituent is capable of
stabilizing either transition state by acting as an electron
source in (46) and an electron sink in (47). The presence of
an electron-releasing substituent (X=o- and p=OCH$_3$) stabilizes
transition state (46) to a greater extent, thereby increasing
the preference for attachment of the boron atom to the terminal
carbon atom. Electron-withdrawing substituents (X=Cℓ or CF$_3$),
however, stablize transition state (47), thus promoting
attachment of the boron atom to the α-position (82).

The formation of triakylboranes comprises at least three
stages as discussed in Sec. 3.2 on the stoichiometry of the

hydroboration reaction. Thus, any discussion of the direction
of addition in hydroboration reactions should consider the
possibility that the distribution of the boron atom between the
carbon atoms of the double bond could differ in each stage.
However, most of the reactions, except those involving acyclic
internal disubstituted alkenes (Sec. 3.1.3), yield products
involving addition of boron to one of the double bond carbons
in excess of 94%, and hence this possibility does not appear to
constitute a major problem.

The proposed four-center transition state is supported by
results obtained in the hydroboration of functionalized alkenes
(Chap. 5). In addition, hydroboration using monochloroborane
has been shown to proceed via a similar transition state (87).
However, use of this transition state model predicts the wrong
configuration for the alcohol formed in the reaction of
cis-1-deuterio-1-butene, using diisopinocampheylborane as
hydroborating reagent (Sec. 3.7) (88). In this case the
formation of a triangular π-complex involving the alkene and
borane has been suggested (88); however, the subsequent triangular
transition state does not provide an entirely satisfactory
explanation of the configuration observed (Sec. 3.7). In this
connection, the formation of an intermediate π-complex which
rearranges to the four-centered transition state discussed above
has also been proposed (83).

A similar four-center transition state also explains the
greater selectivity exhibited by bulky reagents such as
disiamylborane.

(48)   (49)

Thus, transition state (48) is greatly destablized with respect to (49), resulting in enhanced formation of the terminal product. In the hydroboration of styrene with disiamylborane, formation of the terminal adduct is favored to the extent of 98% compared to 81% using borane (Table 3.4) (2); this indicates that, in this case, steric factors are dominant in determining the direction of addition.

The effect of steric factors on the rate of hydroboration is illustrated by the fact that alkenes undergo relatively rapid hydroboration with 9-borabicyclo[3.3.1]nonane compared to the corresponding reactions with disiamylborane (Sec. 3.3) (3). Comparison of the structures of disiamylborane and 9-borabicyclo[3.3.1]nonane reveals that the latter compound is considerably less hindered, and the boron atom is consequently much more exposed (3). However, it is significant that the directive selectivity achieved with the two reagents is equally high. Tetraethyldiborane and other relatively unhindered dialkylboranes also show greatly enhanced selectivity compared to borane itself (89), thereby indicating that bulky alkyl substituents are not essential for selective hydroboration by dialkylboranes.

$PhCH=CH_2$
81 (BH_3);   98 (Sia_2BH);   98 (9-BBN);   92 (Et_2BH)

It has been shown that alkyl- and dialkylboranes usually exist as dimers in ether solution; kinetic studies of the reaction of disiamylborane with cyclopentene in tetrahydrofuran indicate that the reaction is first order with respect to both cyclopentene and the dimeric reagent (90). This observation requires that the dimer reacts with one cyclopentene molecule to form the product trialkylborane and one molecule of monomer; the tetrahydrofuran probably plays an important role in coordinating the monomer molecule, thereby promoting its role as a "leaving group" in the reaction.

R₂B ··· H

H ··· BR₂ (rendered: R_2B ···H, H··· BR_2)

R_2B + R_2BH

In summary, the hydroboration of alkenes proceeds via a four-center transition state having partial charges on the participating atoms. The direction of addition is influenced by both electronic and steric factors, with steric factors being dominant in the reaction with selective hydroborating reagents. However, the presence of vinyl substituents in the form of electron-withdrawing or electron-releasing groups causes the electronic factor to become dominant due to stabilization or destabilization of the adjacent partial charge in the transition state (Sec. 5.1).

3.5 ASYMMETRIC HYDROBORATION

The preparation and use of the asymmetric hydroborating reagents, diisopinocampheylborane, triisopinocampheyldiborane, and monoisopinocampheylborane have been discussed in Sec. 2.2.1.d.

These optically active hydroborating reagents have been widely used in the asymmetric synthesis of organoboranes from alkenes. In many cases a high degree of optical purity is obtained, though, obviously, the degree of optical purity depends on the optical purity of the α-pinene used in the synthesis of the reagents. Since organoboranes may be converted into a variety of other derivatives with retention of configuration, these reagents are of particular value in synthetic organic chemistry.

3.6   SYNTHESIS OF OPTICALLY ACTIVE ALCOHOLS

3.6.1   From Disubstituted cis-alkenes

Reaction of diisopinocampheylborane with acyclic disubstituted cis-alkenes proceeds readily to give the corresponding trialkylboranes (91); these trialkylboranes are capable of maintaining their asymmetry without significant racemization over periods of up to 48 hr at 25°. Oxidation of the organoborane with alkaline hydrogen peroxide gives isopinocampheol and the corresponding alcohol in a state of high optical purity. Thus, reaction of cis-2-butene with (-)-diisopinocampheylborane in diglyme at 0° is complete in 2 to 4 hr; oxidation gives (R)-(-)-2-butanol having an optical purity of 87% in 90% yield (91). Use of (+)-diisopinocampheyl-borane, followed by oxidation, gives (S)-(+)-2-butanol of 86% optical purity. Use of tetrahydrofuran as solvent results in products of lower optical purity (78% compared to 86% for the synthesis of (S)-(+)-2-butanol).

In general, all alcohols derived from the hydroboration of acyclic cis-alkenes by (-)-diisopinocampheylborane possess the R-configuration, with the opposite configuration resulting from use of (+)-diisopinocampheylborane (91).

Reaction of norbornene with (-)-diisopinocampheylborane, followed by oxidation, gives (1S:2S)-(-)-exonorborneol, while use of (+)-diisopinocampheylborane gives 62% (1R:2R)-alcohol. In both cases the optical purity of the alcohols was 65-70%, using α-pinene of 93-95% purity for the preparation of the reagents (91); however, use of 100% optically pure (-)-α-pinene for the preparation of (+)-diisopinocampheylborane results in the preparation of the (1R:2R)-alcohol having 95% optical purity (92). Similarly, reaction of benznorbornadiene with (-)-diisopinocampheyl-borane (prepared from 75% optically pure (+)-α-pinene) gives (1R:2S)-(+)-exobenznorborneol of 46% optical purity (93,94a). In the reaction of 5α-cholest-2-ene with (-)-diisopinocampheyl-

borane the 3α-alcohol is the main product, while with the
(+)-reagent the 2α-alcohol is predominantly formed (94b).

The preparation of some optical active alcohols via
hydroboration of cis-alkenes with diisopinocampheylborane in
diglyme at 0° is summarized in Table 3.5.

3.6.2 From Disubstituted Trans- and Trisubstituted-Alkenes

In contrast to the fast, highly stereoselective and
asymmetric reactions of diisopinocampheylborane with cis-alkenes,
the corresponding reactions with trans- and trisubstituted alkenes
are slow, and proceed with displacement of α-pinene from the
reagent (95). Oxidation with alkaline hydrogen peroxide gives
alcohols having low optical purities (13 to 22%), and possessing
configurations opposite to those observed for alcohols derived
from cis-alkenes.

The fact that the reaction proceeds with displacement of
α-pinene indicates that the hydroborating reagent in these
reactions is triisopinocampheyldiborane. In the case of trans and
moderately hindered alkenes, one mole of α-pinene is displaced
per two moles of alkene hydroborated. Thus the hydroboration of
trans-2-butene may be represented as follows:

TABLE 3.5

Hydroboration of Cis-Alkenes with Diisopinocampheylborane (91)

(IPC)₂BH[a] Used	Alkene	Alcohol formed	Yield %	Optical Purity, %
(-)	Cis-2-butene	(R)-(-)-2-butanol	90	87
(+)	Cis-2-butene	(S)-(+)-2-butanol	--	86 (78% in tetrahydrofuran)
(+)	Cis-2-pentene	(S)-(+)-2-pentanol;	76	82
		3-pentanol	24	
(-)	Cis-3-hexene	(R)-(-)-3-hexanol	--	91
(+)	Cis-4-methyl-2-pentene	(S)-(+)-4-methyl-2-pentanol	92	76
(-)	Norbornene	(1S;2S)-(-)-exo-norborneol	--	67-70
(+)	Norbornene	(1R;2R)-(+)-exo-norborneol	62	65-68
(+)[b]	Norbornene	(1R;2R)-(+)-exo-norborneol	59	95
(-)[c]	Benznorbornadiene	(1R;2S)-(+)-exo-2-benznorborneol	--	46
(-)[d]	5α-Cholest-2-ene	5α-cholestan-2α-ol	8	
		5α-cholestan-3α-ol	88	
(+)[d]	5α-Cholest-2-ene	5α-cholestan-2α-ol	55	
		5α-cholestan-3α-ol	26	
		5α-cholestan-3β-ol	18	

Unless otherwise stated, prepared from α-pinene of 93-95% optical purity: Ref. 91

[b] Prepared from α-pinene of 100% optical purity: Ref. 92

[c] Prepared from α-pinene of 75% optical purity: Ref. 93

[d] Reference 94b

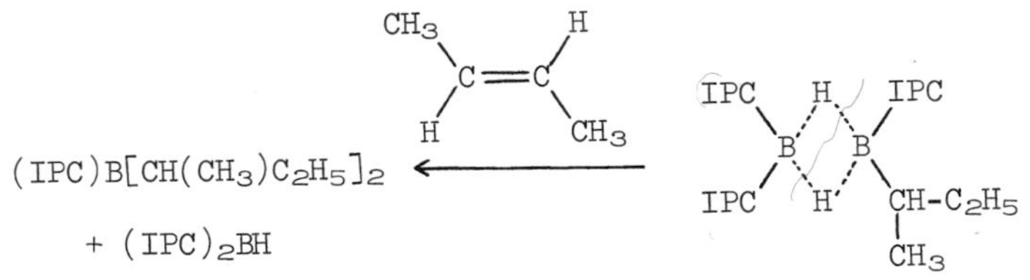

$(IPC)B[CH(CH_3)C_2H_5]_2$

$+ (IPC)_2BH$

With highly hindered alkenes, such as 1-methylcyclopentene, reaction of the corresponding triisopinocampheylmonoalkyldiborane (50) with a second mole of alkene is prevented by steric factors; instead a second mole of α-pinene is eliminated, and a second mole of the alkene then reacts. The process results in the appearance of one mole of α-pinene per one mole of alkene hydroborated.

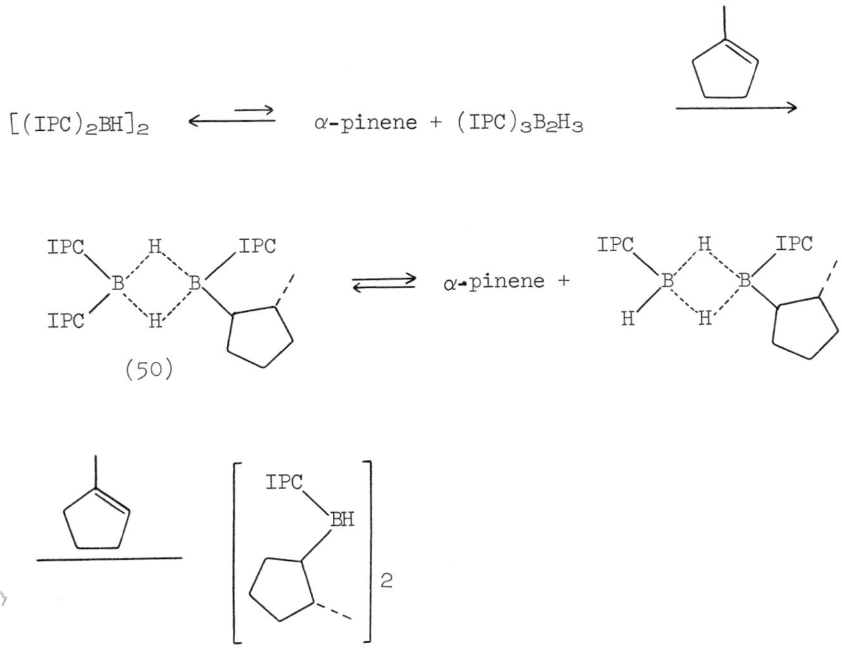

$[(IPC)_2BH]_2 \quad \longleftarrow\!\!\!\longrightarrow \quad α\text{-pinene} + (IPC)_3B_2H_3$

(50)

It is also significant that a marked difference in directive effect is observed in the hydroboration of cis- and trans-4-methyl-2-pentene with diisopinocampheylborane (95). Reaction with the cis-isomer proceeds as expected with 96% of the boron

becoming attached to the less hindered carbon of the double bond
to form the 2-derivative: with the trans-isomer, however, only
62% of the 2-derivative is formed. This latter distribution is
similar to that obtained in the hydroboration of trans-4-methyl-
2-pentene with thexylborane (79); it thus supports the
interpretation that triisopinocampheyldiborane, rather than
tetraisopinocampheyldiborane, is the hydroborating agent in the
reaction with hindered alkenes, since no significant difference
in directive effects is exhibited in the hydroboration of cis-
and trans-alkenes with dialkylboranes such as disiamylborane (2).

The slow step in the hydroboration of hindered alkenes with
tetraisopinocampheyldiborane is probably the elimination of
α-pinene to form triisopinocampheyldiborane, while, in the case
of highly hindered alkenes, the elimination of the second mole
of α-pinene from the initial hydroboration product (50) is even
slower. Reaction of triisopinocampheyldiborane in a 1:1 mole
ratio with the hindered alkene avoids both these steps, and
results in a considerably faster rate of hydroboration; the
products obtained after oxidation compare well in optical purity
with those obtained using diisopinocampheylborane (95). Thus,
reaction of 2-methyl-2-butene with (+)-tetraisopinocampheyldi-
borane in diglyme at 0° for 60 hr gives (R)-(-)-3-methyl-2-
butanol having 14% optical purity, while reaction with
triisopinocampheyldiborane for 3 hr at 0° gives the same alcohol
having 17% optical purity (95).

The preparation of various optically active alcohols via
hydroboration of hindered alkenes with diisopinocampheylborane
or triisopinocampheyldiborane in diglyme at 0° is summarized in
Table 3.6.

The low reactivity of trisubstituted alkenes with
diisopinocampheylborane has led to the use of monoisopinocampheylborane as an asymmetric hydroborating reagent. Thus, reaction of
racemic 3-p-menthene with (+)-monoisopinocampheylborane, followed

TABLE 3.6

Hydroboration of Hindered Alkenes
with Diisopinocampheylborane and Triisopinocampheyldiborane (95)

Reagent Used[a]	Alkene	Reaction time (hr)	Alcohol formed	Optical purity, %
D	Trans-2-butene	24	(S)-(+)-2-butanol	13
D	2-Methyl-2-butene	60	(S)-(+)-3-methyl-2-butanol	14
T	2-Methyl-2-butene	3	(S)-(+)-3-methyl-2-butanol	17
D	1-Methylcyclopentene	60[b]	(1S,2S)-(+)-trans-2-methylcyclopentanol	22
T	1-Methylcyclopentene	3	(1S,2S)-(+)-trans-2-methylcyclopentanol	17.5
T	1-Methylcyclohexene[c]	4	(1S,2S)-(+)-trans-2-methylcyclohexanol	18
T[d]	2-Phenylnorbornene	3	(2S,3S)-(-)-3-endo-phenyl-2-exonorborneol	26

[a]Except where otherwise stated D refers to diisopinocampheylborane and T to triisopinocampheyldiborane, both prepared from (+)-α-pinene of 93-95% optical purity. Solvent is diglyme, and reaction temperature 0° (Ref. 95).

[b]Reaction temperature, 25° (Ref. 95).

[c]Reaction with D is extremely slow (Ref. 95).

[d]Reagent prepared from (-)-α-pinene of 75% optical purity (Ref. 96).

by oxidation, gives 24% (-)-menthol and 44% (-)-isomenthol having optical purities of 14 and 56%, respectively (97).

(1S + 1R)   44%   24%

3.6.3   From Disubstituted Terminal Alkenes

In the hydroboration of internal alkenes the boron atom is directly attached to the new asymmetric center; in the case of disubstituted 1-alkenes, however, the new asymmetric center is developed on the carbon atom immediately adjacent to that bearing the boron atom. The extent of asymmetric induction in the latter case is, therefore, lower than that obtained in the case of cis-alkenes (98). Thus, hydroboration of 2-methyl-1-butene with (+)-diisopinocampheylborane, followed by alkaline peroxide oxidation, gives (R)-(+)-2-methyl-1-butanol having an optical purity of 21%. 2,3-Dimethyl-1-butene and 2-phenylpropene similarly give alcohols of the R-configuration having optical purities of 30% and 5%, respectively; in the case of the hindered 2,3,3-trimethyl-1-butene, the reaction is slow and proceeds with considerable displacement of α-pinene, indicating that triisopinocampheyldiborane also plays a role in the reaction.

Reaction of cholesta-5,25-dien-3β-ol-3-tetrahydropyranyl ether with (+)-diisopinocampheylborane, followed by oxidation, gives the 25-(R)-26-hydroxy derivative, while similar reaction with (-)-diisopinocampheylborane gives the 25-(S)-26-hydroxy derivative (99). In the former case the specific rotation recorded agrees closely with that observed for the microbiologically prepared compound.

1) (+)-(IPC)$_2$BH

2) [O]

$\text{H} \quad \text{CH}_3$

CH$_2$OH

25-(R)

THPO

1) (-)-(IPC)$_2$BH

2) [O]

CH$_3$ H

CH$_2$OH

25-(S)

3.7 MECHANISM OF ASYMMETRIC HYDROBORATION

The configurations of the optically active alcohols prepared
via the hydroboration of alkenes with diisopinocampheylborane
exhibit a definite steric relationship to the configuration of
the particular reagent used. The absolute configuration of these
alcohols have been rationalized in terms of an optimal steric
fit of the alkene and the reagent in the transition state.

Various models have been proposed to account for the
particular configurations observed. Since diisopinocampheylborane
exists as the dimer, sym-tetraisopinocampheyldiborane, it is
probably this form which is the actual hydroborating agent. The
reaction can therefore be envisaged as proceeding via a transition
state similar to that proposed for the hydroboration of alkenes
using reagents such as disiamylborane (Sec. 3.4). One of the
models proposed is based on the preferred conformation of the
dimeric (-)-tetraisopinocampheyldiborane (100), but in the
following discussion only those models which are based on the
monomeric form of the reagent are dealt with; thus, the correlation
rules based on these models do not necessarily have mechanistic
implications, but provide a valuable means for predicting the

structures of products which have not yet been assigned absolute configurations.

The hydroboration-oxidation of a number of acyclic cis-alkenes has been shown to give alcohols of exceptionally high optical purity. Use of (-)-diisopinocampheylborane leads to alcohols having the (R)-configuration, and a model which correctly predicts this configuration in all cases has been proposed by Brown (91); this model is based on the most stable conformation of (-)-diisopinocampheylborane which is shown below (51).

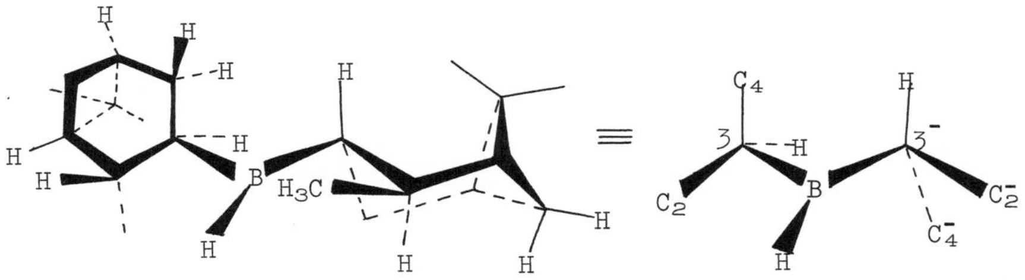

(51)

In the above conformation, the borane group and adjacent methyl groups (at C2 and C2') in each pinane moiety bear a trans-diequatorial relationship to one another; the pinane moieties are oriented so as to place the two methyls (C2 and C2') at an angle of approximately 150° with respect to one another.[†]

In the above model it is assumed that the alkene approaches the borane from above the plane of the B-H bond to form a four-center transition state. Two possible transition states can be formed, and are illustrated for the reaction with cis-2-butene (52 and 53).

[†]Dreiding stereomodels with hydrogen Van der Waals radii have been used throughout this discussion.

(52)   (53)

The major nonbonded interactions in these transition state
models are those between the methyl group and hydrogen atom of the
alkene, and the hydrogen atom at C_3' and the C_4 methylene group.
In transition state model (52), the alkene methyl and C_4 methylene
groups are positioned away from each other, whereas in (53) these
two groups are in close proximity. Thus, model (52) will be
sterically favored; oxidation of the corresponding organoborane
gives R-(-)-2-butanol, in agreement with experimental observation.
The same reasoning applies to unsymmetrical cis-alkenes, with
the boron atom becoming attached to the double bond carbon atom
bearing the smaller substituent.

Similar transition state models can be used to account for
the absolute configuration of alcohols formed via hydroboration
of terminal alkenes with (-)-diisopinocampheylborane (98). Once
again, two possible transition states can be formed, and are
illustrated for the reaction with 2-methyl-1-butene (54 and 55).

(54)   (55)

In these cases the asymmetric center is developed on the carbon atom immediately adjacent to that attached to the boron atom. The major nonbonded interactions in (54) and (55) do not involve the C_3' hydrogen atom and the C_4 methylene group, but rather the alkene alkyl substituents and the methyl group attached to C_2'. In transition state model (54), this interaction involves two methyl groups compared to an ethyl and a methyl group in (55). Thus model (54) will be sterically favored; oxidation of the corresponding organoborane gives (S)-(-)-2-methyl-1-butanol, in agreement with experimental observation. Since the reagent must only discriminate between a methyl and an ethyl group the optical purity of the alcohol formed is understandably low (21%) relative to that obtained in the case of cis-alkenes. Replacement of the ethyl by an isopropyl group, as in 2,3-dimethyl-1-butene, results in a more pronounced discrimination leading to a product of higher optical purity (30%) (98).

The above models devised for cis- and terminal alkenes have been found not to apply to trans- and hindered alkenes (95). For example, on the basis of transition state model (52), it would be anticipated that hydroboration-oxidation of trans-2-butene using (-)-diisopinocampheylborane would give (R)-(-)-2-butanol (i.e., the same as that obtained from cis-2-butene); (S)-(+)-2-butanol is, however, obtained experimentally. The same findings apply to the reaction of trisubstituted alkenes such as 2-methyl-2-butene. In fact, in all the cases of hydroboration-oxidation of trans- and hindered alkenes thus far reported, the absolute configurations of the alcohols formed are the opposite of those predicted on the basis of the transition-state model (52). Moreover, the optical purities of the alcohols formed are considerably lower (13-26%) than those observed in the case of cis-alkenes. This discrepancy between the behavior of cis- and trans-alkenes, as well as other hindered alkenes, has been ascribed to the operation of a different mechanism. Thus, triisopinocampheyldiborane has been proposed as the actual hydroborating agent in the hydroboration of trans- and

hindered alkenes (95) (Sec. 3.6.2); however, no model using this
reagent has been devised. Nevertheless, the consistency of the
results reported has led to the following generalization: "Whenever
displacement of α-pinene occurs in stoichiometric amounts in the
hydroboration of alkenes with diisopinocampheylborane, the alcohol
obtained will possess the configuration opposite to that predicted
on the basis of the simple addition model (52). For such alkenes
the use of triisopinocampheyldiborane will yield the same results"
(95).

    Despite the success of the Brown model in predicting the
absolute configuration of alcohols formed from cis- and terminal
alkenes, a noteworthy discrepancy has been reported in the
hydroboration of cis-1-butene-1-d, where reaction with
(-)-diisopinocampheylborane, followed by oxidation, gives
R-(-)-1-butanol-1-d of 56% optical purity (88).

$$\underset{H_5C_2}{\overset{H}{\diagdown}}C=C\underset{D}{\overset{H}{\diagup}} \quad \xrightarrow[\text{2) [O]}]{\text{1) (-)-(IPC)}_2\text{BH}} \quad (R)-(-)-C_3H_7CHDOH$$

Since the differences in the nonbonded interactions between
deuterium and hydrogen are negligibly small, it would be
anticipated that transition state model (54) would correctly
predict the absolute configuration of the 1-butanol formed;
however, on the basis of model (54), (S)-(+)-1-butanol-1-d should
be formed. In order to account for this discrepancy in predicted
and observed results a model involving the initial formation of
a triangular alkene-borane π-complex has been proposed (88)
(Sec. 3.4).

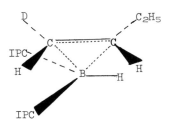

As before two possible complexes can be formed (56 and 57).

(56)   (57)

In the case of the above triangular π-complex models it is suggested that the transition state involves a relatively small perturbation from the model as shown.

Comparison of model (56) with the corresponding Brown model (54) shows that (56) would be relatively less stable as a result of the closer proximity of the alkene ethyl and C_4 methylene groups. In model (57) the alkene ethyl group interacts with the C_3' hydrogen atom, and hence, on the basis of these interactions alone, model (57) would be sterically favored over (56). This model (57) predicts the formation of (R)-(-)-1-butanol-1-d as observed experimentally. However, model (57) is destabilized relative to (56) by a marked steric interaction between the alkene ethyl and the C_2' methyl group. Hence neither model, (56) nor (57), would appear to represent satisfactory models of the transition state; on the basis of steric interactions the Brown model (54) still seems more favorable. However, it has been claimed that use of Fisher-Hirschfelder models enables the construction of a transition state model similar to the Brown model (55), in which the boron atom can bend toward a tetrahedral configuration, thus permitting the more ready formation of the required transition state (101).

A further model based on the monomeric form of (-)-diisopinocampheylborane has been proposed (102), but the

conformation of the borane chosen as the basis for this model
appears to be destabilized by nonbonded interactions. The
model leads, however, to the correct prediction of configuration
of the products of reaction with cis and terminal alkenes.

In summary, no model has, to date, been proposed which
satisfactorily accounts for the absolute configurations of the
alcohols formed in the hydroboration-oxidation of all types of
alkenes with diisopinocampheylborane. In the case of cis- and
terminal alkenes, models (52) and (54), respectively, provide a
satisfactory means for predicting the structures of the alcohols
formed using (-)-diisopinocampheylborane. With trans and
hindered alkenes use of model (52) consistently predicts
configurations opposite to those experimentally observed; however,
provided this discrepancy is borne in mind, model (52) can provide
a useful means for predicting the configurations of the alcohols
formed in these cases.

However, caution must be exercised in the assignment of
absolute configurations on the basis of the models for asymmetric
hydroborations. It has been found that reaction of a given alkene
with (-)-diisopinocampheyldiborane [(-)-monoisopinocampheylborane]
in tetrahydrofuran leads to either enantiomer of the same substance
depending on the time elapsed between the preparation of the
reagent and the addition of the alkene (93). Thus, in the case of
benznorbornadiene, immediate addition of the alkene gives, after
oxidation and acetylation, (1R,2S)-(+)-exo-2-benznorbornenyl
acetate of optical purity 7.4%, while addition of the alkene
after 21 hr, followed by the same reaction sequence, gives the
(1S,2R)-enantiomer having an optical purity of 7.2%. It has been
suggested that these results might be due to the rearrangement
of the unsymmetrical 1,1-adduct (58) to the cis- and/or trans-
1,2-adduct (59); these two different hydroborating reagents could
produce the enantiomers obtained as shown below (93).

(Opt. purity: 7.4%)   (Opt. purity: 7.2%)

3.8   USE OF DIISOPINOCAMPHEYLBORANE IN THE RESOLUTION OF ALKENE RACEMATES

In view of the successful application of diisopinocampheyl-borane to the asymmetric hydroboration of alkenes its use in the separation of enantiomeric alkenes was investigated. Thus, hydroboration of a cis-alkene racemate with a deficient quantity of diisopinocampheylborane converts the more reactive enantiomer into the organoborane, leaving the less reactive enantiomer free to be recovered in its optically active form (91).

In general, the reaction is carried out using a mole ratio of alkene to borane of two to one; the configuration of the enantiomer isolated depends on the configuration of the diisopinocampheylborane used, and optical yields of up to 65% have been obtained. Since both (+)- and (-)-diisopinocampheylborane are readily prepared, it is possible to prepare both optical isomers of a given alkene with equal ease. In addition, the

reactive enantiomer can often be recovered from the organoborane
by displacement with another alkene (Sec. 2.3.3).

   The configuration of the enantiomer which reacts with the
diisopinocampheylborane can be predicted on the basis of
transition state models similar to those discussed in Sec. 3.7.
Thus, it is possible to predict the configuration of the recovered
enantiomer; in all the cases where such models have been applied,
the predicted and observed configurations agree (91). For example,
treatment of the 3-methylcyclopentene racemate with 50 mole % of
(-)-diisopinocampheylborane affords 75% (R)-(+)-3-methylcyclo-
pentene, having an optical purity of 45%. Using the stable
conformation of (-)-diisopinocampheylborane proposed in Sec. 3.7
[structure (51)] the following transition states can be formed.

(60) S-Enantiomer   (61) R-Enantiomer

   (Two other transition states are possible, but in these
models the asymmetric center of the cyclopentene molecule is too
remote from the pinane moieties of the reagent to have any
significant effect on the reaction.)

   Consideration of the above two models shows that model (60)
is definitely more favorable, and this results in the isolation
of the free R-enantiomer as experimentally observed. In the above
case it has been found that use of 80 mole % of the reagent gives
residual alkene of 65% optical purity (91).

   The preparation of optically active alkenes by reaction of
the racemates with diisopinocampheylborane is summarized in
Table 3.7.

TABLE 3.7

Resolution of Alkene Racemates (91)

Alkene	Alkene (mmoles)	$(-)-(IPC)_2BH^a$ (mmoles)	Config. of recovered alkene	Yield, %	Optical purity, %
3-Methylcyclopentene	62	31	(R)	75	45
3-Ethylcyclopentene	150	120	(R)		65
3-Ethylcyclopentene	100	50	(R)	80	37
1-Methylnorbornene	100	50	(1S,4R)	67	--
4-Methylcyclohexene	700	300^b	(R)	--	--
Trans-cyclooctene	50	18.4^c	d	40	20

^{a}Prepared from (+)-α-pinene of 93-95% optical purity (Ref. 91).

^{b}Prepared from (+)-α-pinene of 41% optical purity (Ref. 103).

^{c}Prepared from (+)-α-pinene of approximately 100% optical purity (Ref. 104).

^{d}(-)-Trans-cyclooctene isolated (Ref. 104).

REFERENCES

1. H. C. Brown and G. Zweifel, J. Am. Chem. Soc., 82, 4708 (1960).

2. H. C. Brown and G. Zweifel, J. Am. Chem. Soc., 83, 1241 (1961).

3. E. F. Knights and H. C. Brown, J. Am. Chem. Soc., 90, 5281 (1968).

4. S. Nishida, I. Moritani, K. Ito, and K. Sakai, J. Org. Chem., 32, 939 (1967).

5. M. S. Newman, A. Arkell, and T. Fukunaga, J. Am. Chem. Soc., 82, 2498 (1960).

6. M. Hanack and H. M. Ensslin, Annalen, 713, 49 (1968).

7. J. Klein and D. Lichtenberg, J. Org. Chem., 35, 2654 (1970).

8. H. C. Brown and J. H. Kawakami, J. Am. Chem. Soc., 92, 1990 (1970).

9. J. Lhomme and G. Ourisson, Tetrahedron, 24, 3201 (1968).

10. H. C. Brown and G. Zweifel, J. Am. Chem. Soc., 83, 2544 (1961).

11. S. P. Acharya, H. C. Brown, A. Suzuki, S. Nozawa, and M. Itoh, J. Org. Chem., 34, 3015 (1969).

12. G. Zweifel, N. R. Ayyangar, and H. C. Brown, J. Am. Chem. Soc., 85, 2072 (1963).

13. H. C. Brown and A. W. Moerikofer, J. Am. Chem. Soc., 85, 2063 (1963).

14. H. C. Brown and B. C. Subba Rao, J. Am. Chem. Soc., 81, 6428 (1959).

15. H. C. Brown and A. W. Moerikofer, J. Am. Chem. Soc., 84, 1478 (1962).

16. A. M. Krubiner and E. P. Oliveto, J. Org. Chem., 31, 24 (1966).

17. A. M. Krubiner, N. Gottfried, and E. P. Oliveto, J. Org. Chem., 33, 1715 (1968).

18. G. Ohloff and G. Uhde, Helv. Chim. Acta, 48, 10 (1965).

19. D. J. Pasto and F. M. Klein, J. Org. Chem., 33, 1468 (1968).

20. J. Klein, E. Dunkelblum, and D. Avrahami, J. Org. Chem., 32, 935 (1967).

21. R. Köster, S. Arora, and P. Binger, Angew. Chem. Intern. Ed., 8, 205 (1969).

22. D. K. Shumway and J. D. Barnhurst, J. Org. Chem., 29, 2320 (1964).

23. D. J. Pasto and F. M. Klein, Terahedron Letters, 1967, 963.

24. W. Cocker, P. V. R. Shannon, and P. A. Staniland, J. Chem. Soc., C1967, 485.

25. S. P. Acharya and H. C. Brown, J. Am. Chem. Soc., 89, 1925 (1967).

26. H. C. Brown and A. Suzuki, J. Am. Chem. Soc., 89, 1933 (1967).

27. I. Tabushi, K. Fujita, and R. Oda, J. Org. Chem., 35, 2383 (1970).

28. F. Sondheimer and S. Wolfe, Can. J. Chem., 37, 1870 (1959).

29. J. A. Marshall and A. R. Hochstetler, J. Am. Chem. Soc., 91, 648 (1969).

30. C. H. Heathcock and T. Ross Kelly, Tetrahedron, 24, 1801 (1968).

31. J. A. Marshall, W. I. Fanta, and G. L. Bundy, Tetrahedron Letters, 1965, 4807.

32. J. A. Marshall, M. T. Pike and R. D. Carroll, J. Org. Chem., 31, 2933 (1966).

33. C. H. Heathcock and R. Ratcliffe, J. Am. Chem. Soc., 93, 1746 (1971).

34. G. De Faye, Compte. Rend. Acad. Sci. Paris, 260, 2843 (1965).

35. R. T. Gray and C. Djerassi, J. Org. Chem., 35, 753 (1970).

36. R. Pesnelle and G. Ourisson, J. Org. Chem., 30, 1744 (1965).

37. B. Rickborn and S. E. Wood, Chem. & Ind. (London), 1966, 162.

38. W. F. Erman, J. Am. Chem. Soc., 86, 2887 (1964).

39. G. W. Oxer and D. Wege, Tetrahedron Letters, 1969, 3513.

40. E. J. Corey and R. S. Glass, J. Am. Chem. Soc., 89, 2600 (1967).

41. R. R. Sauers, S. B. Schlosberg, and P. E. Pfeffer, J. Org. Chem., 33, 2175 (1968).

42. (a) L. Borowiecki and Y. Chretien-Bessiere, Bull. Soc. Chim. France, 1967, 2364; (b) J. M. Coxon, M. P. Hartshorn, and A. J. Lewis, Australian J. Chem., 24, 1017 (1971).

43. S. P. Jindal and T. T. Tidwell, Tetrahedron Letters, 1971, 783.

44. P. G. Gassmann and J. L. Marshall, J. Am. Chem. Soc., 88, 2822 (1966).

45. G. Zweifel and H. C. Brown, J. Am. Chem. Soc., 86, 393 (1964).

46. E. Klein and W. Rojahn, Chem. Ber., 100, 1902 (1967).

47. M. Barthelemy and Y. Bessiere-Chretien, Tetrahedron Letters, 1970, 4265.

48. S. P. Acharya and H. C. Brown, J. Org. Chem., 35, 196 (1970).

49. G. Büchi, W. D. Macleod, and J. Padilla, J. Am. Chem. Soc., 86, 4438 (1964).

50. J. P. Schaefer, J. C. Lark, C. A. Flegal, and L. M. Honig, J. Org. Chem., 32, 1372 (1967).

51. W. D. K. Macrosson, J. Martin, and W. Parker, Tetrahedron Letters, 1965, 2589.

52. J. A. Marshall and H. Faubl, J. Am. Chem. Soc., 92, 948 (1970).

53. F. Y. Edamura and A. Nickon, J. Org. Chem., 35, 1509 (1970).

54. (a) S. P. Acharya and H. C. Brown, J. Org. Chem., 35, 3874 (1970); (b) F. Fringuelli and A. Taticchi, J. Chem. Soc., C1971, 2011.

55. G. Defaye-Duchateau, Bull. Soc. Chim. France, 1964, 1469.

56. J. C. Sircar and G. S. Fisher, J. Org. Chem., 34, 404 (1969).

57. D. K. Black and G. W. Hedrick, J. Org. Chem., 32, 3758 (1967).

58. W. Herz and J. J. Schmid, J. Org. Chem., 34, 3464 (1969).

59. J. W. ApSimon, P. V. Demarco, and J. Lemke, Can. J. Chem., 43, 2793 (1965).

60. R. E. Ireland and L. N. Mander, J. Org. Chem., 34, 142 (1969).

61. M. Nussim, Y. Mazur, and F. Sondheimer, J. Org. Chem., 29, 1120 (1964).

62. C. W. Shoppee, J. C. Coll, and R. E. Lack, J. Chem. Soc., C1970, 1893.

63. A. Hassner and C. Pillar, J. Org. Chem., 27, 2914 (1962).

64. G. Cainelli and A. Selva, Gazz. Chim. Ital., 97, 720 (1967).

65. B. Matkovics, Zs. Tegyey, and Gy. Göndös, Steroids, 5, 117 (1965).

66. R. R. Sobti and S. Dev, Tetrahedron Letters, 1966, 3939.

67. H. Nakata, Bull. Chem. Soc. Japan, 38, 378 (1965).

68. C. W. Shoppee, R. Lack, and B. McLean, J. Chem. Soc., 1964, 4996.

69. G. Quinkert, B. Wegemund, F. Homburg, and G. Cimbollek, Chem. Ber., 97, 958 (1964).

70. R. Boutarel, Bull. Soc. Chim. France, 1964, 1665.

71. J. M. Midgley, J. E. Parkin, and W. B. Whalley, Chem. Commun. 1970, 789.

72. R. W. Kelly and P. J. Sykes, J. Chem. Soc., C1968, 416.

73. Y. Tsuda, T. Sano and Y. Inubushi, Tetrahedron, 26, 751 (1970).

74. H. C. Brown and G. J. Klender, Inorg. Chem., 1, 204 (1962).

75. W. J. Wechter, Chem. & Ind. (London), 1959, 294.

76. T. J. Logan and T. J. Flautt, J. Am. Chem. Soc., 82, 3446 (1960).

77. R. Fellous and R. Luft, Tetrahedron Letters, 1970, 1505; R. Fellous, R. Luft, and A. Puill, ibid., 1972, 245.

78. J. E. Dubois and G. Mouvier, Tetrahedron Letters, 1965, 1629.

79. G. Zweifel and H. C. Brown, J. Am. Chem. Soc., 85, 2066 (1963).

80. K. Sisido, K. Inomata, T. Kageyema, and K. Utimoto, J. Org. Chem., 33, 3149 (1968).

81. R. A. Benkeser and E. M. Kaiser, J. Org. Chem., 29, 955 (1964); Org. Syn., 50, 88 (1970).

82. H. C. Brown and R. L. Sharp, J. Am. Chem. Soc., 88, 5851 (1966).

83. J. Klein, E. Dunkelblum, and M. A. Wolff, J. Organometal. Chem., 7, 377 (1967).

84. G. S. Hammond, J. Am. Chem. Soc., 77, 334 (1955).

85. B. R. Ree and J. C. Martin, J. Am. Chem. Soc., 92, 1660 (1970), and references cited therein.

86. J. D. Roberts and V. C. Chambers, J. Am. Chem. Soc., 73, 5030 (1951).

87. D. J. Pasto and S. Z. Kang, J. Am. Chem. Soc., 90, 3797 (1968).

88. A. Streitwieser, L. Verbit, and R. Bittman, J. Org. Chem., 32, 1530 (1967).

89. R. Köster, Angew. Chem. Intern. Ed., 3, 174 (1964); R. Köster, G. Griasnow, W. Larbig, and P. Binger, Annalen, 672, 1 (1964).

90. H. C. Brown and A. W. Moerikofer, J. Am. Chem. Soc., 83, 3417 (1961).

91. H. C. Brown, N. R. Ayyangar, and G. Zweifel, J. Am. Chem. Soc., 86, 397 (1964).

92. R. N. McDonald and R. N. Steppel, J. Am. Chem. Soc., 92, 5664 (1970).

93. D. J. Sandman, K. Mislow, W. P. Giddings, J. Dirlam, and G. C. Hanson, J. Am. Chem. Soc., 90, 4877 (1968).

94. (a) H. Tanida, H. Ishitobi, T. Irie, and T. Tsushima, J. Am. Chem. Soc., 91, 4512 (1969); (b) J. E. Herz and L. A. Marquez, J. Chem. Soc., C1971, 3504.

95. H. C. Brown, N. R. Ayyangar, and G. Zweifel, J. Am. Chem. Soc., 86, 1071 (1964).

96. H. T. Thomas and K. Mislow, J. Am. Chem. Soc., 92, 6292 (1970).

97. J. Katsuhara, H. Watanabe, K. Hashimoto and M. Kobayashi, Bull. Chem. Soc. Japan, 39, 617 (1966).

98. G. Zweifel, N. R. Ayyangar, T. Munekata, and H. C. Brown, J. Am. Chem. Soc., 86, 1076 (1964).

99. E. Caspi, M. Galli Kienle, K. R. Varma, and L. J. Mulheirn, J. Am. Chem. Soc., 92, 2161 (1970).

100. D. R. Brown, S. F. A. Kettle, J. McKenna, and J. M. McKenna, Chem. Commun., 1967, 667.

101. D. S. Matteson, Organometal. Chem. Rev., B6, 323-399 (1970), p. 359.

102. K. R. Varma and E. Caspi, Tetrahedron, 24, 6365 (1968).

103. S. I. Goldberg and F. L. Lam, J. Org. Chem., 31, 240 (1966).

104. W. L. Waters, J. Org. Chem., 36, 1569 (1971).

Chapter 4

OXIDATION OF ORGANOBORANES

In this chapter the various methods for the oxidation of organoboranes are discussed from the point of view of the mechanism of the reactions and their synthetic utility.

In general, hydroboration proceeds by a cis-Markownikoff addition of the boron-hydrogen bond to the alkene double bond, the reaction proceeding preferentially from the less hindered direction. Oxidation with either alkaline hydrogen peroxide (Sec. 4.3) or amine-N-oxides (Sec. 4.4) converts the organoboranes to the corresponding alcohols. Since both these methods of oxidation proceed with retention of configuration, and without any skeletal rearrangement, these hydroboration-oxidation sequences provide very useful methods for the stereoselective cis-anti-Markownikoff hydration of alkenes. The full synthetic utility of the method has been discussed in Sec. 3.1 on the scope of the hydroboration reaction as applied to alkenes. Autoxidation of organoboranes (Sec. 4.1.1) has also been applied to the synthesis of alcohols, though this method lacks the stereoselectivity of the other two methods mentioned above.

Oxidation of organoboranes with chromic acid gives ketones (Sec. 4.6), thus providing a useful method for the direct conversion of alkenes into ketones via hydroboration.

4.1 AUTOXIDATION OF ALKYLBORANES

Lower alkylboranes are spontaneously inflammable in air and react violently with oxygen, while higher homologs, though being sensitive to oxygen, are not inflammable. Solutions of alkylboranes should therefore be stored under nitrogen; an exception is the selective hydroborating reagent, 9-borabicyclo [3.3.1]nonane, which is very stable (1). Arylboranes are less sensitive to oxygen, and derivatives containing bulky groups, such as trimesitylborane, are relatively stable (2).

Examination of both the aerial (3) and autoxidation (4) of alkylboranes reveals that the reaction is initially very fast, with the first boron-carbon bond being oxidized faster than the second which, in turn, is oxidized much faster than the third. The autoxidation proceeds via the initial formation of alkylperoxyboranes which have been isolated in a number of cases (5). In view of the fact that many of the usual radical scavengers, such as hydroquinone, fail to inhibit the reaction, it was initially thought that the reaction proceeds by an ionic rather than a radical mechanism (6). Later work has, however, shown that galvinoxyl exhibits a marked inhibiting effect (7-10). This evidence, together with the observations that the autoxidation of (+)-dihydroxy-1-phenylethylborane [(+)-1-phenylethylboronic acid] (8), diisopinocampheylbutylborane (9), and the epimeric trinorbornylboranes (10) proceeds with considerable racemization, indicates the involvement of intermediate free-radical species in the reaction.

On the basis of the above studies and other studies involving the autoxidation of various butylboranes the following homolytic chain mechanism has been proposed (11,12).

$$
\begin{array}{lll}
\text{Initiation} & R_3B + O_2 \longrightarrow R\cdot & (1) \\
\text{Propagation} & R\cdot + O_2 \longrightarrow RO_2\cdot & (2) \\
& RO_2\cdot + R_3B \longrightarrow RO_2BR_2 + R\cdot & (3) \\
\text{Termination} & 2RO_2\cdot \ (\text{or } 2R\cdot) \longrightarrow \text{inactive products} & (4)
\end{array}
$$

Iodine has been found to be a powerful inhibitor of radical chain reactions of organoboranes, and this property has been used in a study of the initiation reaction (13). Initiation has been shown to involve the slow attack of oxygen on the boron atom, the rate of the reaction being retarded by the presence of bulky substituents on the boron. Thus, the rate of initiation of autoxidation of tri-n-butylborane is found to be considerably greater than that of tri-sec-butylborane, which, in turn, is greater than that of tri-isobutylborane (13b). The overall rate of autoxidation is, however, found to be greater for tri-sec-alkylboranes than for primary alkyl derivatives (4), indicating that the chain propagation steps are more favorable in the former cases.

The proposed propagation step involving rapid attack by an alkylperoxy radical on boron with displacement of an alkyl radical [Eq. (3)] is similar to reactions that occur between tert-butoxy radicals and trialkylboranes. Thus, the photolysis of di-tert-butylperoxide in the presence of trialkylboranes results in the formation of alkyl radicals that have been detected by ESR spectroscopy (14).

$$\text{tert-BuO} \cdot + R_3B \longrightarrow \text{tert-BuOBR}_2 + R \cdot$$

The alkylperoxyborane produced in the initial oxidation [Eq. (3)] may either react with a second mole of oxygen to give the corresponding dialkylperoxyborane [Eq. (5)] (4), or may undergo an intermolecular redox reaction to produce an alkoxyborane [Eq. (6)] (15).

$$RO_2BR_2 + O_2 \longrightarrow (RO_2)_2BR \qquad (5)$$
$$RO_2BR_2 + R_3B \longrightarrow 2ROBR_2 \qquad (6)$$

The autoxidation of trialkylboranes under controlled conditions has found application in the synthesis of alcohols and hydroperoxides.

4.1.1 Synthesis of Alcohols

Early attempts to apply autoxidation of alkylboranes to the
synthesis of alcohols made use of long reaction times and elevated
reaction temperatures; the reaction was reported to be complex and
proved to be unsatisfactory (3,15). However, use of mild
conditions and the theoretical amount of oxygen (1.5 moles O_2 per
mole of R_3B), followed by treatment of the product [possibly
$ROB(O_2R)R$] with aqueous alkali, gives the corresponding alcohols
in high yield (4).

$$R_3B \quad \xrightarrow[\text{2) NaOH, } H_2O]{\text{1) 1.5 } O_2\text{, THF, } 0^\circ} \quad 3\,ROH$$

$$>80\%$$

This method of synthesis of alcohols has a disadvantage in that
it lacks the stereospecificity of the oxidation with alkaline
hydrogen peroxide or amine N-oxides (Secs. 4.3 and 4.4) as
illustrated below (4).

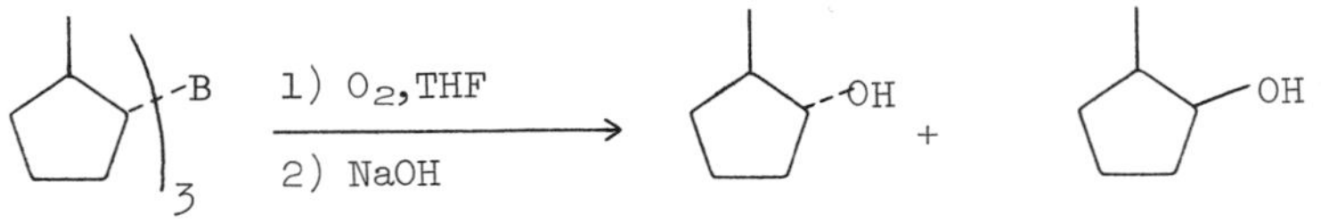

(81%)   (19%)

The formation of the alternate isomers is due to the radical
nature of the reaction. In cases where stereochemical factors
are not involved the method provides a convenient route to
alcohols.

4.1.2 Synthesis of Hydroperoxides

Minimization of the intermolecular redox reaction [Eq. (6)]
results in preferential uptake of a second mole of oxygen
[Eq. (5)], thus providing a useful route to alkyl hydroperoxides.
Such minimization has been achieved by use of very dilute
(0.01-0.05 M) solutions of the alkylborane (16). Oxidation of

the remaining carbon-boron bond with peracids or hydrogen peroxide
gives alkyl hydroperoxides in 30-60% yields (16).

A more convenient synthesis involves initial autoxidation at
-78° which results in a rapid uptake of the first mole of oxygen;
warming the reaction mixture to 0° leads to rapid absorption of
the second mole of oxygen. Addition of an aqueous solution of
hydrogen peroxide at 0° gives the alkyl hydroperoxide in high
yields (17).

$$R_3B + O_2 \xrightarrow[-78°]{THF} R_2B(O_2R) \xrightarrow[0°]{O_2,\ THF} RB(O_2R)_2 \xrightarrow[H_2O]{H_2O_2} 2RO_2H + ROH$$
$$>80\%$$

The alkyl hydroperoxide may be separated from the alcohol by
conversion to the water soluble potassium salt, followed by
isolation of the aqueous phase and acidification (17).

4.2   NEUTRAL HYDROGEN PEROXIDE

The reaction of equimolar quantities of aqueous hydrogen
peroxide and tri-n-hexylborane in tetrahydrofuran results in the
cleavage of two of the three boron-carbon bonds to give a mixture
of hydrocarbons and n-hexanol (18).

$$(n\text{-}C_6H_{13})_3B \xrightarrow[0°,\ 24\ hr]{H_2O_2,THF} n\text{-}C_6H_{14} + n\text{-}C_{12}H_{26} + C_4H_9\underset{\underset{CH_3}{|}}{C}HC_6H_{13} +$$

$$\qquad\qquad\qquad\qquad 6.5\% \qquad 28.6\% \qquad 8.1\%$$

n-C$_6$H$_{13}$OH

5.4%

Subsequent treatment of the mixture with alkaline hydrogen
peroxide gives a further 34% n-hexanol and 2.3% hexan-2-ol.

The formation of 5-methylundecane is the result of the
reaction of 1-methylpentyl groups formed in the initial
hydroboration of 1-hexene. When the reaction is carried out

in carbon tetrachloride solution, 27.3% of 1-chlorohexane is
formed at the expense of dimer and monomer production which
drops to 9.6 and 3.1% respectively. In addition, 24% n-hexanol
is formed (18).

In the case of alkylboranes containing secondary or tertiary
carbon-boron bonds, some cleavage of the third boron-carbon bond
also occurs (18). There is a marked increase in the amounts of
alcohols and monomeric hydrocarbons formed, and a sharp decrease
in dimer formation.

The above findings suggest that radicals are formed during
these reactions. Further evidence for the formation of
intermediate radicals has been obtained from studies of the
reactions of three epimeric mixtures of trinorbornylborane with
aqueous hydrogen peroxide (19). The epimeric composition of
the 2,2'-bisnorbornane dimer formed in each reaction was found
to be independent of the epimeric compositions of trinorbornylborane used; in addition, when the reactions were carried out in
carbon tetrachloride, the epimeric composition of the norbornyl
chloride produced was likewise independent of the epimeric
composition of the trinorbornylborane used. On the basis of
the above results the following mechanism has been proposed (18).

$$R_2B\text{-}O\text{-}OH + R_3B \longrightarrow R_2B\text{-}O\cdot + R\cdot + R_2B\text{-}OH \qquad (7)$$

$$R_2B\text{-}O\cdot + H\text{-}O\text{-}O\text{-}H \longrightarrow R_2B\text{-}OH + \cdot O_2H \qquad (8)$$

$$R_2B\text{-}O\cdot + H\text{-}O\text{-}O\text{-}H \longrightarrow R_2B\text{-}O_2H + \cdot OH \qquad (8')$$

$$R_3B + \cdot O_2H \longrightarrow R_2B\text{-}O_2H + R\cdot \qquad (9)$$

It has been shown that the monomeric hydrocarbons are formed
by abstraction of hydrogen from the solvent. In the case of
alkylboranes containing secondary and tertiary carbon-boron bonds
the reduced formation of dimers is attributed to steric factors.

Alcohol production can arise from the combination of alkyl
and hydroxyl radicals (or hydrogen peroxide); however, the
epimeric composition of the norbornanols produced in the above
reactions of trinorbornylborane with aqueous hydrogen peroxide is

close to that of the trinorbornylborane used, indicating that
most of the alcohol is formed via a polar process involving
retention of configuration (see Sec. 4.3).

$$R_2\overset{\overset{\displaystyle R}{|}}{\underset{\ominus}{B}}\text{-O-OH} \longrightarrow R_2BOR + OH^{\ominus}$$

4.3 ALKALINE HYDROGEN PEROXIDE

The oxidation of organoboranes with alkaline hydrogen
peroxide is extensively used in the synthesis of alcohols via
the hydroboration of alkenes (see Sec. 3.1). An early report of
the reaction involved the complete dealkylation of tri-n-butylborane to form boric acid and n-butyl alcohol (20), and the
reaction was later developed as an analytical procedure for
boron estimation (21).

Early applications (20) of this method involved treatment
of the organoborane with excess hydrogen peroxide and concentrated
sodium hydroxide under reflux, but it was later found that milder
conditions gave better yields of the desired alcohols. Detailed
studies of the reaction showed that hydrogen peroxide concentration,
base concentration, and oxidation temperature can be varied widely
without greatly affecting the yield of alcohol (22).

The general procedure may be illustrated by the oxidation
of tri-n-hexylborane (22). The trialkylborane (16.6 mmoles) in
diglyme (40 ml) is treated with sodium hydroxide (15 mmoles; 5 ml
of a 3 M solution), followed by the dropwise addition of a 20%
excess of hydrogen peroxide (60 mmoles; 6 ml of a 30% solution) at
25 to 30°. A vigorous reaction occurs, and the mixture is then
allowed to stand at room temperature for an additional hour. The
alcohol is isolated by extraction with ether. Where milder basic
conditions are required sodium acetate can be used in place of
sodium hydroxide (23a).

The reaction proceeds equally well in tetrahydrofuran, but,
if the initial hydroboration is carried out in diethyl ether,

efficient oxidation requires the addition of ethanol as cosolvent, presumably to aid mixing of the solvents ($\underline{22}$).

Unlike chromic acid oxidations discussed in Sec. 4.6, oxidation with alkaline hydrogen peroxide appears to be unaffected by the steric requirements of the alkyl groups of the organoborane; the application of the reaction to a wide variety of organoboranes is illustrated in Sec. 3.1. In addition, the reaction is unaffected by the presence of other functional groups, such as alkenes, alkynes, aldehydes, ketones, esters, halides, and nitriles which remain unchanged at the end of the oxidation ($\underline{22}$).

The oxidation proceeds without skeletal rearrangements of the alkyl groups, and with predominant retention of configuration. Thus, in the oxidation of (+)-tris[(R)-2-methylbutyl]borane and (+)-tris[(S)-3-methylpentyl]borane the extent of racemization of the alkyl groups was calculated as 7 and 0.3%, respectively ($\underline{23b}$). Overall retention of configuration is observed in the hydroboration-oxidation of a wide variety of cyclic alkenes (Secs 3.1.5 to 3.1.8), and has been discussed in detail in the case of norbornene ($\underline{24}$).

No detailed kinetic studies of the reaction with trialkylboranes have been reported. However, studies of the reaction of alkyl- ($\underline{25}$) and aryldihydroxyboranes ($\underline{26}$, $\underline{27}$) (boronic acids) with alkaline hydrogen peroxide indicate that the reaction proceeds by an S_E2 mechanism. The following mechanism has been proposed for the base-catalyzed reaction ($\underline{25}$).

$$H_2O_2 + OH^{\ominus} \rightleftarrows HO_2^{\ominus} + H_2O$$

$$\underset{\overset{|}{OH}}{\overset{R}{\underset{|}{HO-B}}} + {}^{\ominus}O_2H \rightleftarrows \left[\underset{\overset{|}{OH}}{\overset{R}{\underset{|}{HO-B-O-OH}}} \right]^{\ominus} \longrightarrow \left[HO-\overset{\overset{\delta-}{R}}{\underset{\overset{|}{OH}}{B}}\cdots\underset{\overset{}{OH}}{\overset{}{O}}\,{}_{\delta-} \right] \longrightarrow$$

$$\underset{\overset{|}{OH}}{HO-B-OR} \overset{H_2O}{\longrightarrow} ROH + H_3BO_3$$

$$+OH^{\ominus}$$

An identical mechanism, in three successive stages, may be suggested for the oxidation of trialkylboranes, and is consistent with the retention of configuration and freedom from rearrangement observed in the oxidation.

It should be noted that oxidation of tri-n-hexylborane with alkaline hydrogen peroxide has been reported to yield small amounts of dimeric products, showing that coupling reactions do compete, even under alkaline conditions (18). Likewise phenylethane-1,2- and 2,2-diboronic esters have been shown to undergo considerable radical cleavage on treatment with alkaline hydrogen peroxide (28).

$$PhCH_2CH \left[-B \underset{O}{\overset{O}{\diagup}} \right]_2 \xrightarrow[OH^\ominus]{H_2O_2} \quad PhCHO + PhCH_2OH \\ \text{(main products)}$$

4.4 AMINE N-OXIDES

Organoboranes containing primary and secondary alkyl, aryl, alkenyl, and cycloalkenyl groups, as well as heterocyclic boron compounds, are readily oxidized to the corresponding boric esters by anhydrous amine N-oxides, such as trimethylamine N-oxide or pyridine N-oxide (29,30). The boron-carbon bond of alkynyl substituents is, however, not affected. Determination of the liberated amine permits the quantitative determination of boron-carbon bonds, and the reaction has been developed as an analytical procedure for the determination of such bonds in organoboranes (31).

The general procedure (29) involves the dropwise addition of a solution of the organoborane (1 mole equivalent) in anhydrous benzene (or toluene) to a boiling suspension of the trimethylamine N-oxide (3 mole equivalents) in the same solvent. The mixture is refluxed for 30 min and the liberated trimethylamine is carried over in a stream of nitrogen into excess 0.5 M sulfuric acid; titration of the excess acid determines the amount of trimethyl-amine formed in the reaction. Filtration of the residual reaction

mixture and evaporation of the solvent gives the borate ester.
The alcohol is obtained by transesterification with methanol.
Alternatively, the reaction mixture may be hydrolyzed with
dilute sulfuric acid, and the alcohol extracted with ether ($\underline{32}$).

The oxidation occurs without the complication of the homolysis
of O-O bonds, and appears to be a milder method than that using
alkaline hydrogen peroxide. Thus, in the case of the oxidation
of tricyclopropylborane, the sensitivity of cyclopropanol to
aqueous hydrogen peroxide precludes the use of alkaline hydrogen
peroxide, whereas trimethylamine N-oxide gives the cyclopropanol
in high yield ($\underline{33}$).

As in the case of the autoxidation of alkylboranes (Sec. 4.1),
the first boron-carbon bond is oxidized faster than the second,
which, in turn, is oxidized faster than the third ($\underline{31}$). Thus,
treatment with one equivalent of N-oxide at room temperature gives
the pure dialkyl-monoalkoxyborane, while use of two equivalents
gives the monoalkyl-dialkoxyborane. The oxidation of the final
boron-carbon bond requires more vigorous conditions, such as
higher temperature. This permits the selective oxidation of
boron-carbon bonds, and is illustrated by the selective oxidation
of divinylalkylboranes to give the corresponding divinylalkoxy-
boranes ($\underline{34}$).

Reaction of (+)-dibutoxy-1-phenylethylborane with
trimethylamine N-oxide proceeds with 98% retention of
configuration to give (-)-1-phenylethanol in 80% yield ($\underline{32}$). The
mechanism of the reaction is probably similar to that proposed
for oxidation with alkaline hydrogen peroxide, involving
1,2-migration of the alkyl group from boron to oxygen.

4.5   SODIUM HYPOCHLORITE

Treatment of alkaline solutions of organoboranes in diglyme with sodium hypochlorite at 25-30° gives the corresponding alcohols in yields exceeding 70% (35).

4.6   CHROMIC ACID OXIDATION

The treatment of organoboranes with aqueous 8 N chromic acid gives ketones, and thus provides a useful method for the conversion of alkenes into ketones via hydroboration (36). While any of the usual ethereal solvents can be used in the initial hydroboration reaction, it has been found that use of ethyl ether greatly facilitates the isolation of the ketone, and hence this solvent is generally preferred. The procedure involves hydroboration of the alkene in ethyl ether (see Sec. 2.2.1); addition of a small quantity of water (to destroy residual hydride) is followed by the slow addition of a 10% excess of aqueous 8 N chromic acid. After refluxing the mixture for 2 hr the product is isolated in yields usually exceeding 60%. The preparation of a variety of ketones using this method is summarized in Table 4.1.

TABLE 4.1

Preparation of Ketones via Hydroboration-Chromic Acid Oxidation

Alkene	Ketone	Yield %	Reference
Cyclohexene	Cyclohexanone	65	36
1-Methylcyclohexene	2-Methylcyclohexanone	87	36
1-Phenylcyclohexene	2-Phenylcyclohexanone	63	36
α-Pinene	Isopinocamphone	72	36
Cis-bicyclo[4.2.0] oct-7-ene	Cis-bicyclo[4.2.0] octan-7-one	60	37
3β,20β-Diacetoxypregn-5-ene	3β,20β-Diacetoxypregnan-6-one	48	38

The method has also been applied to the synthesis of the
C_{16} musk compound, 8-cyclohexadecen-1-one, in 44% yield, via
monohydroboration of 1,9-cyclohexadecadiene (39).

In a number of cases the direct chromic acid oxidation of
alkylboranes has been found to lead to rearranged products (40).
Thus, monohydroboration of norbornadiene, followed by oxidation,
gives nortricyclanone and bicyclo[2.2.1]hept-5-en-2-one
(dehydronorcamphor) in 51 and 49% yields, respectively.

51%   49%

Oxidation of the intermediate borane with alkaline hydrogen
peroxide, followed by chromic acid oxidation of the alcohol,
proceeds without rearrangement. However, the application of the
hydroboration-chromic acid oxidation procedure to the synthesis
of ketones from norbornene derivatives, as opposed to norbornadiene, has been reported to proceed without rearrangement (41).

While no mechanistic studies of the chromic acid oxidation
of alkylboranes have been reported, the kinetics of the oxidation
of a number of alkyldihydroxyboranes (alkylboronic acids) $RB(OH)_2$
has been investigated (42). The results indicate that oxidation
at pH values between 3 and 7 (alcohols are relatively stable to
Cr^{VI} in this pH range) leads to alcohols, while at lower pH
(>2 N acid) the products are ketones. The following tentative
mechanism has been proposed (42).

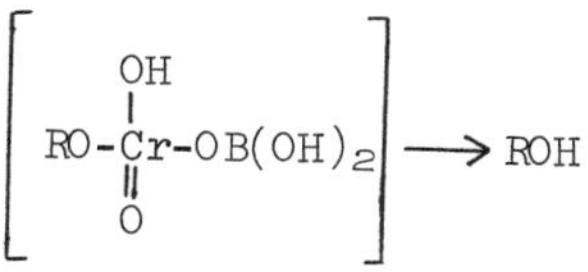

$$\left[\begin{array}{c} \text{OH} \\ | \\ \text{RO-Cr-OB(OH)}_2 \\ \| \\ \text{O} \end{array} \right] \longrightarrow \text{ROH}$$

Moreover, the rate of oxidation has been found to be more sensitive to the structure of R than in the case of alkaline hydrogen peroxide oxidation; thus, the second order rate constants in 0.114 M perchloric acid at $30°$ are 7.5×10^{-2} (R=tert-butyl), 6.6×10^{-4} (R=ethyl), and 2.4×10^{-7} liter mole^{-1} sec^{-1} (R=methyl). This should make possible the selective cleavage of alkylborane bonds (42).

The occurrence of rearrangement observed in the oxidation of the borane derived from norbornadiene (40) (see discussion above) may possibly arise from C-O fission of the intermediate chromate ester as follows:

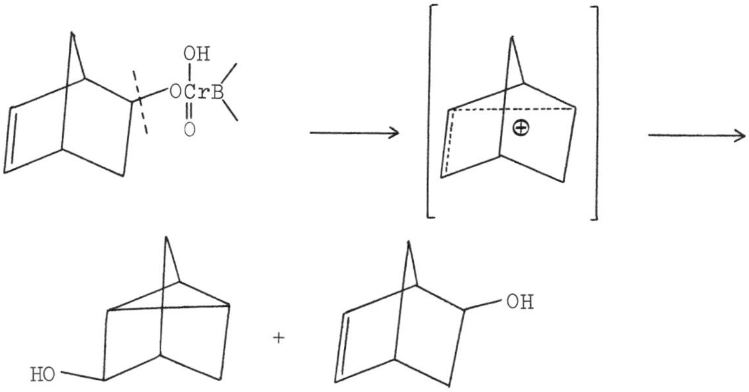

The method has been applied to synthesis of optically active ketones using diisopinocampheylborane (Sec. 3.5) as hydroborating reagent. Thus, reaction of norbornene with (-)-diisopinocampheyl-borane, followed by chromic acid oxidation, gives (1S, 4R)-(+)-norcamphor of 21% optical purity in 27% yield (43).

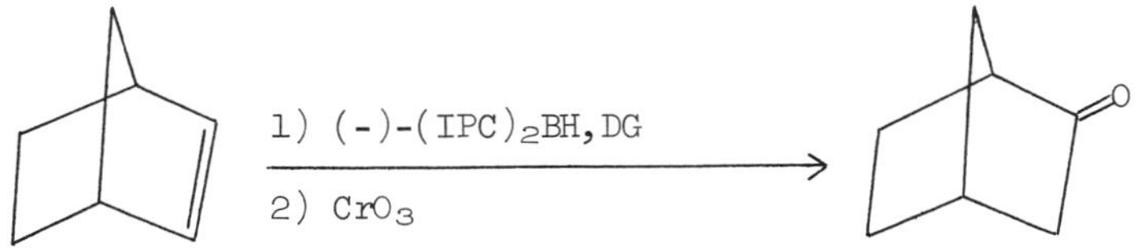

$$\xrightarrow[\text{2) CrO}_3]{\text{1) }(-)\text{-(IPC)}_2\text{BH,DG}}$$

27%

(21% opt. purity)

REFERENCES

1. E. F. Knights and H. C. Brown, J. Am. Chem. Soc., 90, 5280 (1968).

2. H. C. Brown and V. H. Dodson, J. Am. Chem. Soc., 79, 2302 (1957).

3. S. B. Mirviss, J. Am. Chem. Soc., 83, 3051 (1961).

4. H. C. Brown, M. M. Midland, and G. W. Kabalka, J. Am. Chem. Soc., 93, 1024 (1971).

5. A. G. Davies, D. G. Hare, and O. R. Khan, J. Chem. Soc., 1963, 1125; and references cited therein.

6. M. H. Abraham and A. G. Davies, J. Chem. Soc., 1959, 429.

7. P. G. Allies and P. B. Brindley, Chem. & Ind. (London), 1968, 1439.

8. A. G. Davies and B. P. Roberts, J. Chem. Soc., B 1967, 17.

9. P. G. Allies and P. B. Brindley, Chem. & Ind. (London), 1967, 319.

10. A. G. Davies and B. P. Roberts, J. Chem. Soc., B 1969, 311.

11. P. B. Allies and P. B. Brindley, J. Chem. Soc., B 1969, 1126.

12. A. G. Davies, K. U. Ingold, B. P. Roberts, and R. Tudor, J. Chem. Soc., B 1971, 698.

13. (a) M. M. Midland and H. C. Brown, J. Am. Chem. Soc., 93, 1506 (1971); (b) H. C. Brown and M. M. Midland, Chem. Commun., 1971, 699.

14. A. G. Davies and B. P. Roberts, J. Organometal. Chem., 19, P17 (1969); Chem. Commun., 1969, 699; A. G. Davies, D. Griller, and B. P. Roberts, J. Chem. Soc., B 1971, 1823; P. J. Krusic and J. K. Kochi, J. Am. Chem. Soc., 91, 3942 (1969).

15. S. B. Mirviss, J. Org. Chem., 32, 1713 (1967).

16. G. Wilke and P. Heimbach, Annalen, 652, 7 (1962).

17. H. C. Brown and M. M. Midland, J. Am. Chem. Soc., 93, 4078 (1971).

18. D. B. Bigley and D. W. Payling, J. Chem. Soc., B 1970, 1811.

19. A. G. Davies and R. Tudor, J. Chem. Soc., B 1970, 1815.

20. H. R. Snyder, J. A. Kuck, and J. R. Johnson, J. Am. Chem. Soc., 60, 105 (1938).

21. R. Belcher, D. Gibbons, and A. Sykes, Mikrochim. Acta, 40, 76 (1952).

22. G. Zweifel and H. C. Brown, Organic Reactions, 13, 1 (1963).

23. (a) E. Negishi and H. C. Brown, Synthesis, 1972, 196; (b) P. Pino, L. Lardicci, and A. Stefani, Ann. Chim. (Rome), 52, 456 (1962).

24. H. C. Brown and G. Zweifel, J. Am. Chem. Soc., 83, 2544 (1961).

25. H. Minato, J. C. Ware, and T. G. Traylor, J. Am. Chem. Soc., 85, 3024 (1963).

26. H. G. Kuivila, J. Am. Chem. Soc., 76, 870 (1954).

27. H. G. Kuivila and A. G. Armour, J. Am. Chem. Soc., 79, 5659 (1957).

28. D. J. Pasto, S. K. Arora, and J. Chow, Tetrahedron, 25, 1571 (1969).

29. R. Köster and Y. Morita, Angew. Chem. Intern. Ed., 5, 580 (1966).

30. R. Köster, German Patent 1,294,964, May 14, 1969; Chem. Abstr., 71, 39139t (1969).

31. R. Köster and Y. Morita, Annalen, 704, 70 (1967).

32. A. G. Davies and B. P. Roberts, J. Chem. Soc., C 1968, 1474.

33. R. Köster, S. Arora, and P. Binger, Angew. Chem. Intern. Ed., 8, 205 (1969).

34. G. Zweifel, N. L. Polston, and C. W. Whitney, J. Am. Chem. Soc., 90, 6243 (1968).

35. H. C. Brown, U. S. Patent 3,439,046, April 15, 1969; Chem. Abstr., 71, 50273c (1969).

36. H. C. Brown and C. P. Garg, J. Am. Chem. Soc., 83, 2951 (1961).

37. W. R. Moore and W. R. Moser, J. Org. Chem., 35, 908 (1970).

38. J. F. Bagli, P. F. Morand, and R. Gaudry, J. Org. Chem., 27, 2938 (1962).

39. L. G. Wideman, J. Org. Chem., 33, 4541 (1968).

40. P. T. Lansbury and E. J. Nienhouse, Chem. Commun., 1966, 273.

41. G. J. Dufresne and M. Blanchard, Bull. Soc. Chim. France, 1968, 385; K. D. Berlin and R. Ranganathan, Tetrahedron, 25, 793 (1969); A. F. Thomas and B. Willhalm, Helv. Chim. Acta., 50, 826 (1967).

42. J. C. Ware and T. G. Traylor, J. Am. Chem. Soc., 85, 3026 (1963).

43. R. K. Hill and A. G. Edwards, Tetrahedron, 21, 1501 (1965).

Chapter 5

HYDROBORATION OF FUNCTIONALIZED ALKENES

The use of the hydroboration-oxidation sequence of reactions
as a means for achieving anti-Markownikoff cis hydration of
alkenes has been discussed in Sec. 3.1. The presence of
functional groups in the alkene molecule can, however, have a
marked effect on the nature of the products formed. Thus, the
occurrence of elimination and transfer reactions, as discussed
in Sec. 5.1, can lead to products not expected on the basis of
earlier discussions in Chap. 3.

However, by use of controlled reaction conditions, it is
possible to prepare a variety of β-substituted alcohols. Thus,
convenient synthetic routes to vicinal diols (Secs. 5.3.2, 5.3.3,
and 5.4), alkoxy alcohols (Secs. 5.2.2 and 5.3.3), amino alcohols
(Sec. 5.2.4), and chlorohydrins (Sec. 5.3.1) have been developed.
Vicinal amino alcohols can be converted into allylic alcohols and
alkenes, and can be used as intermediates in the transposition
of keto groups to neighboring positions (Sec. 5.2.4), while
γ-chloroorganoboranes, derived from allylic chlorides, are readily
converted into cyclopropanes (Sec. 5.3.1). In addition, allylic
alcohols and α,β-unsaturated carbonyl compounds can be converted
into alkenes (Secs. 5.3.2 and 5.4), and fragmentation of suitably
substituted cyclic boronate derivatives leads to 1,5- and 1,6-dienes
(Sec. 5.8). Finally, hydroboration-elimination of Δ^2-dihydropyrans
and 2,3-dihydrofurans provides a means of synthesis of a variety
of unsaturated alcohols (Sec. 5.9.1).

137

5.1 MECHANISTIC CONSIDERATIONS

The direction of addition of the boron-hydrogen moiety to
substituted alkenes has been shown to depend on both steric and
electronic factors (Sec. 3.4). The hydroboration of vinyl
derivatives can proceed via transition states (1) and (2) to
give α- and β-heterosubstituted organoboranes, respectively.

α-substituted   β-substituted

The relative stability of the transition states (1) and (2),
and hence the relative ease of formation of the α- and β-heterosubstituted organoboranes, depends on the nature of the
substituent, X. Thus, electron-withdrawing substituents favor
formation of transition state (1), while formation of (2) is
favored by substituents capable of stabilizing the adjacent
partial positive charge.

α- and β-Heterosubstituted organoboranes can react further
depending on the conditions of hydroboration used, and the nature
of X. α-Substituted organoboranes tend to undergo rearrangements
in which the substituent, X, is replaced by a group attached to
the boron atom. Such a replacement reaction is referred to as
an α-transfer process (1).

The process has been shown to occur with complete inversion of
configuration at the terminal carbon, and is catalyzed by excess
borane or Lewis acids (2,3). A mechanism involving an intramolecular migration of R from boron to carbon via transition
state (3) has been proposed (4).

(3)

It is probable that the ether solvent facilitates the migration
of R (5). A study of the kinetics of the α-transfer reaction of
the α-chloroorganoborane derived from 1-chloro-2-methylpropene
has provided strong supporting evidence for the intramolecular
mechanism (4). Thus, the rate of the reaction is first order
in the borane dimer, indicating that dissociation of the dimer
occurs and one of the monomer molecules then acts as a catalyst
in promoting the rearrangement (4).

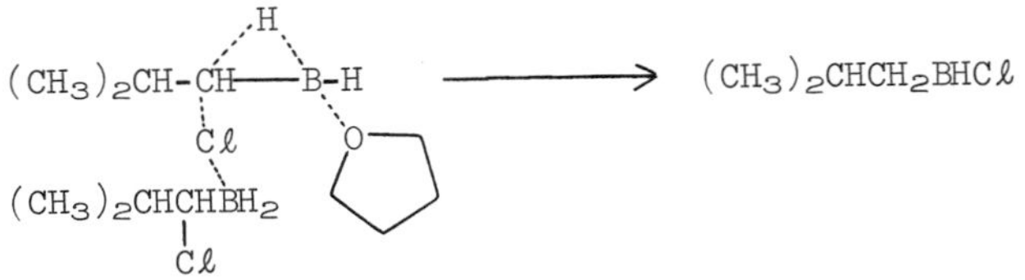

$$(CH_3)_2CH\text{-}\overset{\overset{H}{\cdots}}{\underset{\underset{Cl}{\cdot}}{CH}}\text{---}\overset{\cdots}{\underset{O}{B}}\text{-}H \longrightarrow (CH_3)_2CHCH_2BHCl$$

$$(CH_3)_2CHCHBH_2$$
$$\underset{Cl}{|}$$

In the case of displacement by hydrogen an alternative
mechanism involving intermolecular nucleophilic displacement of
X has been proposed (Sec. 5.2.3) (3).

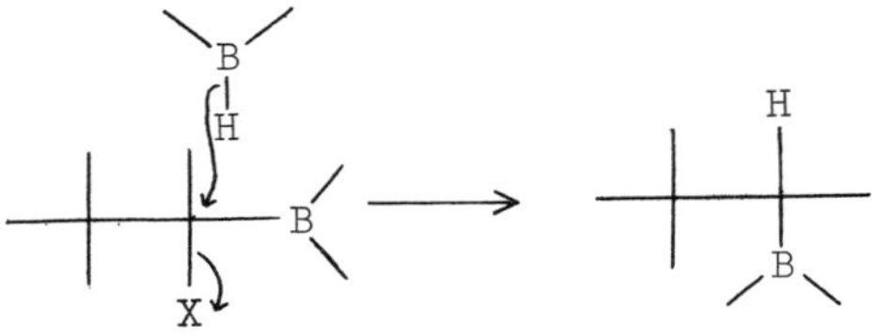

The source of hydride could be a second borane molecule, or could
possibly arise from $BH_4^{\ominus}$ or $B_2H_7^{\ominus}$ ions formed in the equilibriums
shown below (Sec. 2.1) (6,7).

$$2BH_3{:}THF \rightleftharpoons H_2B(THF)_2^{\oplus} + BH_4^{\ominus}$$

or

$$3BH_3{:}THF \rightleftharpoons H_2B(THF)_2^{\oplus} + B_2H_7^{\ominus} + THF$$

β-Heterosubstituted organoboranes tend to undergo elimination
reactions to form alkenes, which subsequently undergo hydroboration.
These eliminations can be uncatalyzed, or acid- or base-catalyzed
depending on the reaction conditions and the substrate. The
uncatalyzed process has been shown to be a cis-process, while
both the acid- and base-catalyzed eliminations proceed in a trans
fashion (8).

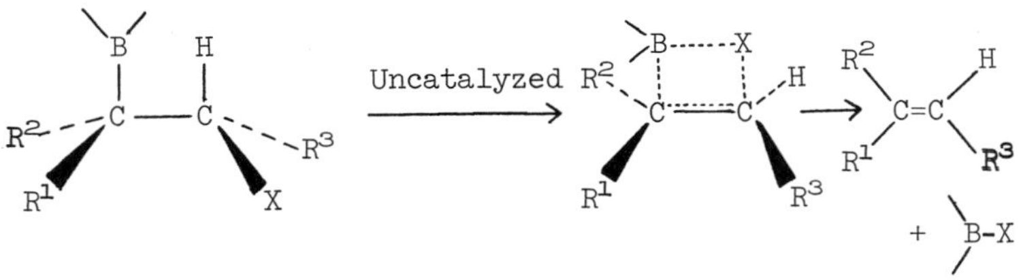

$$\text{(structure)} \quad \xrightarrow[\text{base catalyzed}]{\text{Acid or}} \quad \text{(structure)} \quad + \quad \text{B-X}$$

In cases where X is a good leaving group the ether solvent is often a sufficiently strong base to effect elimination.

Another reaction possibly undergone by β-heterosubstituted organoboranes is that of **β-transfer**, whereby the group X is displaced by a group attached to the boron atom ($\underline{9}$).

$$\text{(structure)} \quad \xrightarrow{\text{β-transfer}} \quad \text{(structure)}$$

α-Heterosubstituted organoboranes have been suggested as ideal systems for the generation of carbenes ($\underline{10}$).

$$RCH_2\overset{|}{\underset{Br}{C}}H-B \longrightarrow \left[RCH_2CH: \ + \ \underset{Br}{B-} \right] \longrightarrow RCH=CH_2$$

The intermediacy of a carbene derived from α-chloroorgano-boranes formed in the reaction of organoboranes with phenyl-(bromodichloromethyl)mercury has been proposed (Sec. 8.8.2)($\underline{11}$).

5.2 HYDROBORATION OF VINYL DERIVATIVES

Hydroboration of vinyl derivatives gives α- and β-substituted organoboranes, with the extent of formation of each being dependent on the nature of the substituent. The α-substituted organoboranes

can undergo α-transfer reactions, while β-substituted organoboranes
can undergo elimination-rehydroboration reactions (Sec. 5.1).

5.2.1 Hydroboration of Vinyl Halides

Hydroboration of 1-halo-1-alkenes places the boron atom
predominantly on the α-carbon atom ($\underline{1},\underline{12}$), while with cyclic
vinyl halides a slight preference for α-attack is observed ($\underline{1},\underline{12}$).

$(CH_3)_2C=CHC\ell$ $C_2H_5CH=CHC\ell$ PhCH=CHBr
      (100)   (85)   (95)

(60)

(40)

(60)

The facile occurrence of α-transfer and elimination-rehydroboration
reactions makes interpretation of the results difficult. Oxidation
of the adducts formed from 1-chloro-2-methylpropene and 1-chloro-
1-butene gives the corresponding aldehydes in 84 and 60% yields,
respectively ($\underline{12}$). The reaction of cis- and trans-2-chloro-2-
butene gives inconclusive results as to the direction of
hydroboration ($\underline{12}$).

When the above results are compared with those obtained in
the hydroboration of the parent alkenes (Sec. 3.1.1), it can be
seen that the halide substituent exerts a small directive influence
favoring α-attack. The slight preference for α-attack can be
ascribed to the inductive effect of the halide exerting a small
stabilizing influence on transition state (1) (Sec. 5.1). It has
been found, however, that attempted alkaline peroxide oxidation
of the organoborane derived from 7-chlorodibenzobicyclo
[2.2.2]octatriene (4) gives dibenzobicyclo[2.2.2]octatriene in 86%
yield ($\underline{13}$); but oxidation with perbenzoic acid in chloroform gives
56% of the trans-7,8-chlorohydrin. This indicates that, in this
case, addition occurs to give mainly the β-chloroorganoborane.

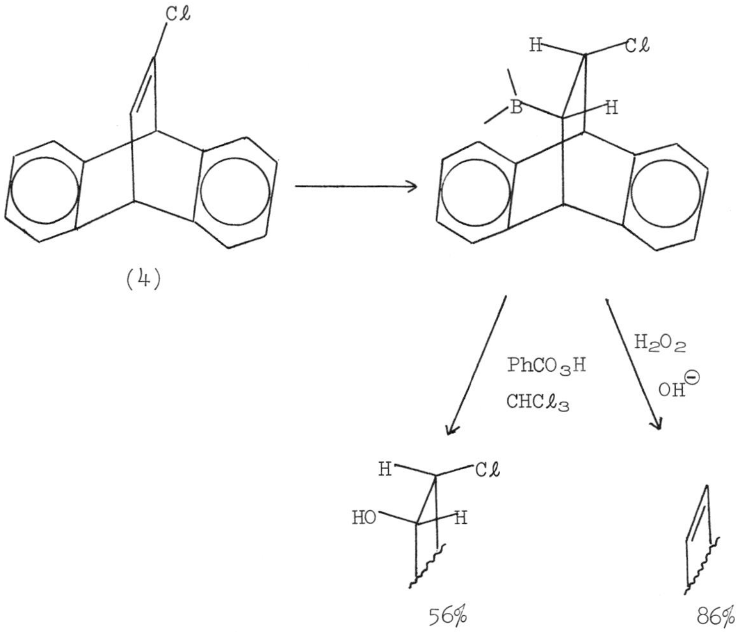
Cℓ
H
Cℓ
B
H
PhCO3H
CHCℓ3
H2O2
OH⊖
H
Cℓ
HO
H
56%
86%

(4)

5.2.2 Hydroboration of Enol Ethers

Alkoxy groups exert a marked directive influence in the hydroboration of enol ethers as illustrated below (12).

$C_2H_5CH{=}CHOC_2H_5$	$(CH_3)_2C{=}CHOC_2H_5$	$PhCH{=}CHOC_2H_5$
↑ (~100%)	↑ (~100%)	↑ 95%

Likewise, hydroboration-oxidation of cyclic enol ethers gives the corresponding trans-2-alkoxy-1-alcohols in high yield (12).

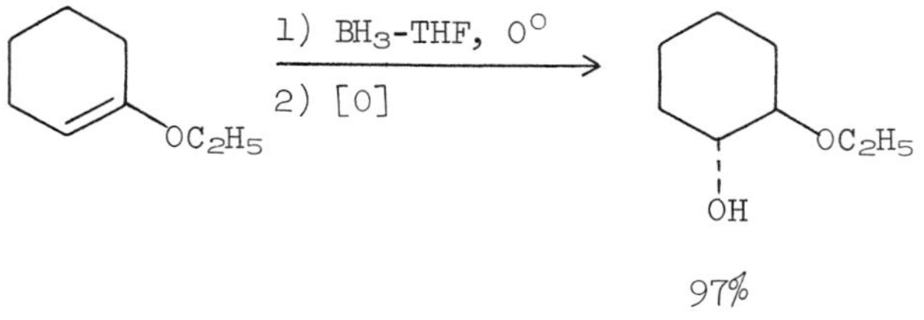
1) BH3-THF, 0°
2) [O]
OC2H5
OC2H5
OH
97%

In the above reactions the hydroboration is carried out using a purified borane-tetrahydrofuran solution and a temperature of $0°$. At $25°$, however, the occurrence of elimination and transfer reactions has been reported (14). Thus, reaction of β-ethoxystyrene with borane-tetrahydrofuran at $25°$, followed by oxidation, produces significant amounts of 1- and 2-phenylethanol. This result is rationalized in terms of a cis-elimination to form styrene (Sec. 5.1) followed by hydroboration (8).

$$PhCH=CHOC_2H_5 \xrightarrow[25°]{BH_3-THF} \left[C_2H_5OCH_2-\underset{Ph}{CH}-BH\right]_2 \longrightarrow PhCH=CH_2 +$$

$$\underset{H\ Ph}{C_2H_5O\,B-CHCH_2OC_2H_5}$$

Similar results have been reported for the reaction of ethylvinylether (15).

$$C_2H_5OCH=CH_2 \xrightarrow[\text{2) r.t.}]{\text{1) } BH_3, Et_2O, -70°} C_2H_5OB(CH_2CH_2OC_2H_5)_2$$
$$(77\%)$$

However, the hydroboration of a large number of alkyl vinyl ethers in hexane as solvent reportedly gives high yields of the corresponding tris(β-alkoxyethyl)boranes (16).

Hydroboration-oxidation of 3-methoxy-5α-cholest-2-ene gives the corresponding 3β-methoxy-2α-hydroxy derivative (17); treatment of the intermediate borane with sodium hydroxide gives 5α-cholest-2-ene.

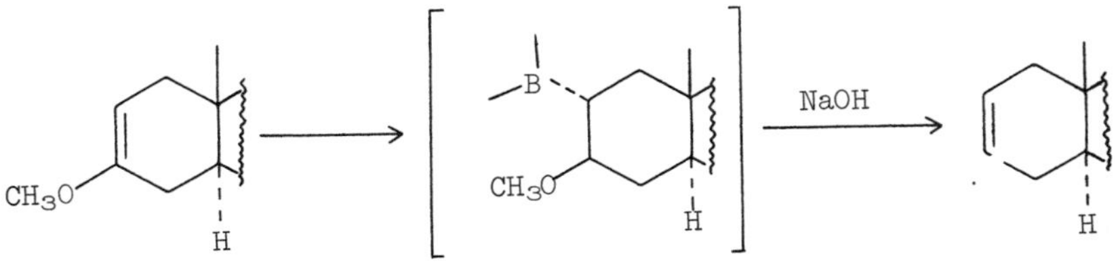

3β-Hydroxy-5α-androst-16-ene has been prepared from
3β-acetoxy-17-ethoxy-5α-androst-16-ene in a similar manner (17).

The application of hydroboration-oxidation to olefinic sugar
derivatives is illustrated by the following example (18).

59%

The exclusive formation of the β-alkoxyorganoborane
derivatives in the hydroboration of enol ethers is attributable
to the stabilization of transition state (2) (Sec. 5.1) by the
mesomeric effect of the adjacent oxygen atom. In view of the
fact that boron trifluoride catalyzes the elimination of
β-alkoxy-organoboranes (14), controlled hydroboration conditions
and rigorous exclusion of boron trifluoride etherate are necessary
for the efficient synthesis of vicinal alkoxy-alcohols from enol
ethers.

5.2.3 Hydroboration of Enol Esters

The hydroboration of enol acetates derived from aldehydes
having two α-hydrogens gives a mixture of α- and β-acetoxyorgano-
boranes. The β-derivative undergoes elimination-rehydroboration
(Sec. 5.1) to give, after oxidation, the corresponding primary
alcohol (12). Reaction with disiamylborane results in exclusive
β-attack followed by elimination-rehydroboration (19).

$$RCH=CHOAc \xrightarrow[\text{2) [o]}]{\text{1) } BH_3\text{-THF}} RCH_2CHO + RCH_2CH_2OH$$

$$\text{30\%} \qquad \text{50\%} \qquad (R=C_2H_5)$$

1) Sia_2BH, THF

2) [o]

$RCH_2CH_2OH \quad \sim 80\%$

Hydroboration of enol acetates derived from dialkyl acetaldehydes with borane-tetrahydrofuran at 0° and using a molar ratio of borane to enol acetate of 1.67 to 1 gives mainly the α-acetoxyorganoborane, which, on oxidation, gives the corresponding aldehyde (12). Use of a large excess of borane (8 equivalents of hydride per mole of enol acetate) and reaction temperatures of 20-40°, however, gives high yields of the corresponding primary alcohol (3). The latter reaction is thought to proceed via intermolecular nucleophilic displacement of the acetoxy group in the α-acetoxyorganoborane by hydride (Sec. 5.1)

$$(CH_3)_2C=CHOAc \xrightarrow[\text{40}^\circ, \text{ 8 hr}]{\text{Excess } BH_3\text{-THF}} \left[(CH_3)_2CHCHB\begin{smallmatrix}\\ | \\ OAc \end{smallmatrix} \right] \xrightarrow{BH_3} (CH_3)_2CHCH_2B\begin{smallmatrix}<\\ \end{smallmatrix}$$

1) BH_3-THF, 0°

2) [o]

$\downarrow$ [o]

$(CH_3)_2CHCHO \qquad 95\%$

$(CH_3)_2CHCH_2OH$

90%

Supporting evidence for the intermolecular displacement of the acetoxy group is provided by the reaction of 1-acetoxy-2-ethyl-1-butene with disiamylborane (3). Only 2-ethylbutanal and 2-ethyl-1-butanol are obtained, while no trace of the alcohol (5) expected to be formed by intramolecular migration of a "siamyl" group, is found.

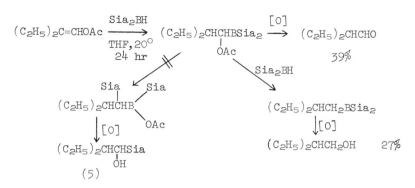

In this respect, it should be noted that intramolecular migrations of bulky groups from boron to carbon are known to be sluggish processes (20).

The hydroboration of enol acetates, derived from acyclic ketones, with borane-tetrahydrofuran (8 equivalents of hydride per mole of enol acetate) at 22° results in a rapid uptake of four equivalents of hydride per mole of enol acetate, to give mainly β-substituted organoboranes (21); elimination rehydro-boration gives a mixture of mono-alcohols. The α-acetoxyorgano-boranes undergo replacement of the acetoxy group by hydrogen to give, after oxidation, the corresponding alcohols (21).

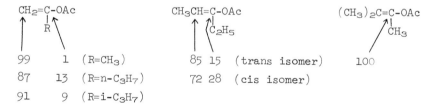

The nature of the products obtained from the hydroboration-oxidation of enol acetates derived from cyclic ketones is dependent on the reaction conditions used in the hydroboration reaction. Thus, hydroboration of 1-acetoxycyclohexene with borane-tetrahydrofuran at 0° using a mole ratio of borane to alkene of 2 to 1, gives 60% trans-cyclohexane-1,2-diol and 20% cyclohexanol (22); use of a mole ratio of borane to alkene of

10 to 1 and in situ reaction conditions gives 56% of the mono-ol
and only 12% of the diol (22).

1) 2BH$_3$-THF, 0°	2) [O]	60%		20%
1) 10BH$_3$ (in situ) 2)[O]		12%		56%

Reaction of 1-acetoxy-2-methylcyclohexene gives essentially
pure cis-2-methylcyclohexanol formed from the α-acetoxyorganoborane via α-transfer of hydrogen (Sec. 5.1) (3).

Hydroboration of 3-acetoxy-5α-cholest-2-ene with a large
excess of borane-tetrahydrofuran at 0°, followed by oxidation
gives 51% of the 2α,3β-diol and 44% of a mixture of the
3-alcohols, consisting mainly of 5α-cholestan-3β-ol (22). Use of
in situ hydroboration conditions gives 80% of the 3-alcohols, and
only 15% of the 2α,3β-diol. The absence of any 2α-alcohol indicates
that the 3-alcohols are formed from the α-acetoxyorganoborane via
α-transfer of hydrogen, since elimination-rehydroboration of the
β-acetoxyborane should result in the formation of some 2α-alcohol
(hydroboration of 5α-cholest-2-ene: Sec. 3.1.8). Refluxing of
the intermediate borane, formed from 3-acetoxy-5α-cholest-2-ene,
with either sodium hydroxide or acetic anhydride is reported to
give 5α-cholest-2-ene (17).

Reaction of 1-acetoxycyclopentene gives only cyclopentanol in
75% yield (22). The reaction probably proceeds via formation of
the α-acetoxyborane followed by α-transfer of hydrogen. In
contrast, hydroboration-oxidation of androst-16-en-3β,17β-diol
diacetate (6) gives the 3β,16α,17β-triol in 80% yield (22).

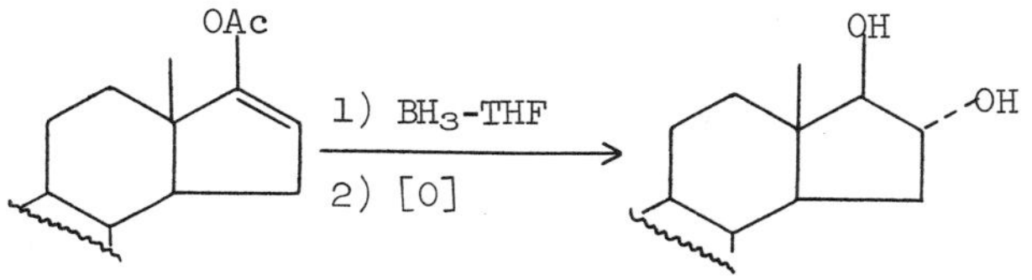

(6)   80%

This result is possibly explained by preclusion of the formation
of the α-acetoxyorganoborane for steric reasons (<u>22</u>).

In contrast to the reaction with borane, reaction of enol
acetates, derived from acyclic ketones, with disiamylborane
proceeds very slowly and fails to give detectable amounts of
alcohols (<u>19</u>). An exception is 2-acetoxypropene (isopropenyl
acetate) which gives 1-propanol in 96% yield (<u>23</u>).

$$CH_2=\underset{\underset{CH_3}{|}}{C}-OAc \quad \xrightarrow[\text{2) [o]}]{\text{1) Sia}_2\text{BH}} \quad CH_3CH_2CH_2OH$$

96%

5.2.4 Hydroboration of Enamines

The hydroboration of enamines gives mainly the
β-aminoorganoborane due to the stabilization of transition state
(2) by the amino group (Sec. 5.1). Attack on the double bond
must occur before coordination of the borane with the nitrogen
atom, since such coordination would favor the formation of the
α-aminoorganoborane.

Hydroboration-oxidation of cyclohexanone enamine derivatives
gives the corresponding <u>trans-amino-alcohols</u> in good yields (<u>24</u>).

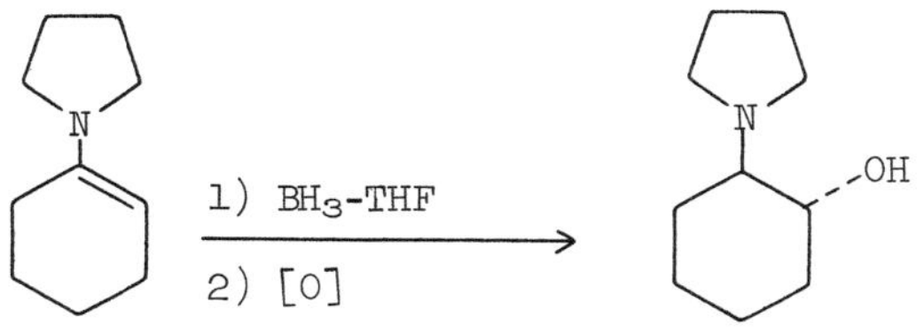

78%

In the case of the pyrrolidine derivative of
2-methylcyclohexanone only the isomeric aminocyclohexanols
formed from the less-substituted enamine are obtained ($\underline{24}$).

40%   24%

Treatment of the trans-amino-alcohols with hydrogen peroxide
in methanol gives the corresponding N-oxides, which, on pyrolysis
at 150°, undergo the Cope elimination to give the 2-cyclohexenol
derivatives ($\underline{25a}$).

60%

Oxidation of the allylic alcohol, followed by reduction of the
double bond, gives the corresponding ketone. This sequence of
reactions thus provides a convenient method for transposing a
ketone group to the neighboring position ($\underline{25b}$).

$$\sim 26\%$$

Hydroboration of enamines derived from acyclic ketones and straight-chain aldehydes gives the corresponding β-aminoorgano-boranes, which, on treatment with propanoic acid in refluxing diglyme, give the corresponding alkenes in high yields (<u>26</u>).

Enamines derived from 2-alkyl substituted aldehydes, however, react to give considerable quantities of saturated hydrocarbons (<u>26</u>).

The formation of saturated hydrocarbons can be explained by either formation of the α-aminoorganoborane followed by α-transfer of hydrogen, or formation of the β-aminoorganoborane followed by β-transfer of hydrogen (Sec. 5.1). In each case the resulting borane can be reduced to the hydrocarbon by propanoic acid (Sec. 10.11) (<u>26</u>).

In the case of β-aminoorganoboranes derived from monocyclic ketones, treatment with propanoic acid in refluxing diglyme gives the corresponding cyclic alkenes in yields exceeding 90% (<u>26</u>).

Since elimination to form the alkene is a trans-process (8), it would be expected that the trans-diaxial arrangement of the pyrrolidino- and boryl groups would lead to the most efficient reaction. However, 4-tert-butyl-1-N-pyrrolidinocyclohexene, which forms equal amounts of both possible isomeric boranes, undergoes efficient elimination to give 90% 4-tert-butylcyclohexene (26).

This indicates that elimination occurs equally easily from the trans-diequatorial aminoborane; however, it is likely that the molecule adopts a suitable twist conformation to enable efficient E2 elimination. Application of the reaction sequence to steroid enamines gives low yields of the corresponding alkenes. Thus, 3-N-pyrrolidino-5α-cholest-2-ene and 3β-N-pyrrolidinocholest-4-ene give 40% 5α-cholest-2-ene and 30% 5α-cholest-3-ene, respectively (26). These low yields have been attributed to the formation of the corresponding diequatorial β-aminoorganoboranes which fail to undergo efficient elimination. However, oxidation of the

β-aminoorganoborane, derived from 3-N-pyrrolidino-5α-androst-2-
en-17β-ol (7), gives 80% of a mixture of the 3-pyrrolidino-
derivatives, with the 3α-epimer (9) forming 80% of the mixture;
this indicates that hydroboration occurs mainly by β-face attack
to give the diaxial 2β-boryl-3α-pyrrolidino derivative (8) (27a).
This unexpected result has been explained in terms of the
pyrrolidino group adopting such a conformation as to hinder α-face
attack (27a).

The formation of the amine rather than the 3-amino-2-alcohol has
been attributed to 1,3-diaxial interaction between the 2β-boryl
group and the 19-angular methyl group promoting internal
displacement of the boryl group by hydride ion (27a). This
explanation is supported by the fact that the 2β-hydroxy-3α-
pyrrolidino derivative is the major product in the hydroboration-
oxidation of 19-norsteroid derivatives (27b). Treatment of the
3-N-pyrrolidino-3,5-diene derivative (10), formed from progesterone,
with diborane in diglyme, followed by acetic acid, gives the
3β-N-pyrrolidino-4-ene in 60% yield (28).

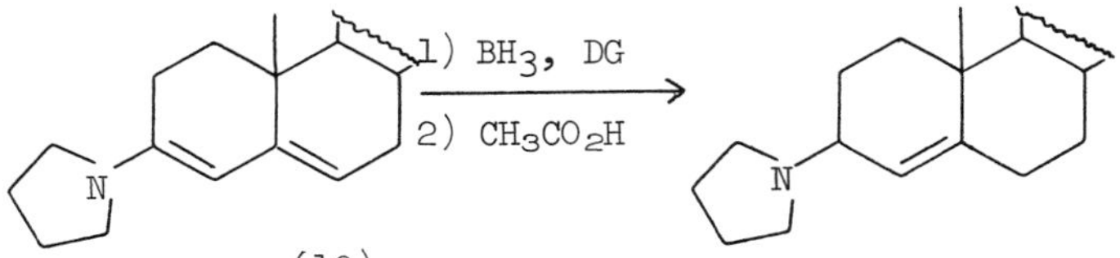

(10)   60%

Hydroboration of N-alkenylureas and N-alkenylcarbamates gives, after treatment with methanol, the corresponding aminoalkyldimethoxyboranes in yields of 30-90% (29).

$$H_2NCONHCH=CH_2 \xrightarrow[\text{2) CH}_3\text{OH}]{\text{1) BH}_3\text{-THF}} H_2NCONHCH_2CH_2B(OCH_3)_2 \qquad (80\%)$$

$$C_2H_5OCONHCH=CH_2 \xrightarrow{} C_2H_5OCONHCH_2CH_2B(OCH_3)_2$$

The failure of the hydroboration to proceed beyond the monoalkylborane stage is accounted for by internal coordination of the borane moiety by the carbonyl oxygen as shown below (29).

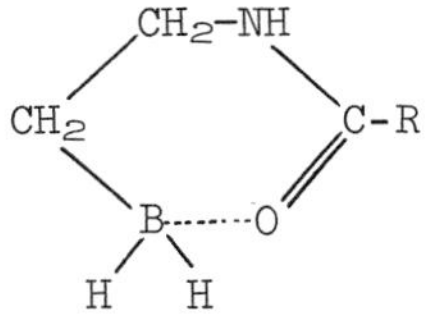

5.2.5 Hydroboration of Miscellaneous Vinyl Derivatives

The hydroboration-oxidation of <u>enethiol ethers</u> gives complex mixtures of products; these arise from the formation of both α- and β-substituted organoboranes which undergo eliminations, and α- and β-transfer processes involving hydrogen and alkyl groups (8,9,30-32).

Hydroboration-oxidation of <u>vinylsilanes</u> gives mixtures of α- and β-hydroxyethylsilanes in yields exceeding 60% (33-35), while hydroboration of the <u>vinyl phosphonate</u> derived from 5α-cholestan-3-one, followed by oxidation with chromic acid, gives the corresponding 3α-hydroxy-3-phosphonate in 44% yield (36).

Hydroboration of <u>1-vinylboron derivatives</u> results mainly
in α-addition as illustrated by the formation of 1,1-diboro
derivatives in the dihydroboration of terminal alkynes (Sec.
7.2.1). Similar results have been obtained in the hydroboration
of dialkoxyvinylboranes ($\underline{37},\underline{38}$), though use of bulky alkoxy
substituents leads to the increasing formation of 1,2-diboro
derivatives ($\underline{39}$).

In the case of <u>arylalkenes</u>, the results of the hydroboration
of substituted styrenes ($\underline{40}$) have been discussed in Sec. 3.4
(Table 3.4). Hydroboration-oxidation of 1-aryl cyclic alkenes
results in the exclusive formation of trans-β-substituted
alcohols ($\underline{41},\underline{42}$).

$$68-80\%$$

$$\begin{bmatrix} Ar=Ph, \quad p-CH_3- \\ p-C\ell-, \quad m-C\ell C_6H_4 \end{bmatrix}$$

5.3 ALLYLIC DERIVATIVES

Hydroboration of allyl derivatives ($CH_2=CHCH_2X$) gives
mixtures of β- and γ-boron derivatives; the extent of formation
of the β-substituted organoboranes decreases with decreasing
electronegativity of the substituent, X ($\underline{43}$).

$CH_2=CHCH_2X$
↑   ↑
45%(X=OTs); 40%(X=Cℓ); 35%(X=OCOCH$_3$); 19%(X=OC$_2$H$_5$)
% balance

Hydroboration of extended allylic systems ($RCH=CHCH_2X$;
R=alkyl), however, results in formation of the β-substituted
organoborane to the extent of 84% or greater in all cases ($\underline{44}$).

$$RCH=CHCH_2X$$

% balance $\overset{\uparrow}{}\overset{\uparrow}{}> 84\%$ (X=Cℓ, OAc, OH, OEt, OPh)

The β-substituted organoboranes can undergo elimination-rehydroboration reactions as discussed in Sec. 5.1. In the case of X being a good leaving group, such as tosylate or chloride, spontaneous elimination occurs by a trans-mechanism; with ester groups rapid elimination occurs by a cis-mechanism (Sec. 5.3.4). With other groups, such as ethers, the extent of elimination can be limited by use of controlled hydroboration reaction conditions, thus providing useful synthetic routes to vicinal diols and their derivatives (Sec. 5.3.3).

5.3.1 Hydroboration of Allylic Halides

Hydroboration of 3-chloro-1-propene (allyl chloride) with borane-tetrahydrofuran at 25° results in the uptake of 1.4 equivalents of hydride; treatment of the product with sodium hydroxide gives 51% cyclopropane, while oxidation of the residual alkaline reaction mixture with hydrogen peroxide gives a mixture of 1- and 2-propanol (45).

$$CH_2=CHCH_2C\ell \xrightarrow[25°, \ 4\ hr]{BH_3 \cdot THF} \diagdown BCH_2CH_2CH_2C\ell + CH_3CHCH_2C\ell \xrightarrow[elim.]{fast}$$

$\downarrow OH^{\ominus}$

$\triangle$

51%

$$CH_3CH=CH_2$$

$\downarrow$ 1) BH$_3$-THF

2) [O]

$$CH_2CH_2CH_2OH + CH_3\underset{\underset{OH}{|}}{C}HCH_3$$

38%   2%

Use of disiamylborane results in the predominant formation of the γ-chloroorganoborane which gives 82% cyclopropane on treatment

with alkali; however, concurrent addition of alkali and hydrogen
peroxide at pH 7-8 gives the γ-chlorohydrin (45).

$$CH_2=CHCH_2C\ell \xrightarrow[\text{THF, }0^\circ]{Sia_2BH} Sia_2BCH_2CH_2CH_2C\ell \xrightarrow[\substack{\text{pH 7-8,}\\5-10^\circ}]{H_2O_2, \ OH^\ominus} HOCH_2CH_2CH_2C\ell \quad 77\%$$

Similar results are obtained with 3-chloro-1-butene (46).

$$\underset{\substack{(49)\ (51)\\(96)\ (4)\ (Sia_2BH)}}{\overset{\uparrow\qquad\uparrow}{CH_2=CH-\underset{|}{\overset{}{C}}HCH_3}}$$
$$Cl$$

The formation of substantial amounts of the β-chloroorganoboranes
is a result of the strong inductive effect of the chloride (46). In
the case of dissymmetric allylic systems containing more β- than
γ-substitution, however, the directive effect of the β-alkyl group
largely overcomes the directive effect of the chloro-group,
resulting in predominant γ-addition (46).

$$\underset{\substack{(88)\ (12)\\(100)\ (0)\ (Sia_2BH)}}{\overset{\uparrow\qquad\uparrow}{CH_2=\underset{\underset{CH_3}{|}}{C}-CH_2C\ell}}$$

The cyclization of γ-chloroorganoboranes on treatment with
sodium hydroxide provides a useful method for the synthesis of
cyclopropanes (49). The reaction proceeds via prior coordination
of the base with the boron atom to produce a quaternary derivative
which then cyclizes.

The usefulness of the method is greatly enhanced by the increased
selectivity exhibited by disiamylborane in the hydroboration of
β-substituted allylic systems; however, the hindered nature of
the intermediate borane results in the slow coordination of the
boron by the base, and leads to competitive direct displacement
of the chloride by base. This problem has been overcome by use
of 9-borabicyclo[3.3.1]nonane(9-BBN) as the hydroborating reagent
(50). The boron atom in the B-substituted-9-BBN derivatives is
far more open to attack by nucleophiles, and is thus readily
coordinated by the base.

$$
CH_2=\overset{\overset{\displaystyle Ph}{|}}{C}CH_2C\ell \quad
\xrightarrow[\text{2) NaOH}]{\begin{array}{c}\text{1) 9-BBN}\\ \text{THF}\end{array}} \quad
\underset{92\%}{\triangle}\!\!-Ph \quad ; \quad
CH_2=CH-\overset{\overset{\displaystyle CH_3}{|}}{\underset{\underset{\displaystyle CH_3}{|}}{C}}-C\ell \quad
\longrightarrow \quad
\underset{CH_3}{\triangle}\!\!\overset{-CH_3}{} \quad 75\%
$$

$$
CH_2=CHCHC\ell_2 \quad \longrightarrow \quad \underset{C\ell}{\triangle} \quad 90\%
$$

In the case of "symmetrical" systems, such as 1-chloro-2-
butene (crotyl chloride), hydroboration with either borane or
disiamylborane gives 100% of the β-adduct (44,46); elimination-
rehydroboration gives a mixture of 1- and 2-butanols. Similar
results have been obtained in the reaction of phytyl bromide (11)
(47).

$$
\underset{\underset{\displaystyle 100}{\uparrow}}{CH_3CH{=}CHCH_2C\ell} \quad
\xrightarrow[\text{2) [O]}]{\text{1) BH}_3\text{-THF, 0}^{\circ}} \quad
CH_3CH_2CH_2CH_2OH \; + \; CH_3CH_2\underset{\underset{\displaystyle OH}{|}}{C}HCH_3
$$

$$\left[\begin{array}{c} \text{CH}_2-\overset{\overset{\displaystyle CH_3}{|}}{C}=CHCH_2Br \\ \end{array} \right]$$

$$[R]$$

$$\xrightarrow[\text{2) [O]}]{\text{1) BH}_3\text{-THF, } 0°} \quad \underset{\overset{\displaystyle |}{OH}}{R\overset{\overset{\displaystyle CH_3}{|}}{C}HCH_2CH_2OH} + R\overset{\overset{\displaystyle CH_3}{|}}{C}H\overset{\overset{\displaystyle }{\underset{OH}{|}}}{C}HCH_3$$

$$90\% \qquad\qquad 8\%$$

[R]

(11)

It has been observed, however, that use of diethyl ether in place of tetrahydrofuran as hydroboration solvent avoids elimination of β-chloroorganoboranes; oxidation of the mixture with m-chloroperbenzoic acid gives the corresponding β-chlorohydrins (48).

Hydroboration of 3-chlorocyclopentene results in practically exclusive formation of the β-organoborane, which undergoes elimination-rehydroboration to form 88% cyclopentanol (51); reaction using disiamylborane, while being slower, gives similar results (51).

$$\xrightarrow[\text{2) [O]}]{\text{1) BH}_3\text{-THF}}$$

88%

(1) (10)

(4) (85)

Reaction of 3-chlorocyclohexene has been reported to give only cyclohexanol (52); detailed studies have, however, shown that all four possible isomeric organoboranes are produced with the β-adducts predominating (48).

The predominance of β-attack in the above two cases, and
also in the case of "symmetrical" acyclic systems, such as
1-chloro-2-butene, has been ascribed to the strong inductive
effect of the chloro group (48); the high percentage of addition
trans to the chloro group in the case of 3-chlorocyclohexene,
as compared to the hydroboration of 3-methylcyclohexene (66% trans;
Sec. 3.1.5), is attributed to the presence of a substantial amount
of the pseudoaxial conformer shown above, in which the chloro
group inhibits cis-attack (48). In tetrahydrofuran solution the
trans-β-derivative undergoes rapid elimination-rehydroboration,
whereas in diethyl ether, it is stable at room temperature (48).

In contrast to the above results, reaction of 1-chloronor-
bornene gives the γ-adduct as the major product (53).

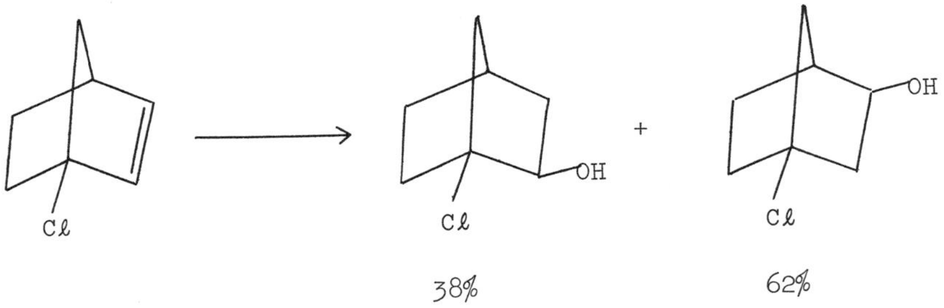

38%   62%

It has been suggested that the predominant formation of the
β-adducts in the case of compounds, such as 3-chlorocyclohexene,
may be due to coordination of the hydroborating reagent by the
chloro group directing the boron to the neighboring β position
(53); in the case of 1-chloronorbornene such an effect could not
operate. The low yield of 1-chloro-exo-2-norbornanol might be
due to inhibition by torsional effects (54); the torsional strain
between the C_1-Cl bridgehead bond and the developing C_2-boron
bond would exceed that existing between the C_4-H bridgehead bond
and the developing C_3-boron bond, thereby favoring formation of
the 3-adduct.

5.3.2 Hydroboration of Allylic Alcohols and Borates

Hydroboration of allylic alcohols occurs via initial consumption of one equivalent of borane to form the corresponding alkoxyboranes, with the formation of the dialkoxyborane generally being favored (44). This intermediate then undergoes rapid hydroboration of the double bond (43).

$$CH_2=CHCH_2OH \xrightarrow[0^\circ]{BH_3-THF} CH_2=CHCH_2OB \xrightarrow[2)~[O]]{1)BH_3-THF} HOCH_2CH_2CH_2OH +$$

$$\underset{72 \quad 24}{} \qquad \underset{72\%}{}$$

$$CH_3\underset{OH}{CHCH_2OH} + CH_3CH_2CH_2OH$$
$$\underset{22\%}{} \qquad \underset{2\%}{}$$

Slightly different directive effects are observed with allyl borate (43).

$$(CH_2=CHCH_2O-)_3B$$
$$\underset{76 \quad 18}{}$$

In the above cases no significant elimination of the β-substituted-organoborane occurs.

With "symmetrical" systems, such as 2-buten-1-ol (crotyl alcohol), the β-substituted organoborane is the major product; in these cases elimination-rehydroboration occurs to a large extent (44).

$$CH_3CH=CHCH_2OH \xrightarrow[2)~[O]]{1)BH_3-THF,0^\circ} CH_3CH_2CH_2CH_2OH + CH_3CH_2\underset{OH}{CHCH_3} +$$
$$\underset{(10) \quad (90)}{} \qquad\qquad\qquad \underset{26\%}{} \qquad \underset{2\%}{}$$

$$CH_3CH_2\underset{OH}{CHCH_2OH} + CH_3\underset{OH}{CHCH_2CH_2OH}$$
$$\underset{38\%}{} \qquad \underset{9\%}{}$$

The elimination reaction can be largely avoided by prior reaction of the alcohols with one equivalent of disiamylborane at 0° to give the disiamylborinate esters (44); subsequent reaction using borane gives the 1,2-diols in high yield.

$$RCH=CHCH_2OH \xrightarrow[\text{THF, } 0^\circ]{Sia_2BH} RCH=CHCH_2OBSia_2 \xrightarrow[\text{2) [O]}]{\text{1)}BH_3\text{-THF}} \underset{\underset{81\text{-}83\%}{OH}}{RCH_2CHCH_2OH}$$

$$(R=CH_3; Ph)$$

$$+ \quad \underset{\underset{OH \quad 9\text{-}13\%}{|}}{RCHCH_2CH_2OH} + RCH_2CH_2CH_2OH \quad (R=CH_3; 5\%)$$

Attempted hydroboration of the intermediate disiamylborinate ester with disiamylborane at 0° is very slow, while at 25° substantial elimination occurs (44). Reaction of the allylic alcohol with two equivalents of thexylborane, however, gives high yield of the 1,2-diol (44).

Hydroboration-oxidation of phytol (12) at 0° results in 48% elimination-rehydroboration occurring; at -20°, however, 75% of the 1,2-diol (13) is obtained (47).

$$RCH=CHCH_2OH \xrightarrow[\text{2) [O]}]{\text{1) } BH_3\text{-THF,} -20^\circ} \underset{\underset{OH \quad 75\%}{|}}{RCH_2CHCH_2OH} + \underset{22\%}{RCH_2CH_2CH_2OH}$$

$$(12) \qquad\qquad\qquad (13) \qquad (14)$$

$(R=C_{16}H_{33}-(11), \text{ Sec. } 5.3.1)$

Reaction of β-methylcinnamyl alcohol gives only the γ-addition product (55).

$$\xrightarrow[\text{2) [O]}]{\text{1)}BH_3\text{-THF}} \text{Threo-}\underset{\underset{OH \quad CH_3 \quad 80\%}{| \quad |}}{PhCH\text{-}CHCH_2OH}$$

The elimination reactions occurring with the β-substituted organoboranes proceed via a cis-mechanism (Sec. 5.1); the fact that elimination is avoided by use of the disiamylborinate ester has been attributed to steric factors preventing the cis-alignment of the disiamylborinate group and the neighboring boron moiety (44). However, it should be noted that hydroboration of phytol using disiamylborane at 0° leads to considerable elimination-rehydroboration, with the formation of 76% 1-phytanol (14) (47).

Elimination from the β-substituted organoborane formed by hydroboration of the disiamylborinate ester can be induced by treatment with methanesulfonic acid in refluxing tetrahydrofuran (44). This sequence thus provides a convenient synthesis of terminal alkenes from the more stable 2-en-1-alcohols.

$$RCH=CHCH_2OH \xrightarrow[\text{2) BH}_3\text{-THF}]{\text{1) Sia}_2\text{BH, THF}} RCH_2CHCH_2OBSia_2 \xrightarrow[\text{THF, }\Delta]{CH_3SO_3H} RCH_2CH=CH_2$$

(R=Ph; 75%)

Reaction of 3-hydroxycyclopentene gives a mixture of products resulting from β- and γ-addition (51). Use of disiamylborane, however, gives the trans-1,2-diol in 80% yield (51).

Reaction of the disiamylborinate ester of 3-hydroxycyclohexene gives mainly the trans 1,2-diol (44). In the case of alkyl-substituted 3-hydroxycyclohexenes, trans-diequatorial 1,2-diols are the major products, even though the formation of cis-diols would be expected for steric reasons.

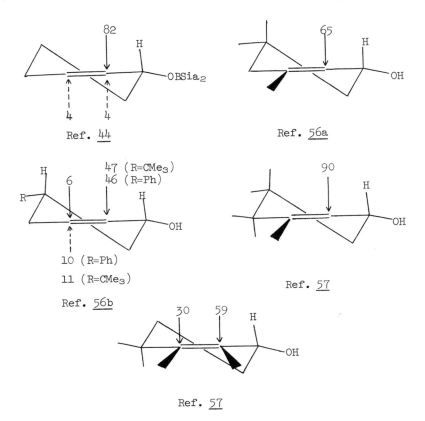

Ref. 44   Ref. 56a

47 (R=CMe₃)
46 (R=Ph)

10 (R=Ph)
11 (R=CMe₃)

Ref. 56b   Ref. 57

Ref. 57

Reaction of bicyclic allylic alcohols related to α-pinene proceeds via attack on the side remote to the gem dimethyl group as expected.

Ref. 58   Ref. 59

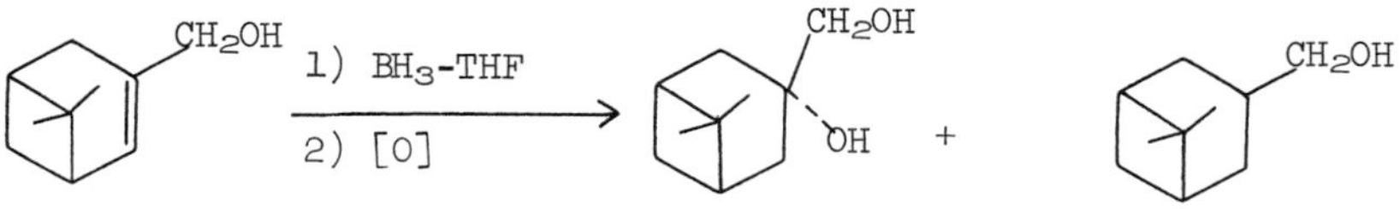

Ref. 60   56%   15%

The formation of some of the mono-ol in the above reaction indicates the occurrence of some elimination-rehydroboration (60).

Hydroboration of bicyclic allylic alcohols related to β-pinene unexpectedly occurs mainly from the side cis to the gem dimethyl group to give 1,3-diols (58,60).

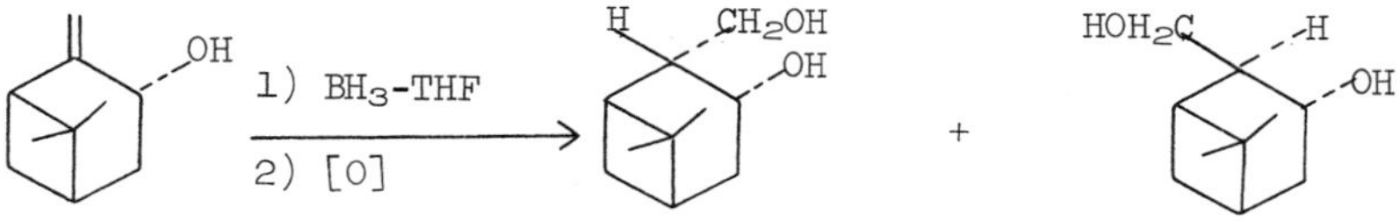

(15) 39%   32%

Use of disiamylborane gives only the cis-1,3-diol (15) in 60% yield (60).

The influence of axial and equatorial hydroxyl groups on the direction of attack is illustrated by the reaction of the epimers of bicyclo[3.3.1]non-3-en-2-ol (16) (61).

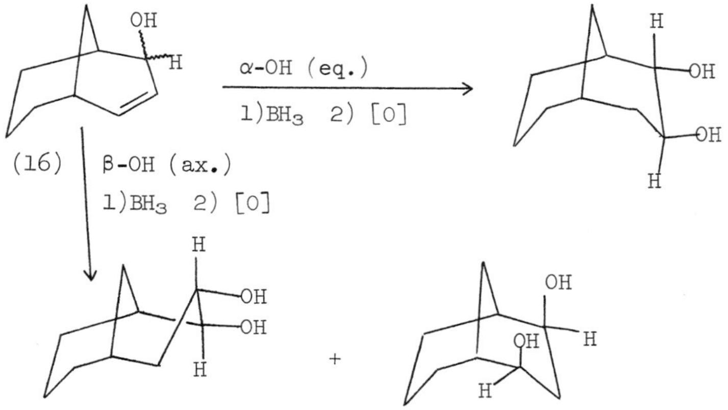

(16)   β-OH (ax.)
       1)BH₃ 2) [o]

(17)   (18)

In the case of the equatorial 2α-hydroxy derivative exo-attack
at position 3 predominates as expected. With the 2β-hydroxy
derivative, however, exo-attack at the 3-position is sterically
hindered, resulting in some endo-attack to give diol (17),
together with some diol (18) (61).

5.3.3 Hydroboration of Allylic Ethers

Hydroboration-oxidation of allyl ethers gives mixtures of
the 1,2- and 1,3-diol monoethers (43).

$CH_2=CHCH_2OC_2H_5$ $CH_2=CHCH_2OPh$
  78   19   58   32

  92   2 (Sia_2BH)

Little elimination occurs under normal hydroboration
conditions; the β-phenoxyorganoborane undergoes base-catalyzed
elimination during oxidation which can be avoided by the
simultaneous addition of the hydrogen peroxide and base (43).

Reaction of the ethyl- and phenyl ethers of 2-buten-1-ol
gives mainly the corresponding vicinal hydroxy ethers (44).

$CH_3CH=CHCH_2OC_2H_5$ $CH_3CH=CHCH_2OPh$
   15 84   13 82

Similar results are obtained with tetrahydropyranyl and
benzyl ethers. The ready conversion of these groups to the
corresponding alcohols provides a convenient route for the
synthesis of 1,2-diols (44).

$PhCH=CHCH_2OTHP$ $\xrightarrow{\begin{array}{c}1)BH_3-THF\\[6pt] 2)\,[O]\\[6pt] 3)pTsH,\,MeOH\end{array}}$ $PhCH_2\underset{\underset{OH}{|}}{C}HCH_2OH$ + $Ph\underset{\underset{OH}{|}}{C}HCH_2CH_2OH$

                                                82%   9%

$CH_3CH=CHCH_2OCH_2Ph$ $\xrightarrow{\begin{array}{c}1)BH_3-THF\\[6pt] 2)\,[O]\\[6pt] 3)H_2,Pd/C\end{array}}$ $CH_3CH_2\underset{\underset{OH}{|}}{C}HCH_2OH$ + $CH_3\underset{\underset{OH}{|}}{C}HCH_2CH_2OH$

                                                   84%   8%

Reaction of 3-ethoxycyclopentene gives trans-2-ethoxycyclo-
pentanol as the major product (51).

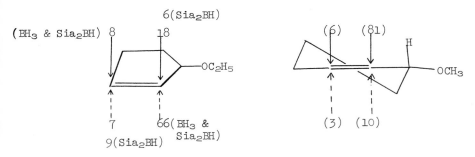

The cis-1,2-adduct undergoes rapid elimination-rehydroboration to
give cyclopentanol (51). With 3-methoxycyclohexene trans-2-
methoxycyclohexanol is the major product (48). The cis-1,2-adduct
undergoes rapid elimination-rehydroboration as in the case of
3-ethoxycyclopentene

The application of the reaction to allylic sugar derivatives
is illustrated by the following examples.

Ref. 62	Ref. 62	Ref. 63

In the case of the last example, 22% acetal cleavage products
were also obtained (63).

5.3.4 Hydroboration of Allylic Esters

Hydroboration of allyl acetate results in preferential
formation of the 1,3-adduct and leads to reduction of the
acetoxy group; use of disiamylborane avoids such reduction and

leads almost exclusively to the 1,3-adduct ($\underline{43}$). In the case of allyl benzoate, reduction of the ester group by borane is relatively slow ($\underline{43}$).

$\text{CH}_2=\text{CHCH}_2\text{OAc}$ $\qquad\qquad$ $\text{CH}_2=\text{CHCH}_2\text{OCOPh}$ $\quad$ $\text{CH}_2=\text{CHCH}_2\text{OTs}$

65 35 $\qquad\qquad\qquad\qquad$ 71 20 $\qquad\qquad\qquad$ 55 45

95 2 (Sia_2BH)

Hydroboration of 1-acetoxy-2-butene (crotyl acetate) ($\underline{44}$) and phytyl acetate ($\underline{47}$) gives almost exclusive β-addition followed by rapid elimination-rehydroboration; disiamylborane gives similar results.

$$\text{RCH=CHCH}_2\text{OAc} \xrightarrow[\text{2) [O]}]{\text{1)BH}_3\text{-THF, 0}^\circ} \text{RCH}_2\text{CH}_2\text{CH}_2\text{OH} + \text{RCH}_2\underset{\underset{\text{OH}}{|}}{\text{CHCH}_3}$$

R=CH₃ $\qquad\qquad\qquad\qquad\qquad$ 83% $\qquad\qquad$ 9%

(R=$\text{C}_{16}\text{H}_{33}$—(11); Sec. 5.3.1) $\qquad$ 92% $\qquad\qquad$ 6%

The very rapid elimination observed with β-acyloxyorgano-boranes probably proceeds via a cyclic mechanism ($\underline{44}$).

With 3-acetoxycyclopentene the rate of hydride consumption is much slower than with the acyclic compounds indicating an overall slower rate of elimination ($\underline{51}$).

Elimination from the trans-β-acetoxyorganoborane proceeds at a
faster rate than that from the cis-isomer; the overall slow rate
of elimination, and the relative stability of the cis-β-
acetoxyorganoborane, have been attributed to the formation of
heterocycles (19) and (20) (51).

(19)   (20)

The reaction with disiamylborane is extremely slow (51).

Hydroboration of cis-verbenyl acetate (21) proceeds by
predominant attack on the side remote from the gem dimethyl
group (64). Disiamylborane gives only 15% of the trans-acetoxy
alcohol (22).

(21)   33%   (22) 27%

                        10%   12%

In the case of trans-verbenyl acetate reaction is very slow and
gives a low yield (6%) of the corresponding trans-acetoxy alcohol
(65). β-Addition is also found to predominate in the hydroboration
of the bicyclo[3.2.1]octyl derivative of cholestane (23) shown
below (66a).

23   13
41   17   (Sia$_2$BH)

(23)

5.3.5 Hydroboration of Miscellaneous Allylic Derivatives

Hydroboration of allyl phenyl thioether results in preferential formation of the γ-adduct (43); reaction of allyl methyl thioether with dimethyl sulfide-borane gives the stable monoalkylborane, 3-(methylthio)propylborane (71%) which exists in the S→B coordinate form (24) (66b). A similar product (25)

(24)   (25)

has been obtained in 25% yield on treatment of dimethylallylamine with trimethylamine-borane (66c). Hydroboration of triallylamine hydrochloride with triethylamine borane, however, gives mainly polymeric material (67), while prolonged reaction times are required for the satisfactory reaction of borane with a number of morphine-type alkaloids (68,69).

Hydroboration of allyltrimethylsilane results in exclusive γ-attack (35), while with allyldimethoxyborane the γ-adduct is the main product (70). Reaction of triallylborane with triethyl-amineborane gives mainly polymeric material (67).

Hydroboration of allylaryl derivatives results in the exclusive formation of the γ-adduct (71,72).

5.4 α,β-UNSATURATED ALDEHYDES AND KETONES

Hydroboration of α,β-unsaturated aldehydes and ketones involves the initial reduction of the carbonyl group to alkoxyboryl derivatives (Sec. 10.1) followed by hydroboration of the double bond. Consequently, the same factors which influence the hydroboration of allylic alcohols (Sec. 5.3.2) operate in the reaction of these compounds.

Saturated monohydric alcohols are reported to be formed in varying yields in the reaction of a variety of α,β-unsaturated carbonyl compounds (73).

$$CH_3CH=CHCHO \xrightarrow[\text{2) [o]}]{\text{1)}BH_3,Et_2O,\ 0^{\circ}} CH_3CH_2CH_2CH_2OH \text{ (major product)}$$

These products are formed by elimination-rehydroboration of the β-substituted organoborane. As with allylic alcohols, prior formation of the disiamylborinate ester largely precludes the subsequent elimination. Thus, reduction of the carbonyl group with disiamylborane, followed by hydroboration-oxidation of the disiamylborinate ester, gives the corresponding <u>vicinal diol</u> (44).

$$RCH=CHCHO \xrightarrow[\text{THF, }0^{\circ}]{Sia_2BH} RCH=CHCH_2OBSia_2 \xrightarrow[\text{2) [o]}]{\text{1)}BH_3\text{-THF}} \underset{\underset{OH}{|}}{RCH_2CHCH_2OH} \quad >80\%$$

$$(CH_3)_2C=CHCOCH_3 \xrightarrow[\text{THF, }0^{\circ}]{Sia_2BH} \underset{\underset{CH_3}{|}}{(CH_3)_2C=CHCHOBSia_2} \xrightarrow[\text{2) [o]}]{\text{1)}BH_3\text{-THF}}$$

$$\underset{\underset{OH\ \ OH}{|\ \ \ |}}{(CH_3)_2CHCH\text{-}CHCH_3}$$

85% (1:1 erythro-threo (74))

Hydroboration-oxidation of conjugated cyclohexenones results in the stereospecific formation of the corresponding <u>trans-diequatorial diols</u> as illustrated below.

R = R′ = H	85%		Ref. 75
R=CH₃; R′=H	70%	17%	Ref. 74
R=R′=CH₃	65%	15%	Ref. 74
R=H; R′=β-Ph, H	45%(+14% 1,3-diols)		Ref. 56b
R=H; R′=β-CMe₃, H	45%(+11% 1,3-diols)		Ref. 56b

85%

Similar results have been reported for the hydroboration-oxidation
of cholest-4-en-3-one (76) and 5α-cholest-1-en-3-one (77), though,
in the case of progesterone, the formation of 70% of the
corresponding 3β-hydroxy-4-ene has been reported (78).

Hydroboration of steroid 4-en-3-ones with a large excess of
borane in diglyme, followed by refluxing with acetic anhydride in
a nitrogen atmosphere, is reported to give the corresponding
3-alkenes in 60-70% yields (79,80).

60-70%

Likewise, the formation of 3β-acetoxy-5α-cholest-6-ene in 80%
yield from the corresponding 5-en-7-one has been reported (81).
However, application of the reaction sequence to androst-4-en-3-one
17β-ol (testosterone) (82) and various bicyclic α,β-unsaturated
ketones (83) gives low yields (20-30%) of the expected alkenes.
Indeed, repetition of the reaction sequence with cholest-4-en-3-one
is reported to give 5α-cholest-3-ene in yields of 30-51% (83b).
3,5,5-Trimethylcyclohexene has been prepared in 52% yield from
3,5,5-trimethyl-2-cyclohexenone by this method (84).

Treatment of the disiamylborinate ester, formed by reduction
of an α,β-unsaturated carbonyl compound with disiamylborane, with
borane-tetrahydrofuran, and followed by treatment with methanesulfonic acid in refluxing tetrahydrofuran, gives the corresponding
alkenes in variable yields (44).

$$RCH=CHCHO \xrightarrow[\text{THF, }0°]{Sia_2BH} RCH=CHCH_2OBSia_2 \xrightarrow[\substack{2)CH_3SO_3H, \\ THF,\Delta}]{1)BH_3-THF} RCH_2CH=CH_2$$

75% (R=Ph)
38% (R=CH$_3$)

$$(CH_3)_2C=CHCOCH_3 \xrightarrow{\substack{1)\ Sia_2BH,\ THF \\ 2)\ BH_3-THF \\ 3)\ CH_3SO_3H,THF,\Delta}} (CH_3)_2CHCH=CHCH_3$$

64%

Application of the reaction sequence to cyclohexenone gives
cyclohexene in 82% yield (44).

1) Sia$_2$BH, THF

2) BH$_3$-THF

3) CH$_3$SO$_3$H

82%

Hydroboration of 1-cyclohexyl-4-benzylidene-2,3-dioxopyrroli-
dine, followed by treatment with ethanol, gives the enol derivative
(25a) (85); similar reaction of trans-1,3-diphenylpropenone
(chalcone) gives significant amounts of 1,3-diphenylpropanone (85).

$$(25a) \quad 58\%$$

$$\text{Trans PhCH=CHCOPH} \xrightarrow[\text{2) } C_2H_5OH]{\text{1) } BH_3\text{-THF}} \text{PhCH}_2\text{CH}_2\text{COPh} + \text{PhCH}_2\text{CH}_2\overset{\overset{\displaystyle OH}{|}}{\text{CHPh}}$$

The possibility of hydroboration occurring via a 1,4-addition
process in these cases has been suggested (85).

5.5   KETENES

Monohydroboration-oxidation of diphenylketene gives
diphenylethanal as the major product, together with various
minor products (86).

$$Ph_2C=C=O \xrightarrow[\text{2) [O]}]{\text{1)}BH_3\text{-THF, } 0°} Ph_2CHCHO + Ph_2C=CH_2 + Ph_2CHCH_2OH + Ph_2CO$$
$$\qquad\qquad\qquad\qquad\qquad\quad 52\% \qquad\quad 6\% \qquad\quad 5\% \qquad\quad 5.5\%$$

The formation of diphenylethanal is rationalized in terms
of reduction of the carbonyl group to give the intermediate

vinyloxyborane which is hydrolyzed; the minor products arise
from elimination and transfer reactions involving the vinyloxyborane (86).

Use of disiamylborane as the hydroborating reagent gives
74% diphenylethanal (86).

5.6 UNSATURATED CARBOXYLIC ACIDS AND DERIVATIVES

Hydroboration-oxidation of α,β-unsaturated carboxylic acids
gives the saturated alcohols as the major products (47,73); thus,
reaction of phytenoic acid with either borane or disiamylborane
at 25° gives mainly 1-phytanol, while at lower temperatures
(0° or -20°) considerable amounts of the 1,2-diol are formed (47).

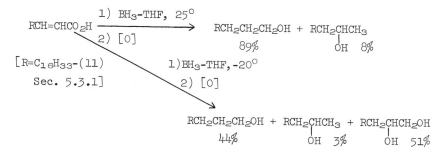

The saturated alcohols are formed via reduction of the
carboxylic acid group and hydroboration of the double bond
followed by elimination-rehydroboration (47).

Hydroboration of α,β-unsaturated acid chlorides has been
reported to proceed with some reduction of the acid chloride group
(87) even though this group is normally unreactive to hydroboration
conditions (Sec. 10.6.3). Similar results have been obtained in
the hydroboration of unsaturated carboxylic esters, where the
presence of an alkene linkage near the ester group greatly enhances
the reactivity of the group toward borane (88). Thus,
ω-unsaturated esters consume 1.3 to 2.8 eq of hydride within 3 hr;
the following order of reactivity has been observed (88).

$$CH_2=CHCO_2C_2H_5 \ (2.8) > CH_2=CHCH_2CO_2C_2H_5 \ (2.15) > CH_2=CHCH_2CH_2CO_2C_2H_5$$

$$(1.8) > CH_2=CH(CH_2)_8CO_2C_2H_5 \ (1.4)$$

Oxidation of the products gives only mixtures of hydroxy-esters and diols indicating that preferential reaction occurs at the double bond.

Competitive hydroboration studies have shown that the reaction of borane with esters is generally considerably slower than that with alkenes (Sec. 10.13) (89). The enhanced reactivity of esters groups in unsaturated carboxylic esters has been attributed to the rapid intramolecular reaction of the ester with the borane intermediate formed by initial hydroboration of the double bond (88).

$$CH_2=CH(CH_2)_nCO_2C_2H_5 \xrightarrow{BH_3} H_2BCH_2CH_2(CH_2)_nCO_2C_2H_5 \longrightarrow$$

In accordance with this explanation is the fact that those esters capable of forming five- or six-membered cyclic intermediates are the most reactive (88).

The problem of reduction of the ester group may be overcome by addition of the theoretical quantity of a dilute borane solution to the neat unsaturated ester at $0°$ (e.g., 14 mmoles 1 M BH_3-THF to 40 mmoles ester); with the exception of ethyl propenoate (ethyl acrylate), high yields of hydroxy-esters are obtained (88).

$CH_2=CHCH_2CO_2C_2H_5$ $CH_2=CHCH_2CH_2CO_2C_2H_5$ $CH_2=CH(CH_2)_8CO_2C_2H_5$

| 63 | 15 | | 85 | 6 | | 80 | 8 |
| 76 | 1 (Sia₂BH) | | 78 | 1 (Sia₂BH) | | 81 | (Sia₂BH) |

The directive influence of the ester group is illustrated by the significant amount of secondary isomer formed in the case of ethyl 3-butenoate. Use of disiamylborane eliminates both the reduction of the ester group and the formation of the secondary isomer (88).

The reaction of ethyl propenoate with excess borane appears to be complex, giving rise to unexpected products (88).

$$CH_2=CHCO_2C_2H_5 \xrightarrow[\text{2) [O]}]{\text{1) } BH_3\text{-THF, } 0^{\circ}} CH_3CHCH_2OH + HOCH_2CH_2CH_2OH +$$

$$\underset{OH}{|}$$

$$30\% \qquad 8\%$$

$$CH_3CH_2CH_2OH + CH_3CHCH_2OC_2H_5$$

$$\underset{OH}{|}$$

$$15\% \qquad 14\%$$

Similar results are obtained with methyl phytenoate (47); at 25° 1-phytanol is the major product, but at -20° the diol and diol monoether are also formed.

$$RCH=CHCO_2CH_3 \xrightarrow[\text{2) [O]}]{\text{1) } BH_3\text{-THF, } -20^{\circ}} RCH_2CH_2CH_2OH + RCH_2CHCH_3 +$$

$$\underset{OH}{|}$$

$$65\% \qquad 5\%$$

[R=$C_{16}H_{33}$-(11),

Sec. 5.3.1]

$$RCH_2CHCH_2OH + RCH_2CHCH_2OCH_3$$

$$\underset{OH}{|} \qquad \underset{OH}{|}$$

$$12\% \qquad 18\%$$

The formation of the diol monoether is attributed to the following reactions (47).

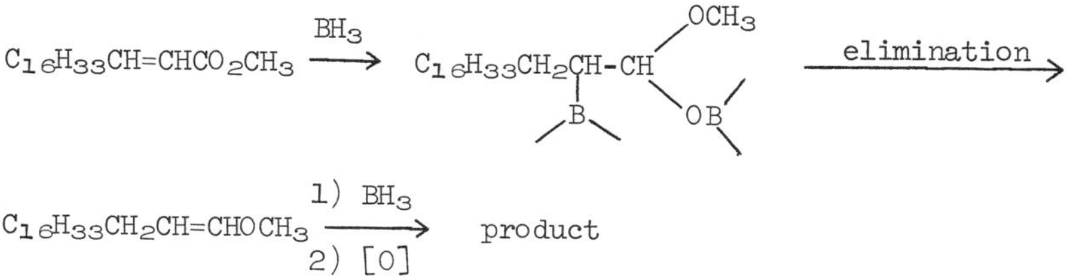

$$C_{16}H_{33}CH=CHCO_2CH_3 \xrightarrow{BH_3} C_{16}H_{33}CH_2CH-CH\overset{OCH_3}{\underset{OB}{\diagup}} \xrightarrow{\text{elimination}}$$

$$\underset{B}{\diagup}$$

$$C_{16}H_{33}CH_2CH=CHOCH_3 \xrightarrow[\text{2) [O]}]{\text{1) } BH_3} \text{product}$$

In the hydroboration of cyclohexene mono- and dicarboxylate derivatives the <u>trans-hydroxy esters</u> are the main products (84).

These results have been rationalized in terms of the polar and steric effects of the carboxylate groups ($\underline{84}$).

The intramolecular interaction of ester groups and intermediate boranes is also encountered in the hydroboration of cyclic unsaturated esters ($\underline{90}$).

The formation of products (26) and (27) has been attributed to
the formation of an intermediate heterocycle (28).

(28)

Reaction of methyl bicyclo[3.1.0]hex-2-ene-6-endo-carboxylate
(29) gives some of the ring-cleaved product (30) (91).

Reaction of the exo-isomer of (29) gives insignificant amounts of
(30), thus providing supporting evidence for the mechanism
proposed above (91).

5.7 THE INFLUENCE OF REMOTE SUBSTITUENTS

The influence of various substituents on the hydroboration of 1-substituted 3-butenes using borane-tetrahydrofuran and disiamylborane is summarized below (yields were determined by vapor phase chromatography)

BH_3			Ref. 92	Sia_2BH		
$CH_2=CHCH_2CH_2X$				$CH_2=CHCH_2CH_2X$		
↑	↑	X		↑	↑	X
73	16	Cℓ		98	1	Cℓ
65	10	OH		89	2	OH
89	11	OCH$_3$		90	2	OCH$_3$
75	11	OPh		86	2	OPh
77	11	SCH$_3$		92	2	SCH$_3$
68	13	OAc		88	1	OAc
62	12	NH$_2$		81	2	NH$_2$

The significant amounts of the secondary isomers formed in the reaction with borane-tetrahydrofuran indicates the operation of appreciable directive influences in all the cases studied. Since 6-7% of the secondary isomer is formed in the hydroboration of 1-alkenes (Sec. 3.1.1), the increase observed in the above cases may be attributed to the directive effect of the substituent. Use of disiamylborane greatly reduces the formation of the secondary isomer. The organoboranes formed are stable and can be readily oxidized or utilized in other reactions (92).

Reaction of 7-dimethylaminoheptene at 25° gives the corresponding 1-alcohol as the major product (93); however, at temperatures exceeding 100° 7-dimethylamino-4-heptanol becomes the major product (94). Formation of this product involves migration of the boron atom in the direction opposite to that normally observed in the high-temperature isomerization of organoboranes (Sec. 2.3.2), and probably proceeds via migration of the boron atom to that position which affords a stable six-membered cyclic amine-borane (31) (94).

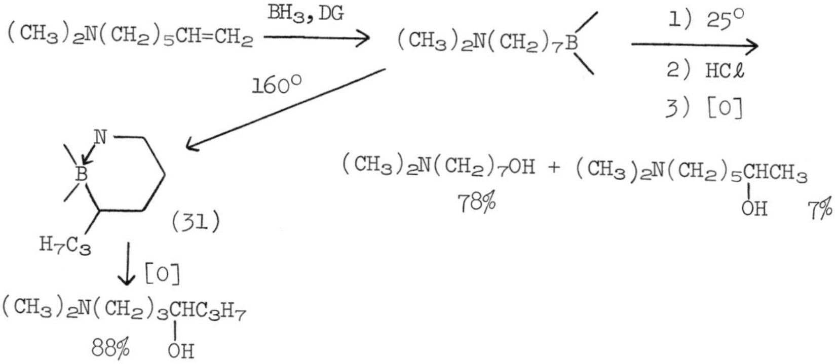

$(CH_3)_2N(CH_2)_5CH=CH_2$ $\xrightarrow[160°]{BH_3, DG}$ $(CH_3)_2N(CH_2)_7B\Big\langle$ $\xrightarrow[\substack{2) \ HC\ell \\ 3) \ [o]}]{1) \ 25°}$

$(CH_3)_2N(CH_2)_7OH$ + $(CH_3)_2N(CH_2)_5CHCH_3$
$\qquad$ 78% $\qquad\qquad\qquad\qquad$ ÓH $\quad$ 7%

(31)

H₇C₃ ↓ [o]

$(CH_3)_2N(CH_2)_3CHC_3H_7$
88%   ÓH

Similar results have been obtained in the hydroboration of
ω-dimethylaminoalkenes with triethylamine-borane in refluxing
xylene (93).

In the hydroboration of 4-methoxycyclohexene with
borane-tetrahydrofuran, a slight preference is shown for the
formation of the trans-3-substituted product (48).

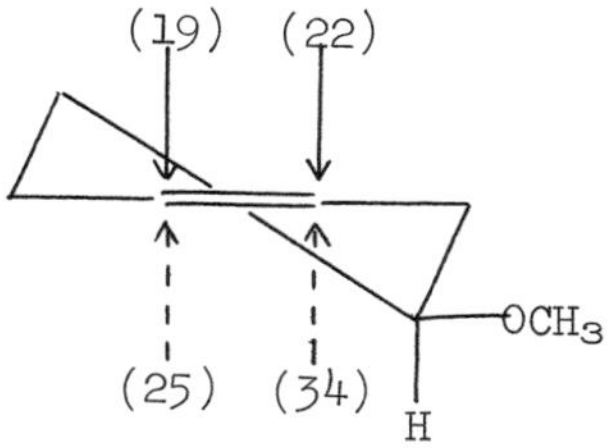

(19) (22)

—OCH₃

(25) (34) H

In the reaction of the diastereoisomeric isopulegols (32)
and (35) the configuration of the 3-hydroxyl group has a marked
influence on the stereochemistry of the diols formed (95).

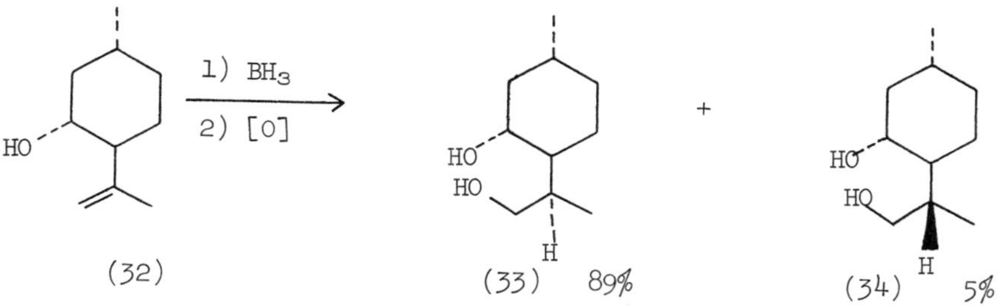

HO   1) BH₃
      ───────→
      2) [o]

HO
HO

+

HO
HQ

(32)

(33)   89%

(34)   5%

(35)   (36)   8%   (37)   82%

The high degree of stereoselectivity observed has been
rationalized in terms of intramolecular interactions between the
isopropenyl double bond and the alkoxyboryl substituent formed by
reaction of the hydroxyl group and borane (95). Thus, in the case
of (-)-isopulegol (32) the interactions result in trans-addition
to the double bond to give heterocycles (38) and (39). The
formation of (38) is favored and gives diol (33).

The results obtained with (+)-neo-isopulegol (35), and with
the other two diastereoisomers containing 4α-isopropylidene
substituents, can be explained in a similar manner (95).

5.8 FRAGMENTATION REACTIONS: SYNTHESIS OF DIENES

A review of this subject has been published (96).

Hydroboration of certain unsaturated cyclic methanesulfonate derivatives, followed by treatment with aqueous sodium hydroxide, gives fragmented products. Thus, the octalin derivative (40) gives 1-methyl-trans,trans-1,6-cyclodecadiene (41) as the major product (97).

(40)   1) BH₃-THF   (41) 77%   +   (42) 13%
         2) NaOH, Δ

The diene (41) is formed via internal carbon-carbon bond cleavage as shown in (43), while formation of (42) proceeds via 1,3-elimination of the tertiary boryl derivative (44) (98).

(43)   (44)

The unsaturated keto mesylate (45) undergoes similar bond cleavage (99), while reaction of (46) gives 1-methyl-trans,trans-1, 5-cyclodecadiene (100).

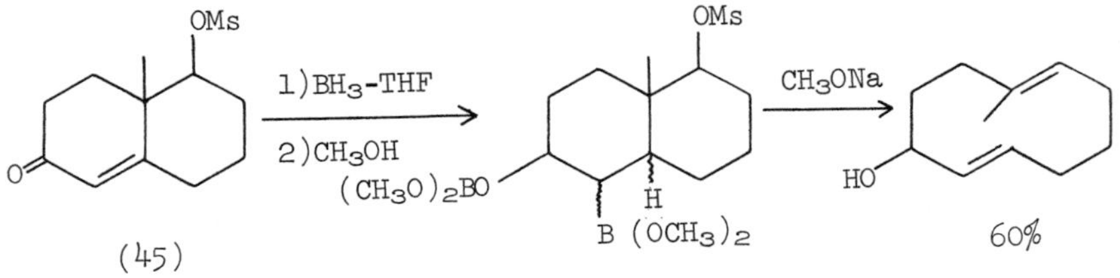

(45)

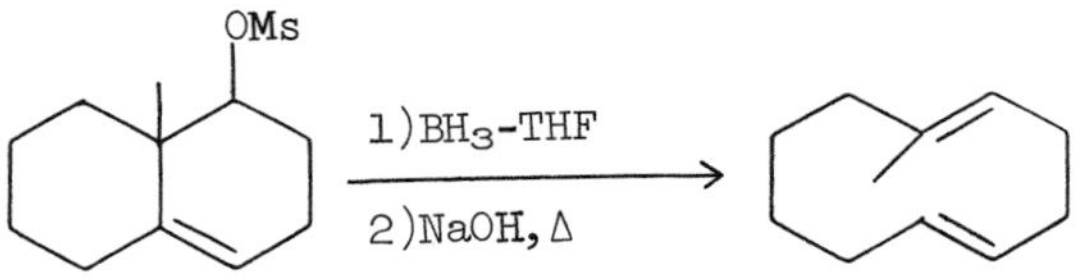

60%

(46)   78%

Application of this reaction sequence to suitable cyclohexenyl methanesulfonate derivatives has led to the development of a useful route to acyclic 1,5-dienes (101).

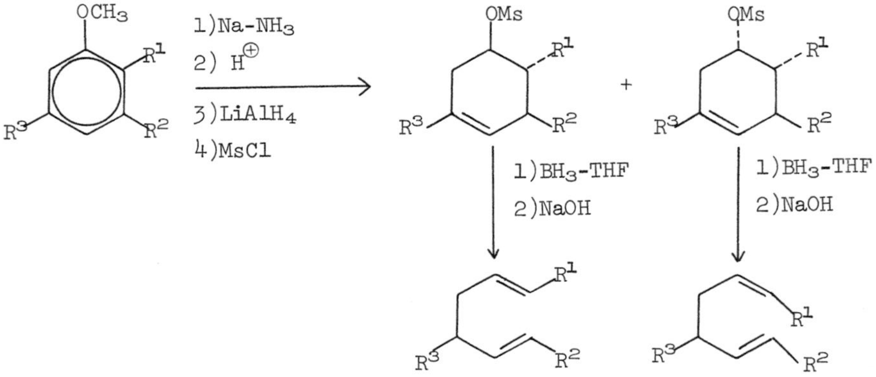

The overall yields of 1,5-dienes are 50-60%. In all cases only those products arising from cleavage of the more substituted carbon-carbon bond are formed (101); in the case of equally substituted bonds both isomeric 1,5-dienes are formed (101).

5.9   HETEROCYCLIC ALKENES

   In the following discussion the hydroboration of heterocyclic compounds containing the hetero atom and the double bond in the same ring is discussed.

5.9.1   Oxygen Heterocycles

   Hydroboration of 2,3- and 2,5-dihydrofurans with borane‑tetrahydrofuran in a mole ratio of 3:1 at 0° readily gives the corresponding trialkylboranes (102).

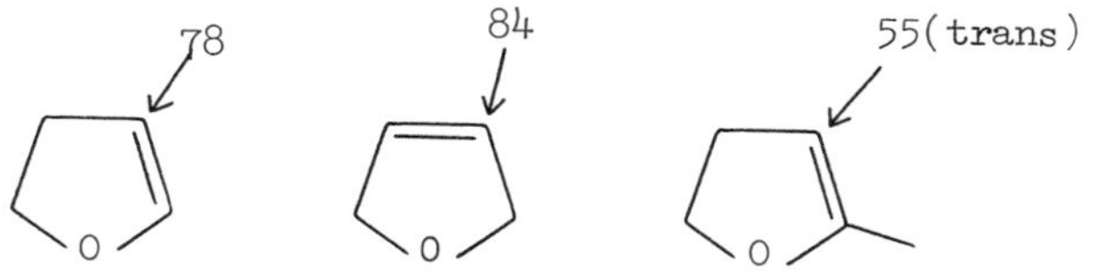

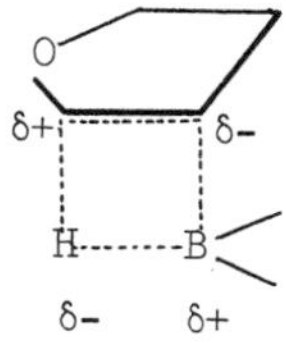

(47)

The rate of reaction of 2,3-dihydrofuran with disiamylborane is markedly enhanced compared to that of cyclopentene, while no such enhancement is observed with 2,5-dihydrofuran (102). This increase in reactivity, as well as the selective attack at position 3, may be explained by the stabilization of the transition state (47) by the neighboring oxygen atom.

   Dihydropyrans likewise react to give the corresponding trialkylboranes (102).

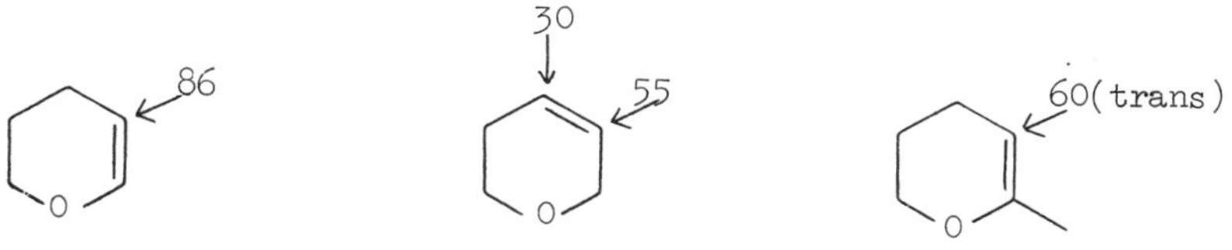

Ref. 103

The selective attack at position 3 of Δ^2-dihydropyran is
attributed to stabilization of the relevant transition state by
the neighboring oxygen atom (102). However, little enhancement
of the reaction rate compared to that of cyclohexene is observed,
indicating that the stabilization in this case is not as efficient
as that observed for 2,3-dihydrofuran (102).

Reaction of 2-alkoxy-3,4-dihydropyran derivatives gives
mixtures of isomeric 2-alkoxy-5-hydroxytetrahydropyrans with the
trans-isomer predominating.

Ref. 102   Ref. 104

The preferential formation of the trans-isomers is attributed to
the axial conformation of the alkoxy group (anomeric effect)
hindering cis-attack (102).

Isoflavene derivatives (48) (105) and chromene derivative
(49) (106a) react as expected to give the corresponding
trans-4-alcohols. Hydroboration-oxidation of 3-flavenes (49a)
gives mainly the trans-4-alcohols (106b).

(48)   (49)   (49a)

It has been observed that the use of excess borane in the hydroboration of dihydropyrans and dihydrofurans leads, after oxidation, to the formation of acyclic diols (102).

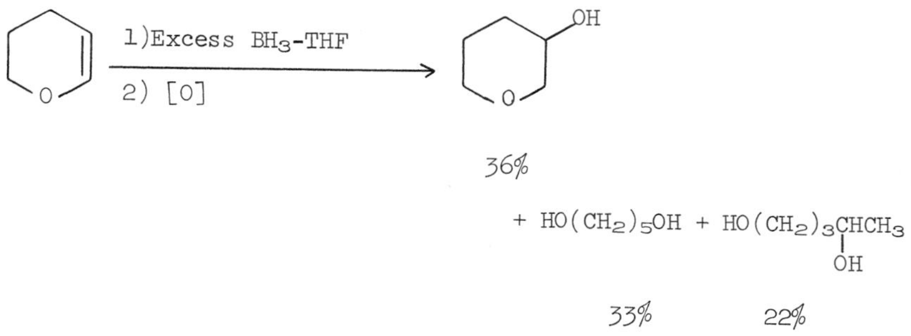

36%

$$+ \ HO(CH_2)_5OH \ + \ HO(CH_2)_3CHCH_3$$
$$\overset{|}{OH}$$

33%   22%

These diols are formed via trans-elimination-rehydroboration and redistribution-β-transfer reactions (Sec. 5.1) (102).

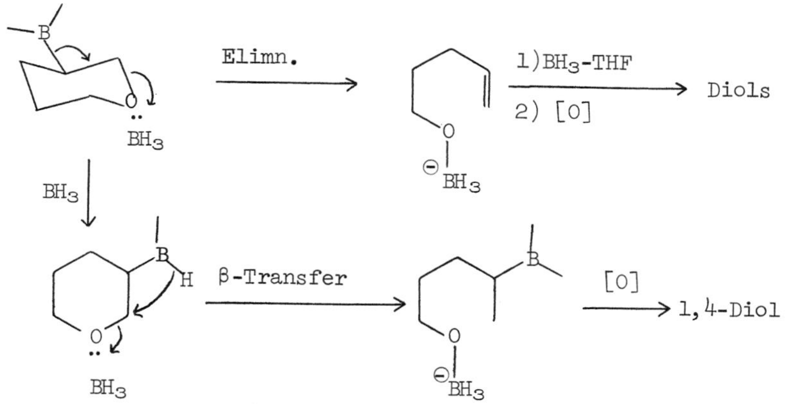

Such a process might also account for the formation of a mixture in the attempted hydroboration of (25R)-5α-spirost-23-ene (50) (107).

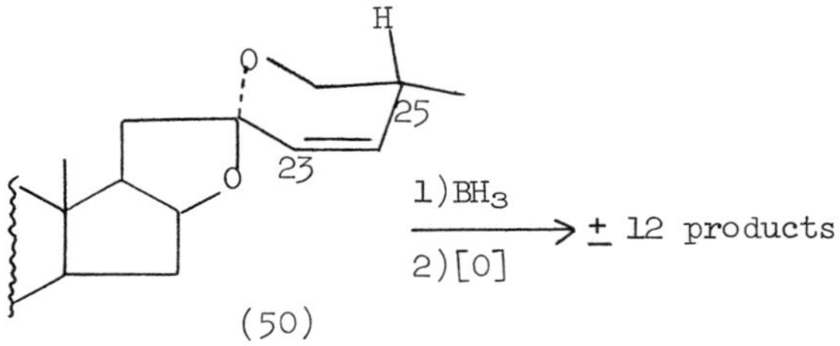

(50)

1) BH$_3$
2) [O] $\rightarrow$ ± 12 products

The above hydroboration-elimination sequence of reactions has been adapted to the synthesis of unsaturated alcohols (102). Thus, hydroboration of Δ^2-dihydropyrans and 2,3-dihydrofurans, followed by treatment with boron trifluoride etherate and hydrolysis gives the corresponding unsaturated alcohols in high yield.

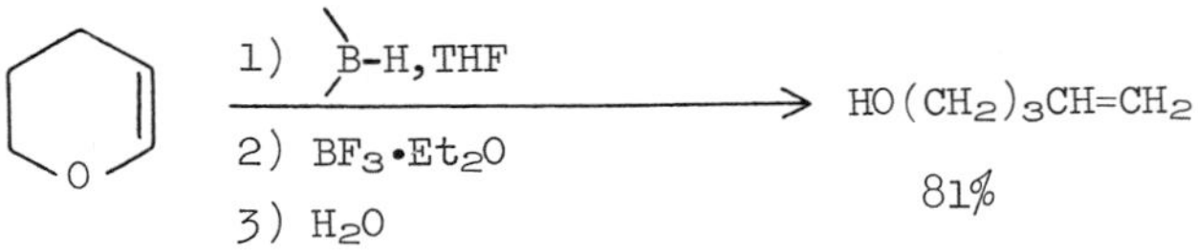

1) $\rangle$B-H, THF
2) BF$_3$·Et$_2$O
3) H$_2$O

$\rightarrow$ HO(CH$_2$)$_3$CH=CH$_2$

81%

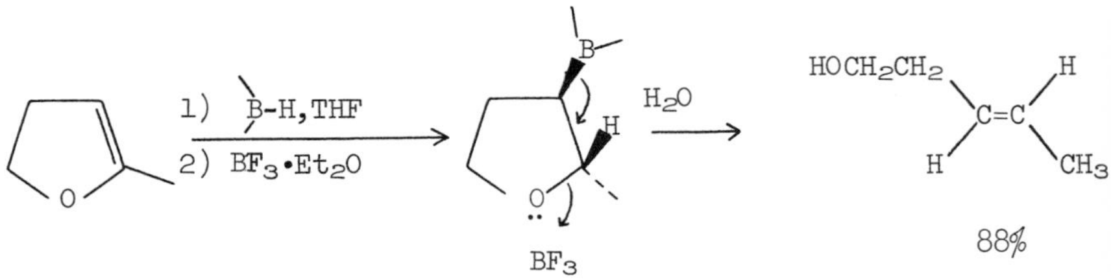

1) $\rangle$B-H, THF
2) BF$_3$·Et$_2$O

$\rightarrow$

H$_2$O $\rightarrow$

HOCH$_2$CH$_2$ $\!\!\!\!$ \ $\quad$ / H
$\qquad\qquad$ C=C
$\qquad$ H / $\quad$ \ CH$_3$

88%

5.9.2 Nitrogen Heterocycles

Hydroboration of 1-alkyl-1,2,3,6-tetrahydropyridines proceeds via the unsaturated amine-borane; attack on this intermediate is directed so as to place boron predominantly at the 3-position (108).

(R=CH$_3$, PhCH$_2$)   53-60%   17-24%

Reaction of the 1-methyl derivative using (-)-diisopino-campheylborane gives (3R)-1-methyl-3-piperidinol (51) (109).

(51) 50%   20%

The configuration of the alcohols is opposite to that predicted by model 52 (Sec. 3.7), and thus the hydroboration of the amine-borane complex gives similar results to those obtained with hindered cyclic alkenes (Sec. 3.6.2).

In the cases of 3- and 4-substituted derivatives of 1-methyl-1,2,3,6-tetrahydropyridine trans-alcohols are formed.

Ref. 110   Ref. 111

Hydroboration of tropidine derivatives (52) occurs predominantly from the α-side due to shielding of the β-side of the molecule by the amine-borane moiety (108).

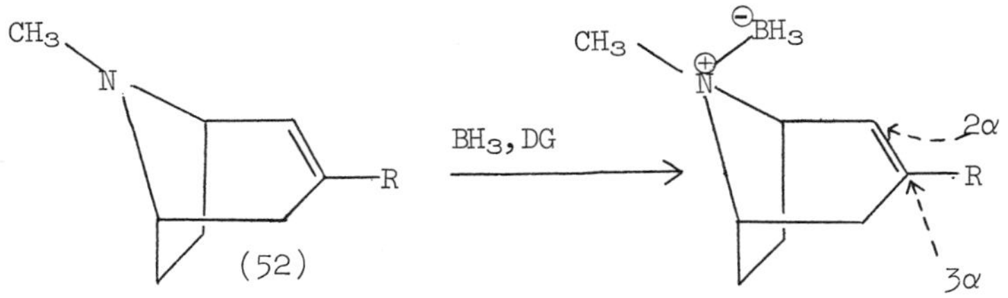

R=H

R=Ph

R=p-CℓC$_6$H$_4$

R=p-CH$_3$OC$_6$H$_4$

2α	3α
30%	34%
55%	18%
45%	19%
60%	15%

The formation of significant amounts of the 3α-alcohols in the reaction of 3-aryl substituted tropidines is surprising in view of results obtained with trisubstituted arylalkenes (Sec. 5.2.5) and substituted tetrahydropyridines discussed earlier in this section. It is possible that destabilization of the transition state leading to the 2α-products by the cis-vicinal ethano bridge promotes some attack at the 3-position, with the amount of attack at this position being influenced by the p-substituent in the aromatic ring (Table 3.4, Sec. 3.4) (108).

Hydroboration of 2,3-dicarbomethoxy-2,3-diazabicyclo[2.2.1]-hept-5-ene (53) gives the exo-dialkylborane which undergoes cleavage on treatment with various bases (112). However, oxidation at 0° gives the corresponding alcohol in high yield (113).

The cleavage reaction occurs via a trans-β-elimination mechanism (Sec. 5.1) (112).

5.9.3 Miscellaneous Heterocycles

The hydroboration-oxidation of various germano-cycloalkenes (114) and sila-cycloalkenes (114-116) proceeds readily to give high yields of alcohols.

REFERENCES

1. D. J. Pasto and R. Snyder, J. Org. Chem., 31, 2773 (1966).

2. D. J. Pasto and J. Hickman, J. Am. Chem. Soc., 89, 5608 (1967).

3. A. Suzuki, K. Ohmori, H. Takenaka, and M. Itoh, Tetrahedron Letters, 1968, 4937.

4. D. J. Pasto, J. Hickman, and T. C. Cheng, J. Am. Chem. Soc., 90, 6259 (1968).

5. D. S. Matteson, Organometal. Chem. Rev., 5, 42 (1969).

6. T. A. Shchegoleva, V. D. Sheludyakov and B. M. Mikhailov, Dokl. Akad. Nauk. SSSR, 152, 888 (1963).

7. O. P. Shitov, S. L. Ioffe, V. A. Tartakovskii, and S. S. Novikov. Russ. Chem. Revs., 1970, 905.

8. D. J. Pasto and R. Snyder, J. Org. Chem., 31, 2777 (1966).

9. D. J. Pasto and J. L. Miesel, J. Am. Chem. Soc., 85, 2118 (1963).

10. H. C. Brown, Hydroboration, Benjamin, New York, 1962, p. 269.

11. D. Seyferth and B. Prokai, J. Am. Chem. Soc., 88, 1834 (1966).

12. H. C. Brown and R. L. Sharp, J. Am. Chem. Soc., 90, 2915 (1968).

13. S. J. Cristol, F. P. Parungo, and D. E. Plorde, J. Am. Chem. Soc., 87, 2870 (1965).

14. D. J. Pasto and C. C. Cumbo, J. Am. Chem. Soc., 86, 4343 (1964).

15. B. M. Mikhailov and A. N. Blokhina, Izv. Akad. Nauk. SSSR. Otdel. Chim. Nauk., 1962, 1373; Chem. Abstr., 58, 5707(e) (1963).

16. G. F. D'Alelio, U. S. Patent 3,115,526, Dec. 24, 1963; Chem. Abstr., 60, 6866(a) (1964).

17. L. Cagliotti, G. Cainelli, G. Maina, and A. Selva, Gazz. Chim. Ital., 92, 309 (1962).

18. H. Arzoumanian, E. M. Acton, and L. Goodman, J. Am. Chem. Soc., 86, 74 (1964).

19. D. B. Bigley and D. W. Payling, J. Chem. Soc., 1965, 3974.

20. C. F. Lane and H. C. Brown, J. Am. Chem. Soc., 93, 1025 (1971).

21. A. Suzuki, K. Ohmori, and M. Itoh, Tetrahedron, 25, 3707 (1969).

22. A. Hassner, R. E. Barnett, P. Catsoulacos, and S. H. Wilen, J. Am. Chem. Soc., 91, 2632 (1969).

23. H. C. Brown, D. B. Bigley, S. K. Arora, and N. M. Yoon, J. Am. Chem. Soc., 92, 7161 (1970).

24. I. J. Borowitz and G. J. Williams, J. Org. Chem., 32, 4157 (1967).

25. (a) J. Gore, J. P. Drouet, and J. J. Barieux, Tetrahedron
Letters, 1969, 9, (b) J. J. Barieux and J. Gore, Bull. Soc. Chim,
France, 1971, 1649, 3978.

26. J. W. Lewis and A. A. Pearce, J. Chem. Soc., B1969, 863.

27. (a)   J. J. Barieux and J. Gore, Tetrahedron, 28, 1537 (1972);
(b) J. J. Barieux and J. Gore, Tetrahedron, 28, 1555 (1972).

28. J. Schmitt, J. J. Panouse, A. Hallot, P. J. Cornu, P. Comoy,
and H. Pluchet, Bull. Soc. Chim. France, 1963, 807.

29. D. N. Butler and A. H. Soloway, J. Am. Chem. Soc., 88, 484
(1966).

30. D. J. Pasto and J. L. Miesel, J. Am. Chem. Soc., 84, 4991
(1962).

31. D. J. Pasto, J. Am. Chem. Soc., 84, 3777 (1962).

32. G. A. Russell and L. A. Ochrymowycz, J. Org. Chem., 35, 764
(1970).

33. D. Seyferth, J. Am. Chem. Soc., 81, 1844 (1959).

34. M. Kumada, N. Imaki, and K. Yamamoto, J. Organometal Chem.,
6, 490 (1966).

35. D. Seyferth, H. Yamazaki, and Y. Sato, Inorg. Chem., 2, 734
(1963).

36. C. Benezra and G. Ourisson, Bull. Soc. Chim. France, 1967,
624.

37. P. M. Aronovich, V. S. Bogdanov, and B. M. Mikhailov, Izv.
Akad. Nauk. SSSR. Ser. Khim., 1969, 362; Chem. Abstr., 71, 13163(w)
(1969).

38. D. J. Pasto, J. Chow, and S. K. Arora, Tetrahedron, 25, 1557
(1969).

39. P. M. Aronovich and B. M. Mikhailov, Izv. Akad. Nauk. SSSR,
Ser Khim., 1968, 2745; Chem. Abstr., 70, 87876(g) (1969).

40. H. C. Brown and R. L. Sharp, J. Am. Chem. Soc., 88, 5851
(1966).

41. C. H. DePuy, G. F. Morris, J. S. Smith, and R. J. Smat,
J. Am. Chem. Soc., 87, 2421 (1965).

42. S. Murahashi and I. Moritani, Bull. Chem. Soc. Japan, 41,
1884 (1968).

43. H. C. Brown and O. J. Cope, J. Am. Chem. Soc., 86, 1801
(1964).

44. H. C. Brown and R. M. Gallivan, J. Am. Chem. Soc., 90, 2906
(1968).

45. H. C. Brown and K. A. Keblys, J. Am. Chem. Soc., 86, 1791
(1964).

46. H. C. Brown and E. F. Knights, Israel J. Chem., 6, 691(1968).

47. Y. Bessiére-Chrétien and J. P. Marion, Bull. Soc. Chim.
France, 1970, 3013.

48. D. J. Pasto and J. Hickman, J. Am. Chem. Soc., 90, 4445
(1968).

49. M. F. Hawthorne, J. Am. Chem. Soc., 82, 1886 (1960).

50. H. C. Brown and S. P. Rhodes, J. Am. Chem. Soc., 91, 2149
(1969).

51. H. C. Brown and E. F. Knights, J. Am. Chem. Soc., 90, 4439
(1968).

52. P. Binger and R. Köster, Tetrahedron Letters, 1961, 156;
R. Köster, G. Griansnow, W. Larbig, and P. Binger, Annalen, 672,
1 (1964).

53. A. J. Fry and W. B. Farnham, J. Org. Chem., 34, 2314 (1969).

54. P. v. R. Schleyer, J. Am. Chem. Soc., 89, 701 (1967).

55. J. Canceill, J. Gabard, and J. Jacques, Bull. Soc. Chim.
France, 1966, 2653.

56. (a) J. Klein and E. Dunkelblum, Tetrahedron Letters, 1966,
6047; (b) E. Dunkelblum, R. Levene, and J. Klein, Tetrahedron,
28, 1009 (1972).

57. A. Uzarewicz, I. Uzarewicz, and W. Zacharewicz, Rocz. Chem.,
39,19 (1965); Chem. Abstr., 62, 16074(b) (1965).

58. Y. Chrétien-Bessiére and G. Boussac, Bull. Soc. Chim. France,
1967, 4728.

59. I. Uzarewicz, A. Uzarewicz, and W. Zacharewicz, Rocz. Chem.,
39, 1051 (1965); Chem. Abstr., 64, 12726 (f) (1966).

60. Y. Bessiére-Chrétien and B. Meklati, Compte Rend. Acad. Sci.
Paris, 271, (C), 318 (1970); Bull. Soc. Chim. France, 1971, 2591.

61. J. M. Davies and S. H. Graham, J. Chem. Soc., C1968, 2040.

62. K. J. Ryan, H. Arzoumanian, E. M. Acton, and L. Goodman,
J. Am. Chem. Soc., 86, 2503, (1964).

63. A. Rosenthal and M. Sprinzl, Can. J. Chem., 47, 4477 (1969).

64. I. Uzarewicz and A. Uzarewicz, Rocz. Chem., 44, 1205 (1970);
Chem. Abstr., 73, 88031(b) (1970).

65. I. Uzarewicz, A. Uzarewicz, J. Krupowicz, and W. Zacharewicz,
Rocz. Chem., 38, 1615 (1964); Chem. Abstr., 62, 9175(e) (1965).

66. (a) W. Nagata, M. Narisada, and T. Wakabayashi, Chem. Pharm.
Bull., 16, 875 (1968);   (b) R. A. Braun, D. C. Brown, and R. M.
Adams, J. Am. Chem. Soc., 93, 2823 (1971);   (c) R. M. Adams and
F. D. Poholsky, Inorg. Chem., 2, 640 (1963).

67. N. N. Greenwood, J. H. Morris and J. C. Wright, J. Chem. Soc.,
1964, 4753.

68. H. Kugita and M. Takeda, Chem. Pharm. Bull., 13, 1422 (1965).

69. M. Takeda, H. Inoue, and H. Kugita, Tetrahedron, 25, 1839
(1969).

70. B. M. Mikhailov, V. F. Pozdnev, and V. G. Kiselev, Dokl.
Akad. Nauk. SSSR, 151, 577 (1963); Chem. Abstr., 59, 12830(b)
(1963).

71. E. N. Marvell, D. Sturmer, and C. Rowell, Tetrahedron, 22,
861 (1966).

72. W. G. Duncan and D. W. Henry, J. Med. Chem., 12, 711 (1969).

73. L. Tai and C. T. Chien, Hua Hsueh Hsueh Pao, 31, 370 (1965);
Chem. Abstr., 64, 8022(b) (1966); C. T. Chien, C. Hsieh, and
L. Tai, Hua Hsueh Hsueh Pao, 31, 376; Chem. Abstr., 64, 8022(c)
(1966).

74. J. Klein and E. Dunkelblum, Tetrahedron, 24, 5701 (1968).

75. M. Zaidlewicz, I. Uzarewicz, and A. Uzarewicz, Rocz. Chem.,
43, 937 (1969); Chem. Abstr., 71, 70167(z) (1969).

76. M. Nussim, Y. Mazur, and F. Sondheimer, J. Org. Chem., 29,
1120 (1964).

77. G. Cainelli and A. Selva, Gazz. Chim. Ital., 97, 720 (1967).

78. M. Stefanović and S. Lajsić, Tetrahedron Letters, 1967, 1777.

79. L. Caglioti, G. Cainelli, G. Maina, and A. Selva, Tetrahedron, 20, 957 (1964).

80. L. Caglioti and G. Cainelli, U. S. Patent 3,083,199, Mar. 26, 1963; Chem. Abstr., 59, 5238(e) (1963).

81. L. Caglioti, G. Cainelli, and A. Selva, Chim. Ind. (Milan), 44, 36 (1962); Chem. Abstr., 60, 12075(d) (1964).

82. H. Tada and Y. K. Sawa, J. Org. Chem., 33, 3347 (1968).

83. (a) K. Bailey and T. G. Halsall, J. Chem. Soc., C1968, 679; (b) F. J. Schmitz and C. A. Peters, ibid., C1971, 1905.

84. J. Klein, E. Dunkelblum, and D. Avrahami, J. Org. Chem., 32, 935 (1967).

85. P. L. Southwick, N. Latif, B. M. Fitzgerald, and N. M. Zaczek, J. Org. Chem., 31, 1 (1966).

86. D. S. Sethi, I. Mehrotra, and D. Devaprabhakara, Tetrahedron Letters, 1970, 2765.

87. K. Kratzl and P. Claus, Monatsch. Chem., 94, 1140 (1963).

88. H. C. Brown and K. A. Keblys, J. Am. Chem. Soc., 86, 1795 (1964).

89. H. C. Brown and W. Korytnyk, J. Am. Chem. Soc., 82, 3866 (1960).

90. D. Varech and J. Jacques, Bull. Soc. Chim. France, 1969, 3505.

91. D. Döpp, Chem. Ber., 102, 1081 (1969).

92. H. C. Brown and M. K. Unni, J. Am. Chem. Soc., 90, 2902 (1968).

93. Z. Polivka, V. Kubelka, N. Holubová, and M. Ferles, Coll. Czech. Chem. Commun., 35, 1131 (1970).

94. Z. Polivka and M. Ferles, Coll. Czech. Chem. Commun., 35, 1147 (1970).

95. K. H. Schulte-Elte and G. Ohloff, Helv. Chim. Acta, 50, 153 (1967).

96. J. A. Marshall, Synthesis, 1971, 229.

97. J. A. Marshall and G. L. Bundy, J. Am. Chem. Soc., 88, 4291 (1966).

98. J. A. Marshall and J. H. Babler, Chem. Commun., 1968, 993.

99. J. A. Marshall and W. F. Huffman, J. Am. Chem. Soc., 92, 6358 (1970).

100. J. A. Marshall and G. L. Bundy. Chem. Commun., 1967, 854.

101. J. A. Marshall and J. H. Babler, Tetrahedron Letters, 1970, 3861.

102. G. Zweifel and J. Plamondon, J. Org. Chem., 35, 898 (1970).

103. J. Gore and F. Guigues, Bull. Soc. Chim. France, 1970, 3521.

104. R. M. Srivastava and R. K. Brown, Can. J. Chem., 48, 2334 (1970).

105. S. Yamaguchi, S. Ito, I. Suzuki, and N. Inoue. Bull. Chem. Soc. Japan, 41, 2073 (1968).

106. (a) K. Hanaya, Bull. Chem. Soc. Japan, 43, 442 (1970); (b) J. Clark-Lewis and M. I. Baig, Aust. J. Chem., 24, 2581 (1971).

107. W. H. Faul, A. Failli, and C. Djerassi, J. Org. Chem., 35, 2571 (1970).

108. R. E. Lyle, K. R. Carle, C. R. Ellefson, and C. K. Spicer, J. Org. Chem., 35, 802 (1970).

109. R. E. Lyle and C. K. Spicer, Tetrahedron Letters, 1970, 1133.

110. Z. Polivka and M. Ferles, Coll. Czech. Chem. Commun., 35, 2392 (1970).

111. R. E. Lyle, D. H. McMahon, W. E. Krueger, and C. K. Spicer, J. Org. Chem., 31, 4164 (1966).

112. E. L. Allred and R. L. Smith, J. Org. Chem., 31, 3498 (1966).

113. E. L. Allred, C. L. Anderson, and R. L. Smith, J. Org. Chem., 31, 3493 (1966).

114. G. Manuel, P. Mazerolles, and J. Florence, Compte. Rend. Acad. Sci. Paris, 269, (C), 1553 (1969).

115. A. G. Brook and J. B. Pierce, J. Org. Chem., 30, 2566 (1965).

116. A. G. Brook, H. W. Kucera, and D. M. MacRae, Can. J. Chem., 48, 818 (1970).

Chapter 6

HYDROBORATION OF DIENES AND POLYENES

The utility of the hydroboration of dienes in the synthesis
of diols and cyclic organoboranes is discussed in this chapter.
In the case of acyclic dienes, use of borane as the hydroborating
reagent often leads to mixtures of diols; use of reagents such as
disiamylborane, however, results in the preferential formation of
a single product, thereby enhancing the synthetic utility of the
reaction (Sec. 6.3). In addition, selective monohydroboration
of dienes is often achieved using disiamylborane, resulting in a
convenient synthetic route to unsaturated alcohols (Sec. 6.5).

The cyclic hydroboration of dienes by reagents such as
thexylborane provides a convenient synthetic route to cyclic
organoboranes, which are useful intermediates in the synthesis of
cyclic ketones by means of carbonylation reactions (Sec. 6.2).
Cyclic organoboranes are also readily synthesized by the
hydroboration of suitable trienes (Sec. 6.8).

The mechanism of the hydroboration of 1,3-butadiene has
been extensively studied, and is therefore discussed separately
in Sec. 6.1.

6.1 HYDROBORATION OF 1,3-BUTADIENE

The mechanism of the hydroboration of 1,3-butadiene has
been investigated by various workers for a number of years and
has only recently been clarified.

Hydroboration of 1,3-butadiene using an equimolar quantity
borane-tetrahydrofuran results in the rapid uptake of hydride to
give almost entirely polymeric material which, on oxidation,
gives a mixture consisting mainly of 1,4- and 1,3-diols ($\underline{1}$).
Distillation of the polymeric material gives a 1:1 dimer (2)
which, for a substance containing residual boron-hydrogen bonds,
exhibits unusual stability toward water, alcohols, and alkenes;
in addition, it is not oxidized by alkaline hydrogen peroxide
at $0°$, while oxidation at room temperature gives only
1,4-butanediol ($\underline{2}$). The structure of the dimer was initially
thought to be that of bisboracyclopentane (1)($\underline{3}$), but later
studies have shown it to be 1,6-diboracyclodecane (2)($\underline{2}$).

Confirmation of structure (2) has been provided by the
preparation of 1,2-tetramethylenediborane (3) which reacts with
1,3-butadiene to give (2)($\underline{4}$).

Hydroboration of 1,3-butadiene with borane-tetrahydrofuran
in the molar ratio of 3:2 gives a mixture of 1,4- and
1,3-bis(1-boracyclopentyl)butane (4 and 5, respectively),which,

on heating at $170°$, gives a mixture of products containing
mainly 1,1-bis(1-boracyclopentyl)butane (6)(5).

$$3CH_2=CH-CH=CH_2 \xrightarrow[0°]{2BH_3-THF}$$

⬠$B-(CH_2)_4-B$⬠ + ⬠$B-CH_2CH_2\overset{\underset{\displaystyle CH_3}{|}}{C}H-B$⬠

(4) (22%)   (5) (56%)

$\downarrow 170°$

+ Polymer (~20%)

(4) + ⬠$B\big[CHCH_2CH_2CH_3\big]_2$

(25%)   (6)   (75%)

The difference in the results obtained using 1:1 and 3:2
molar ratios of diene to borane indicates that the bis(1-
boracyclopentyl)butanes (4) and (5) must be susceptible to the
action of borane. Supporting evidence is provided by the
observation that treatment of B-alkylboracyclopentanes with an
equimolar quantity of borane-tetrahydrofuran leads to the facile
opening of the boracyclopentyl ring to give 1-alkyl-1,2-tetra-
methylenediboranes (e.g., 7) (6).

⬠$B-C_4H_9$ $\xrightarrow[25°]{BH_3-THF}$ (7) structure

(7)

Treatment of a mixture of (4) and (5) with an equimolar quantity
of borane at $25°$ gives the polymeric product (8), thus confirming
the above proposal (6a). Reaction of 1,3-butadiene with
borane-tetrahydrofuran in a ratio of 3:4 gives mainly a mixture
of bis(1,2-tetramethylenediboryl)butanes (9), which are
converted into (8) on addition of an equimolar quantity of a
mixture of (4) and (5) (6a).

$$3(CH_2=CH)_2 \xrightarrow[\text{THF}]{4BH_3-} \quad (9) \quad \xrightarrow{\quad} (4) + (5)$$

$$\left[\quad \right]_n \quad (8) \quad \xleftarrow[\text{THF} \quad (5)]{BH_3 \quad (4) \quad +}$$

The 1:1 hydroboration product (8) reacts readily with water and
methanol but only slowly with alkenes at 25° (6a). Refluxing the
reaction mixture in tetrahydrofuran greatly accelerates the
reaction with alkenes and gives B-alkylboracyclopentanes in
55-60% yields (6a), thereby representing one of the simplest
methods for the synthesis of these compounds.

On the basis of the above evidence the following mechanism
has been proposed for the hydroboration of 1,3-butadiene (5,6a).

$$(CH_2=CH-)_2 \xrightarrow{BH_3-THF} \underset{BH_2}{CH_2CH_2CH=CH_2} \quad + \quad \underset{BH_2}{CH_3CHCH=CH_2}$$

$$\downarrow \text{BH}_3\text{-THF (fast)} \qquad \downarrow \text{BH}_3\text{-THF (fast)}$$

$$\underset{BH_2 \quad BH_2}{CH_2CH_2CH_2CH_2} \qquad \underset{BH_2 \quad BH_2}{CH_3CHCH_2CH_2}$$

(2)

$$\downarrow (CH_2=CH)_2 \qquad \downarrow 2(CH_2=CH)_2 \qquad \downarrow 2(CH_2=CH)_2$$

$$(4) \underline{\qquad\qquad} (5)$$

$$\downarrow \text{BH}_3\text{-THF}$$

(8)

The preference for attack by boron at the internal 3 position of
the diene can possibly be associated with stabilization of the

relevant transition state by the remaining double bond, while the
fast reaction of the monohydroborated species with a second borane
molecule may be attributed to the greater reactivity of the
isolated double bonds compared to the conjugated diene system ($\underline{1}$).

6.2   SYNTHESIS OF CYCLIC ORGANOBORANES BY THE HYDROBORATION OF ACYCLIC DIENES

The synthesis of B-alkylboracyclopentanes by the reaction of
alkenes and the 1:1 hydroboration product (8), formed from
1,3-butadiene and borane-tetrahydrofuran, has been discussed in
Sec. 6.1.

Hydroboration of a number of dienes capable of forming five-,
six-, or seven-membered boracyclanes, using a molar ratio of diene
to borane of 3:2, gives the corresponding bis(1-boracycloalkyl)
alkanes.   Thus, 2-methyl-1,3-butadiene ($\underline{7}$) and 1,3-pentadiene (10)
($\underline{8}$) give the corresponding bis(1-boracyclopentyl)butane **derivatives**,
while 1,4-pentadiene gives mainly bis(1-boracyclohexyl)pentane
($\underline{9}$)and 1,5-hexadiene (11) forms 1,6-bis(1-boracycloheptyl)hexane
($\underline{8}$).

These products, on heating at $170°$, isomerize to form the
thermodynamically more stable boracyclohexane derivatives
(Sec. 2.3.2)($\underline{8}$).

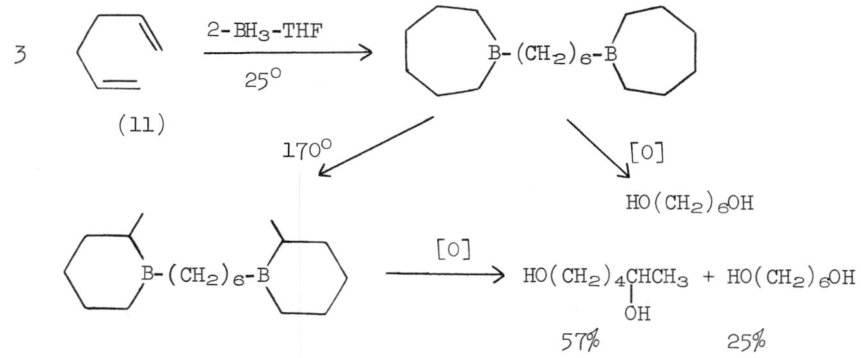

3   (11) $\xrightarrow[25°]{\text{2-BH}_3\text{-THF}}$ B-$(CH_2)_6$-B

170°   [o]

HO$(CH_2)_6$OH

B-$(CH_2)_6$-B $\xrightarrow{[o]}$ HO$(CH_2)_4$CHCH$_3$ + HO$(CH_2)_6$OH
                                                           |
                                                           OH
                                      57%   25%

In contrast to the behavior of B-alkylboracyclopentanes which are cleaved by borane (Sec. 6.1), reaction of B-alkylboracyclohexanes with borane results in a rapid exchange reaction to form bisboracyclohexanes (6,9).

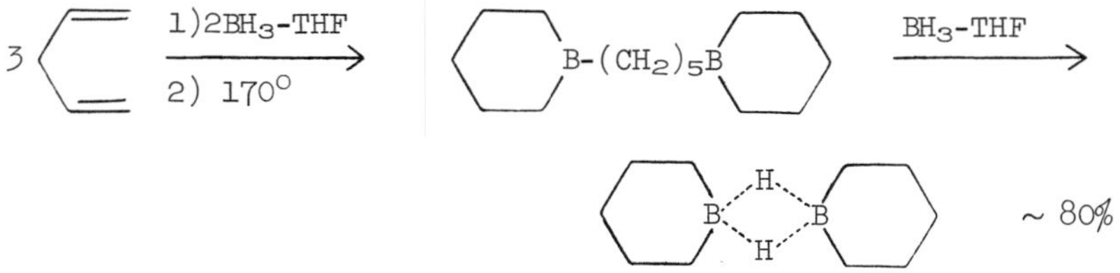

3   $\xrightarrow[\text{2) 170°}]{\text{1)2BH}_3\text{-THF}}$ B-$(CH_2)_5$B $\xrightarrow{\text{BH}_3\text{-THF}}$

B$\cdots$H$\cdots$B   ~ 80%
    H

The bisboracyclohexane reacts readily with alkenes thus providing a convenient general synthesis of B-alkylboracyclohexanes (9). The necessity for the isomerization step (some boracyclopentyl derivative is formed at 25°) is avoided by utilizing dienes containing disubstituted terminal alkene moieties which direct hydroboration nearly exclusively to the terminal position (10a).

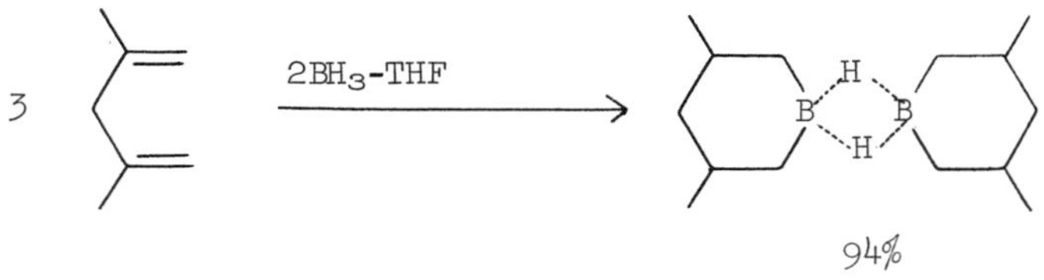

3   $\xrightarrow{\text{2BH}_3\text{-THF}}$ B$\cdots$H$\cdots$B

94%

Bis(3,5-dimethylboracyclohexane) and bis(3,6-dimethylbora-cycloheptane), prepared in a similar manner to that described above, are highly selective hydroborating reagents allowing quantitative conversion of alkenes into alcohols or the corresponding B-alkylboracyclanes (10a).

Hydroboration of acyclic dienes with borane-tetrahydrofuran in a molar ratio of 1:1 in many cases proceeds via extensive formation of the fully substituted bis(1-boracycloalkyl)alkanes, which then react further with the unreacted borane as discussed above (10b). Boracyclopentane derivatives are readily cleaved to give 1,2-tetramethylenediborane derivatives (Sec. 6.1), while six- and seven-membered rings give the corresponding boracyclanes. No evidence for the formation of four-, eight-, or nine-membered rings has been obtained (10b).

Cyclic organoboranes may also be prepared by the reaction of dienes with mono- or dialkylboranes. Thus, diethylborane converts 1,3-butadiene derivatives into boracyclopentanes in high yields (11).

$(R, R^1 = H, CH_3)$ $(>80\%)$

Thexylborane has proved to be a most useful reagent for the conversion of dienes to cyclic and bicyclic organoboranes (12). In the competitive formation of five- and six-membered rings the five-membered boracyclane is strongly favored under the usual hydroboration conditions, while in the case of six- and seven-membered boracyclanes neither isomer is strongly favored over the other (12b). Four-membered boracyclanes are not formed (12b). The B-thexylboracyclanes formed are useful intermediates in the synthesis of cyclic ketones by carbonylation (Sec. 8.5.1) (13,14).

The application of this reaction sequence to the synthesis of
monocyclic ketones is illustrated below (13).

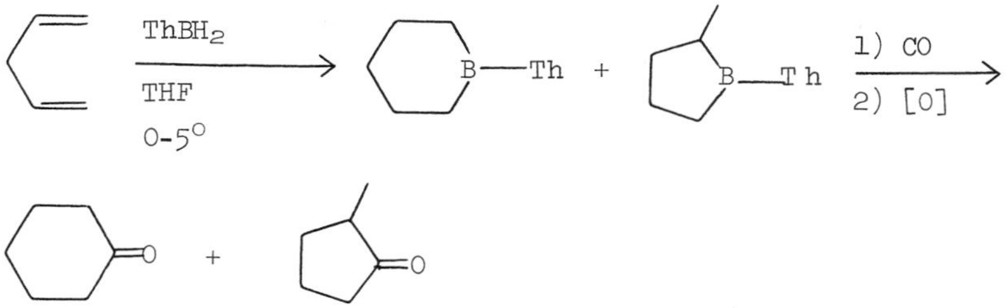

(14%)   (61%)

Cyclic hydroboration of (R)-(+)-limonene (12) with
thexylborane, followed by oxidation, gives essentially pure
(-)-(1S,2R,4R)-limonene-2,9-diol (13)(12); selective protonolysis
of the primary carbon-boron bond with acetic acid, followed by
oxidation, gives (-)-carvomenthol (14)(12).

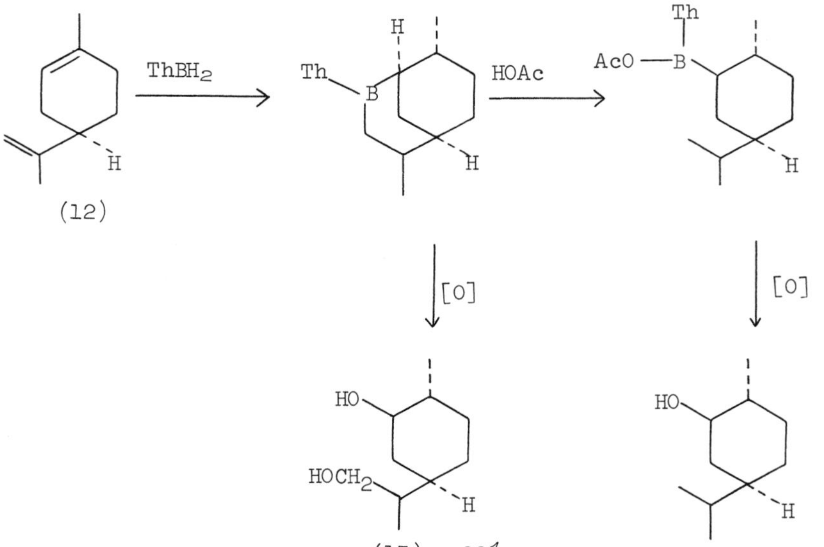

(12)

(13)   92%

(14)   75%

6.3 SYNTHESIS OF ACYCLIC DIOLS

Dihydroboration of acyclic dienes, followed by oxidation, usually gives mixtures of diols, which are the result of the addition of boron to the terminal positions of the diene system, and intramolecular addition of a second boron-hydrogen bond of the monohydroboration product to give cyclic organoboranes (1). The latter process can, however, largely be overcome by use of disiamylborane as the hydroborating reagent (15). Some results are summarized in Table 6.1.

TABLE 6.1

Dihydroboration of Acyclic Dienes : Synthesis of Diols

Diene	Products	Yield	
		BH$_3$-THF (1)	Sia$_2$BH (15)
1,3-Butadiene	1,3-diol	18	8
	1,4-diol	56	70
2-Methyl-1,3-butadiene	1,3-diol	9	--
	1,4-diol	58	--
2,3-Dimethyl-1,3-butadiene	1,4-diol	66	--
	1,3-diol	--	--
1,3-Pentadiene	1,3-diol	8[a]	--
	1,4-diol	75[a]	--
1,4-Pentadiene	1,4-diol	(62)[b]	13
	1,5-diol	(38)	71
1,5-Hexadiene	1,5-diol	(22)	5
	1,6-diol	(69)[c]	67

[a]Ref. 8

[b]41% 1,4-diol isolated.

[c]58% 1,6-diol isolated.

6.4 HYDROBORATION OF CYCLIC DIENES

The results of the hydroboration-oxidation of some monocyclic dienes are summarized in Table 6.2.

In general, mixtures of diols are obtained under the normal reaction conditions. In the case of cyclooctadienes and larger ring dienes hydroboration proceeds via the formation of boron-bridged bicyclo organoboranes, which isomerize in refluxing tetrahydrofuran to give the more stable borabicyclo systems containing a six-membered ring (17,18).

It has been found that cis- and trans-cyclic diols are readily separated by reaction of the diol mixture with n-butyldihydroxyborane (16); the cis-diol forms a cyclic derivative which is distilled from the polymeric products resulting from reaction with the trans-diol. Trans esterification of the separate fractions with ethylene glycol regenerates the pure diols.

In the steroid series hydroboration of cholesta-3,5- or 4,6-dienes gives 5α-cholestane-4α,6α-diol (19).

TABLE 6.2

Hydroboration of Monocyclic Dienes : Synthesis of Cyclic Diols

Diene	Temperature of reaction using BH_3-THF	Products & yield	Reference
Cyclopentadiene	0°	trans- + cis-1,3-diol (ratio 85:15)	16
1,3-Cyclohexadiene	0°	Diol mixture; 61	1
1,3-Cyclooctadiene	0°	Cis-1,3-diol; (23)	17
		Cis-1,4-diol; (77)	
	70°	Cis-1,5-diol; 75	17
1,5-Cyclooctadiene	0°	Cis-1,4-diol; (28)	18
		Cis-1,5-diol; (72)	
	65°	Cis-1,5-diol; 99	18
Cis,cis-1,5-Cyclo-nonadiene	0°	Cis-1,4- + cis-1,5-diols (ratio 1:4)	17
	70°	Cis-1,5-diol; 80	17
Cis,cis-1,6-Cyclo-decadiene	70°	Cis-1,5-diol; 82	17

Reaction of 7,9(11)-dienes in the 5α-steroid series results in exclusive attack of the 9(11)-double bond due to the unreactivity of the 7-double bond (Sec. 3.1.8) (19).

Hydroboration of cholesta-5,7-dien-3β-ol in diglyme using diborane gas gives, after oxidation, 5α-cholest-7-en-3β,6α-diol (20); however, in situ hydroboration using lithium aluminium hydride and boron trifluoride etherate gives 5α-cholest-6-en-3β-ol as the only crystalline product (19). This product has also been shown to arise from acid hydrolysis of the intermediate 7-ene-6α-borane (15) formed using external hydroboration conditions (20).

The 5α-cholest-6-en-3β-ol is probably formed by rearrangement of an intermediate allylic carbanion formed under the protonolysis conditions (Sec. 10.11) (20).

6.5 SELECTIVE HYDROBORATION OF DIENES: SYNTHESIS OF UNSATURATED ALCOHOLS

6.5.1 Acyclic Dienes

Acyclic conjugated dienes are relatively unreactive to hydroboration compared to alkenes. Thus, competitive hydroboration of 1,3-butadiene and 1-hexene results in the preferential reaction of 1-hexene (1). Therefore, in the hydroboration of acyclic conjugated dienes using borane, the unsaturated organoborane initially formed undergoes further reaction in preference to the less reactive diene, resulting in dihydroboration. In the case of nonconjugated dienes the results obtained are more favorable, but preferential dihydroboration still occurs, due mainly to cyclization of the initially formed unsaturated organoborane (1).

Use of disiamylborane as the hydroborating reagent still leads to preferential dihydroboration in the case of symmetrical conjugated dienes (15). However, with dienes in which the double bonds differ markedly in their reactivity toward disiamylborane (Sec. 3.3), the extent of monohydroboration is significantly increased due to preferential reaction at the more reactive site (15). In such cases satisfactory yields of the monohydroboration product are obtained; use of a 100% excess of diene favors the formation of the monohydroboration product to an even greater extent and provides a convenient method for the synthesis of unsaturated alcohols (15). Some results are summarized in Table 6.3.

Diethylborane has been used in the synthesis of citronellol from 3,7-dimethyl-1,6-octadiene (16) by selective hydroboration of the terminal double bond (11).

$$\text{(16)} \quad \xrightarrow[\text{2) [O]}]{\text{1) Et}_2\text{BH, -5}^\circ} \quad \text{CH}_2\text{OH} \quad 81\%$$

6.5.2 Cyclic Dienes

In contrast to the results obtained with acyclic dienes (Sec. 6.5.1), the reaction of cyclic dienes with borane can be

TABLE 6.3

Monohydroboration of Acyclic Dienes: Synthesis of Unsaturated Alcohols (15)

Diene	Monohydroboration,[a] % by vpc		Unsaturated alcohol isolated using Sia_2BH[b]
	BH_3	Sia_2BH	
1,3-Butadiene	4	8	--
2-Methyl-1,3-butadiene	4	14	--
1,3-Pentadiene	12	76	3-Pentenol (74%)[c]
1,4-Pentadiene	10	30	--
1,5-Hexadiene	35	52	5-Hexenol; 51%
2-Methyl-1,5-hexadiene	--	68	5-Methyl-5-hexenol; 68%

[a]25 mmoles diene + 8.4 mmoles BH_3-THF at 0-5° (Ref. 1).
25 mmoles diene + 25 mmoles Sia_2BH in THF at 0-5° (Ref. 15).

[b]0.2 mole diene + 0.1 mole Sia_2BH in THF at 0-5° (Ref. 15).

[c]Cis and trans isomers are formed.

controlled to achieve a degree of monohydroboration (1). Thus, the reaction results in the formation of unsaturated alcohols in 30-45% yields. Use of selective hydroborating agents, however, increases the yields considerably, and provides a useful method for the synthesis of unsaturated cyclic alcohols (15).

Ref. 21   Refs. 1 and 15   Ref. 15

In the above cases only the major products are indicated; minor amounts of isomeric mono-ols and diols are also formed.

The predominant formation of 2-cyclohexenol in the reaction of 1,3-cyclohexadiene with disiamylborane has been ascribed to the steric influence of the methylene hydrogen atoms adjacent to the diene system directing the reagent to the more remote 2-position (15).

Hydroboration-oxidation of substituted cyclohexadienes using borane gives satisfactory yields of the corresponding unsaturated alcohols in which the less substituted double bond has reacted.

(+ 19% Diols)

Ref. 22   Ref. 23   Ref. 24   Ref. 25

Reaction of 1,5-cyclooctadiene with borane (1), disiamylborane (15), and diethylborane (11) results in only dihydroboration
products being formed. 1,3-Cyclooctadiene, however, undergoes in
situ monohydroboration to give the allylic unsaturated organoborane
which is hydrolyzed by oxygen-free water to give cis-cyclooctene
in 35% yield (26).

2,5-Bicyclo[2.2.1]heptadiene (17) undergoes monohydroboration
to give, after oxidation, mainly the unsaturated exo-alcohol
(1,15). In the case of N-benzyl-6-aza-2,4-cholestadien-7-one (18)
the 2-double bond reacts selectively to give a mixture of
unsaturated 2- and 3-alcohols; the lactam function is surprisingly
unaffected (Sec. 10.6.7) (27).

In the reaction of thebaine (19), the amine borane (20) is
initially formed (28); reaction with a further equivalent of
borane gives the unsaturated alcohol (21), while use of excess
borane gives a mixture of products arising from elimination-
rehydroboration reactions (Sec. 5.1) (28).

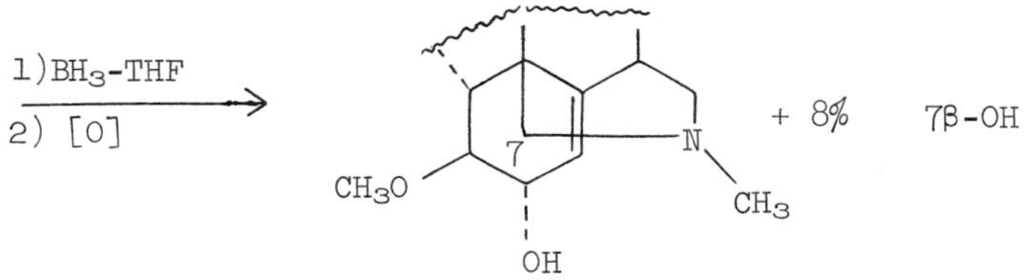

1)BH$_3$-THF
2) [O]

+ 8%   7β-OH

CH$_3$O

7

OH

N

CH$_3$

(21)   46%

6.5.3 "Mixed Acyclic-Cyclic" Dienes

In the case of dienes containing one acyclic and one cyclic double bond the acyclic double bond usually undergoes selective hydroboration.

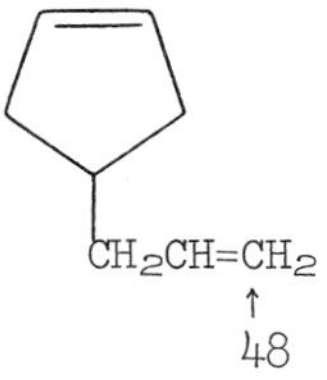

CH$_2$CH=CH$_2$
↑
48

Ref. 29

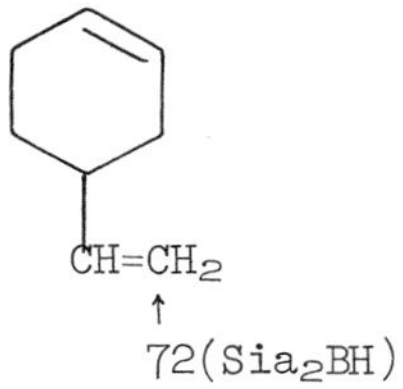

CH=CH$_2$
↑
72(Sia$_2$BH)

Ref. 30

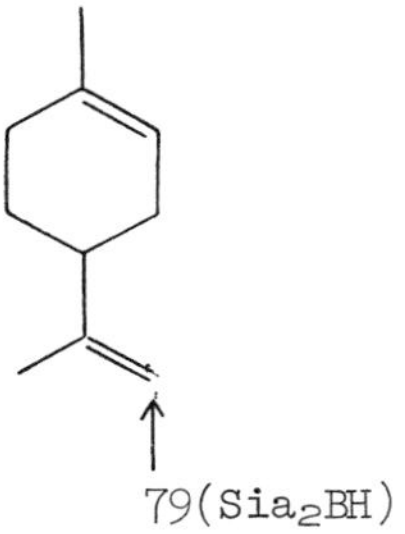

79(Sia$_2$BH)

Ref. 30

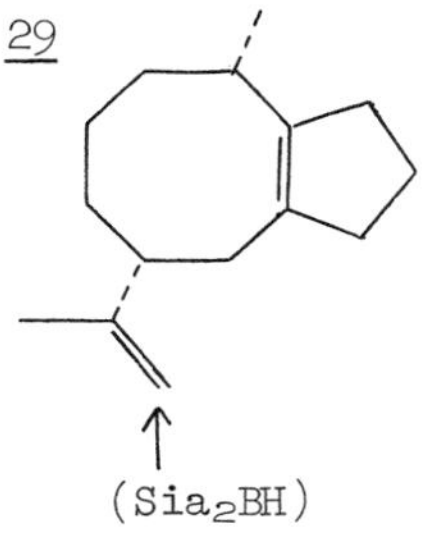

(Sia$_2$BH)

Ref. 31

Selective hydroboration of the exocyclic double bond of verbenene has been used in the synthesis of cis- and trans-δ-pinene (32).

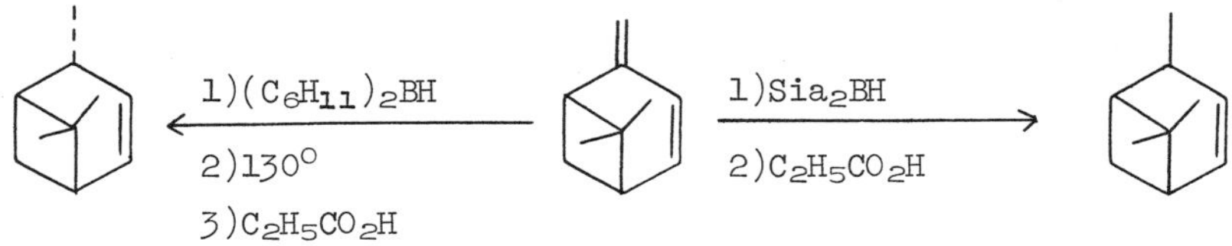

$$\xleftarrow[\substack{2)130^\circ \\ 3)C_2H_5CO_2H}]{1)(C_6H_{11})_2BH}$$
$$\xrightarrow[2)C_2H_5CO_2H]{1)Sia_2BH}$$

In the case of caryophyllene (22), however, selective hydroboration of the very reactive trans cyclic double bond occurs in preference to the exocyclic double bond (33).

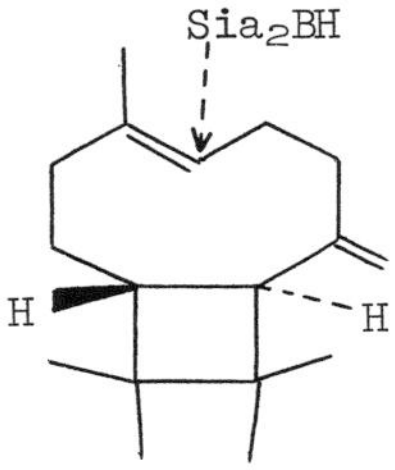

(22)

6.6 HYDROBORATION OF ALLENES

6.6.1 Acyclic Allenes

Gas phase hydroboration of allene leads to the formation of 1,2-trimethylenediborane (23) and n-propyldiborane (24); the former product polymerizes on liquefying but reforms on heating at 60° in vacuum (34).

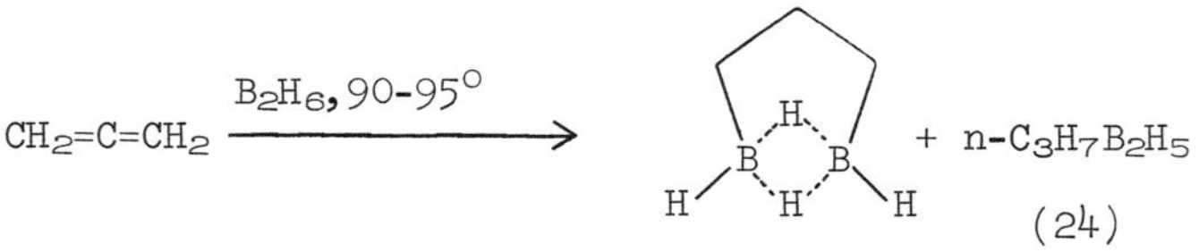

$$CH_2=C=CH_2 \xrightarrow{B_2H_6, 90-95^\circ}$$

$+ \ n\text{-}C_3H_7B_2H_5$

(24)

(23)

Hydroboration-oxidation of acyclic allenes gives 1,3-diols as the major products (35).

$$RCH=C=CH_2 \xrightarrow[\text{2) [o]}]{\text{1)}BH_3} RCHCH_2CH_2OH \; + \; RCH_2CHCH_2OH$$
$$\qquad\qquad\qquad\qquad\qquad\quad |\qquad\qquad\qquad\;\; |$$
$$\qquad\qquad\qquad\qquad\qquad\;\; OH\qquad\qquad\qquad OH$$

(R=H; C_4H_9; Ph)   Major product   (2-20%)

In the monohydroboration of allenes using disiamylborane attack occurs preferentially at the terminal carbon atom in the case of monosubstituted terminal allenes (36). With 1,3-disubstituted, tri-, and tetrasubstituted allenes, however, attack occurs mainly at the central carbon atom (36). The monohydroboration-oxidation of acyclic allenes results in the formation of a variety of products as illustrated in Table 6.4.

The formation of alkenes is attributed to the hydrolytic cleavage of the allylic organoboranes under the oxidation conditions (26). The high yields of alkenes obtained in the case of phenyl-substituted allenes are probably due to the added stabilization of the allylic carbanions formed during hydrolysis (36). The formation of the products are rationalized in the following reaction scheme:

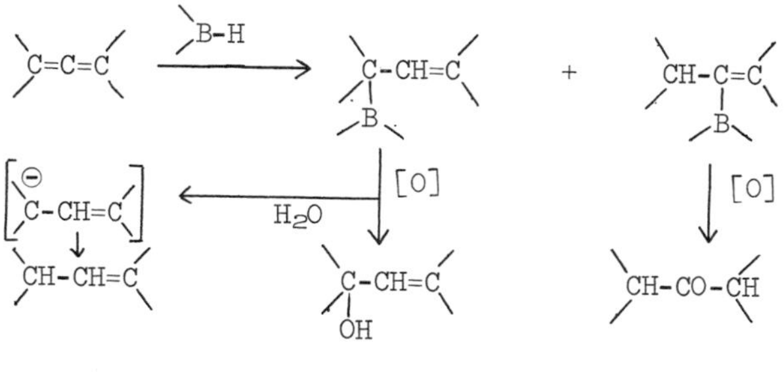

Similar results have been obtained using 4,4,6-trimethyl-1,3,2-dioxaborinane (TMDB; 25) as hydroborating reagent (37). The reactions were carried out in sealed ampules at 130° for 25 to 50 hr, and the organoboranes formed were separated by preparative vapor phase chromatography.

TABLE 6.4

Monohydroboration-Oxidation of Acyclic Allenes Using Disiamylborane (36)

Allene	Products; % yield		
	Ketones	Alcohols	Alkenes
n-C6H13CH=C=CH2	n-C7H15COCH3; 17	n-C6H13CH=CHCH2OH; 39	n-C7H15CH=CH2; 5 n-C6H13CH=CHCH3; 16 (cis and trans)
PhCH=C=CH2	PhCH2COCH3; 17	--	PhCH2CH=CH2; 23 PhCH=CHCH3; 44 (cis and trans)
CH3 PhC=C=CH2	CH3 PhCHCOCH3; 11	--	CH3 PhCHCH=CH2; 19 CH3 PhC=CHCH3; 60 (cis and trans)
C3H7CH=C=CHC3H7	C4H9COC4H9; 60	OH C3H7CHCH=CHC3H7; 20	C3H7CH=CHC4H9; 9 (cis and trans)
(CH3)2C=C=C(CH3)2	(CH3)2CHCOCH(CH3)2; 22	--	--

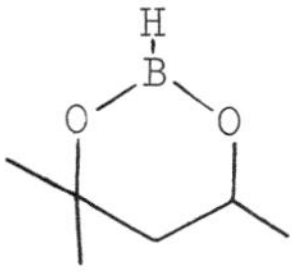

TMDB (25)

$CH_2=C=CH_2$
↑ ↑
3 22   (TMDB)

$(CH_3)_2C=C=CH_2$
↑ ↑
16 56   (TMDB)

$CH_3CH=C=CHCH_3$
↑ ↑
73 19   (TMDB)

$(CH_3)_2C=C=C(CH_3)_2$
↑
38   (TMDB)

The ratio of the attack at the terminal and central carbon atoms of the allenic system has been explained in terms of steric and electronic factors (37). In the case of 2,3-pentadiene attack at the central carbon atom leads mainly to the cis-vinylborane due to nonbonded interactions between the reagent and the substituents on the other orthogonal double bond in the transition state (37).

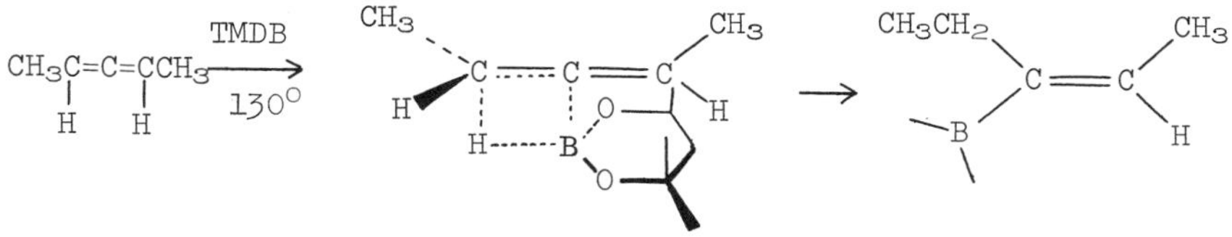

6.6.2 Cyclic Allenes

Monohydroboration of cyclic allenes with either borane (38) or disiamylborane (39) results in predominant attack at the central carbon atom of the allenic system; oxidation of the intermediate organoboranes gives the corresponding cyclic ketones.

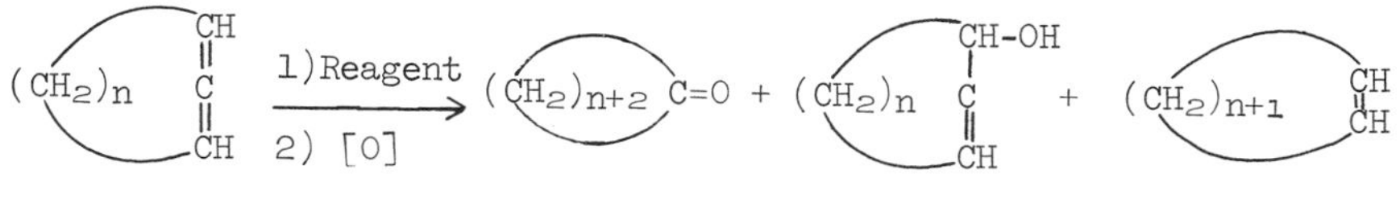

	Reagent			
n=6	BH_3, DG (38)	62%	--	12%
	Sia_2BH, DG (39)	46%	7%	3%
n=7	BH_3, DG (38)	44%	--	7%
n=10	Sia_2BH, DG (39)	59%	29%	7% trans
				2% cis

In the reaction using borane low yields of the saturated alcohols are also formed (38). Under dihydroboration conditions the saturated alcohols are the major products, and probably arise via formation of 1,1-diboryl derivatives (26)(38).

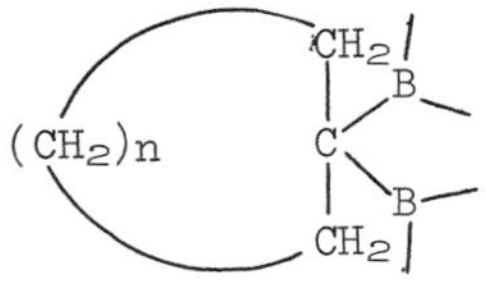

(26)

Protonolysis of the intermediate organoboranes gives high yields of the corresponding alkenes (40). Thus, monohydroboration of 1,2-cyclodecadiene with disiamylborane, followed by protonolysis, gives 81% cis-cyclodecene, while similar treatment of 1,2-cyclotridecadiene gives 74% cis and 26% trans-cyclotridecene (40). In the case of 1,2-cyclononadiene, hydroboration using borane, followed by protonolysis, gives cis-cyclononene in 70% yield (38).

6.7 RESOLUTION OF DIENE RACEMATES VIA ASYMMETRIC HYDROBORATION

Reaction of racemic diene mixtures with a deficient quantity of diisopinocampheylborane results in partial resolution of the racemates (see Sec. 3.8). Thus, the racemic diene mixture formed by Cope rearrangement of trans,trans-2,8-trans-bicyclo[8.4.0] - tetradecadiene (27) is partially resolved using (-)-diisopino-campheylborane (41).

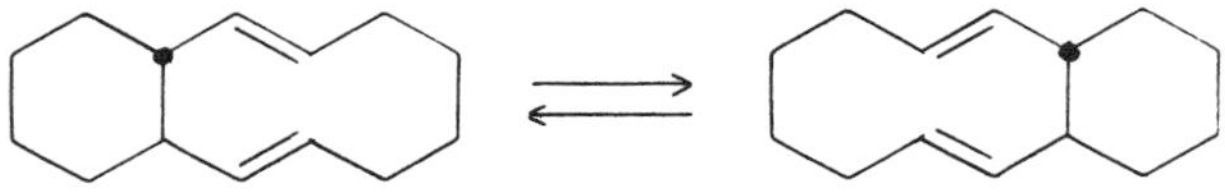

(27)

Racemic 2,3-pentadiene has also been partially resolved using (+)-diisopinocampheylborane (42). The degree of resolution has been found to be dependent on the reaction conditions, being highest when the reaction is carried out in diglyme at $0°$. Use of tetrahydrofuran as solvent gives poor resolution (42).

$$2(\pm)CH_3CH=C=CHCH_3 \xrightarrow[\text{DG},0°,\ 4\ \text{hr}]{+(IPC)_2BH} (R)-(-)-CH_3CH=C=CHCH_3$$
$$100\% \ (25\% \text{ optically pure})$$

Since α-pinene is displaced during the reaction, triisopinocampheyldiborane is probably involved in the hydroboration (Sec. 2.2.1.d). Thus far no attempts to relate the configuration of the allene to a transition state model have been made (Sec. 3.7).

6.8 HYDROBORATION OF POLYALKENES

Hydroboration of 1,4,8-nonatrienes (28) with borane in diglyme gives polymeric boranes which, on heating, undergo isomerization to give bicyclic boranes (43). The bicyclic boranes are carbonylated to give the corresponding bicyclic alcohols (Sec. 8.3.3) (43).

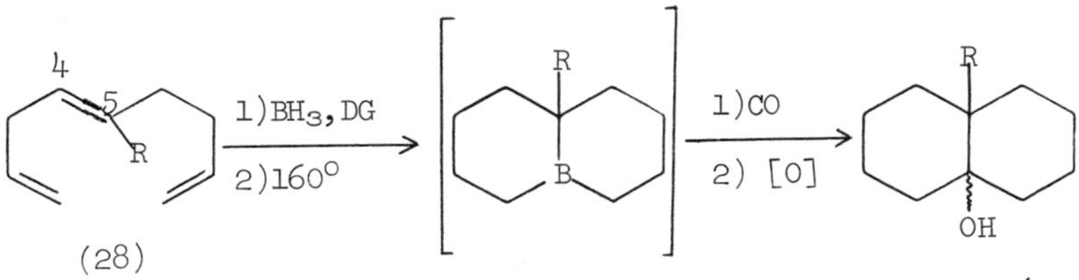

R=H; 73%
(cis/trans = 4/1)
R=CH$_3$; 62% (cis)

The formation of the bicyclic borane involves attack by boron at the tertiary 5 position instead of at the alternative secondary

4 position. This unexpected result probably arises from the
strong tendency of the system to form two fused six-membered
rings (43). Similar results have been obtained with
1,3,7-octatriene (29) (43).

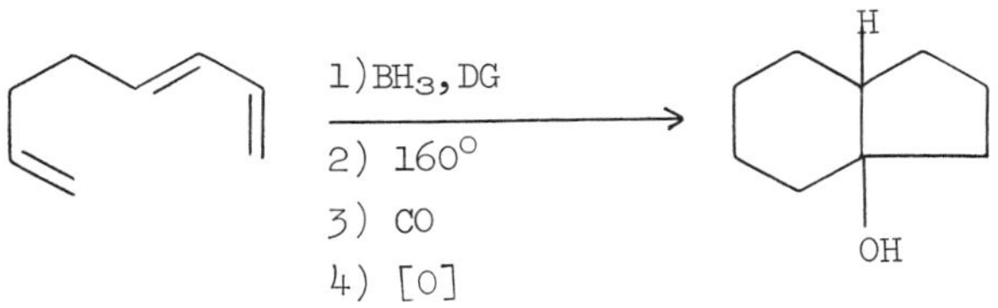

1)BH$_3$,DG

2) 160°

3) CO

4) [O]

(29)   33%

Monohydroboration of myrcene (7-methyl-3-methylene-1,6-
octadiene; 30) using disiamylborane, followed by oxidation, gives
myrcenol (7-methyl-3-methylene-6-octen-1-ol; 31) in 68% yield
(44). Dihydroboration with disiamylborane, followed by oxidation,
gives 7-methyl-3-hydroxymethyl-6-octen-1-ol (32) in 78% yield (44).

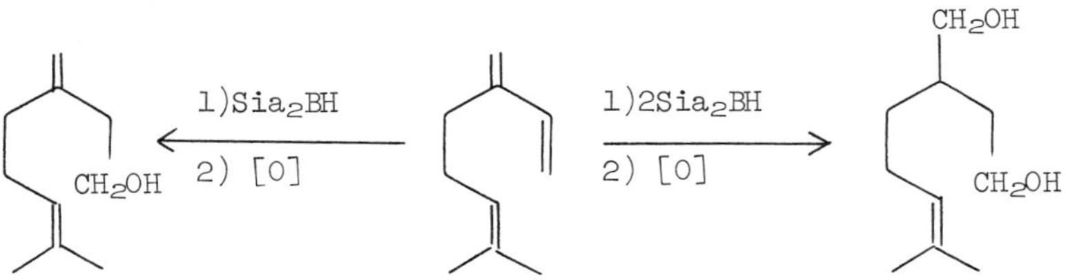

1)Sia$_2$BH   1)2Sia$_2$BH

2) [O]   2) [O]

68%   (30)   78%

(31)   (32)

Hydroboration of the cyclododecatrienes (33) and (34) using
triethylamine-borane gives the isomeric perhydro-9b-boraphenalenes
(35) and (36), respectively (45). The assignments of the
structures of these polycyclic boranes indicated below are based
on the nuclear magnetic resonance spectra of the corresponding
carbonylation products formed by treatment with carbon monoxide

(Sec. 8.3.3) (46), and are the reverse of the assignments originally made (45); originally structure (36) was proposed for the product formed from (33), while (35) was thought to correspond to the product formed from (34).

(33)   Et₃N:BH₃   140°   →   H (35) 97%

(34)   1)Et₃N:BH₃   140°   2) 220°   →   (36) 65%

REFERENCES

1. G. Zweifel, K. Nagase, and H. C. Brown, J. Am. Chem. Soc., 84, 183 (1962).

2. E. Breuer and H. C. Brown, J. Am. Chem. Soc., 91, 4164 (1969).

3. R. Köster, Angew. Chem., 72, 626 (1960).

4. D. E. Young and S. G. Shore, J. Am. Chem. Soc., 91, 3497 (1969).

5. H. C. Brown, E. Negishi, and S. K. Gupta, J. Am. Chem. Soc., 92, 2460 (1970).

6. (a) H. C. Brown, E. Negishi, and P. L. Burke, J. Am. Chem. Soc., 93, 3400 (1971); (b) H. C. Brown, E. Negishi, and P. L. Burke, ibid, 92, 6649 (1970).

7. L. I. Zakharkin and A. I. Kovredov, Izv. Akad. Nauk. SSSR, Ser. Khim., 1969, 106; Chem. Abstr., 70, 115200(c) (1969).

8. K. A. Saegebarth, J. Am. Chem. Soc., 82, 2081 (1960).

9. H. C. Brown and E. Negishi, J. Organometal. Chem., 26, C67 (1971).

10. (a) H. C. Brown and E. Negishi, J. Organometal Chem., 28, C1 (1971); (b) H. C. Brown, E. Negishi, and P. L. Burke, J. Am. Chem. Soc., 94, 3561 (1972).

11. R. Köster, G. Griasnow, W. Larbig, and P. Binger, Annalen, 672, 1 (1964).

12. (a) H. C. Brown and C. D. Pfaffenberger, J. Am. Chem. Soc., 89, 5475 (1967); (b) H. C. Brown and E. Negishi, ibid, 94, 3567 (1972).

13. H. C. Brown and E. Negishi, J. Am. Chem. Soc., 89, 5477 (1967).

14. H. C. Brown and E. Negishi, Chem. Commun., 1968, 594.

15. G. Zweifel, K. Nagase, and H. C. Brown, J. Am. Chem. Soc., 84, 190 (1962).

16. H. C. Brown and G. Zweifel, J. Org. Chem., 27, 4708 (1962).

17. I. Mehrotra and D. Devaprabhakara, Tetrahedron Letters, 1970 4493.

18. E. F. Knights and H. C. Brown, J. Am. Chem. Soc., 90, 5280 (1968).

19. M. Nussim, Y. Mazur, and F. Sondheimer, J. Org. Chem., 29, 1131 (1964).

20. L. Caglioti, G. Cainelli, and G. Maina, Tetrahedron, 19, 1057 (1963).

21. H. M. Hess and H. C. Brown, J. Org. Chem., 32, 4138 (1967).

22. A. Uzarewicz, Rocz. Chem., 38, 599 (1964); Chem. Abstr., 61, 10600 (c) (1964).

23. A. Uzarewicz, Rocz. Chem., 38, 385 (1964); Chem. Abstr., 61, 1771 (b) (1964).

24. W. Zacharewicz and A. Uzarewicz, Rocz. Chem., 38, 591 (1964); Chem. Abstr., 61, 10598 (h) (1964).

25. I. Uzarewicz, M. Zaidlewicz, and A. Uzarewicz, Rocz. Chem., 44, 1403 (1970); Chem. Abstr., 74, 3730 (b) (1971).

26. D. S. Sethi and D. Devaprabhakara, Can. J. Chem., 46, 1165 (1968).

27. J. P. Kutney, G. Eigendorf, and J. E. Hall, Tetrahedron, 24, 845 (1968).

28. M. Takeda, H. Inoue, and H. Kugita, Tetrahedron, 25, 1839 (1969).

29. P. D. Bartlett, W. S. Trahanovsky, D. A. Bolon, and G. Schmid, J. Am. Chem. Soc., 87 1314 (1965).

30. H. C. Brown and G. Zweifel, J. Am. Chem. Soc., 83, 1241 (1961).

31. J. A. Marshall, A. E. Greene, and R. Ruden, Tetrahedron Letters 1971, 855.

32. G. Zweifel and C. C. Whitney, J. Org. Chem., 31, 4178 (1966).

33. H. C. Brown, Hydroboration, Benjamin, New York, 1962, p.226.

34. H. H. Lindner and T. Onak, J. Am. Chem. Soc., 88, 1886 (1966).

35. S. Corsano, Atti. Accad. Nazl. Lincei, Rend., Classe Sci. Fis., Mat. Nat., 34, 430 (1963); Chem. Abstr., 60, 3993 (a) (1964).

36. D. S. Sethi, G. C. Joshi, and D. Devaprabhakara, Can. J. Chem., 47, 1083 (1969).

37. R. H. Fish, J. Am. Chem. Soc., 90, 4435 (1968).

38. D. Devaprabhakara and P. D. Gardner, J. Am. Chem. Soc., 85, 1458 (1963).

39. D. S. Sethi, G. C. Joshi, and D. Devaprabhakara, Can. J. Chem., 46, 2632 (1968).

40. D. S. Sethi, G. C. Joshi, and D. Devaprabhakara, Indian J. Chem., 6, 402 (1968).

41. P. S. Wharton and R. A. Kretchmer, J. Org. Chem., 33, 4258 (1968).

42. W. L. Waters, W. S. Linn, and M. C. Caserio, J. Am. Chem. Soc., 90, 6741 (1968).

43. H. C. Brown and E. Negishi, J. Am. Chem. Soc., 91, 1224 (1969).

44.   H. C. Brown, K. P. Singh, and B. J. Garner, J. Organometal. Chem., 1, 2 (1963).

45.   G. W. Rotermund and R. Köster, Annalen, 686, 153 (1965).

46.   H. C. Brown and E. Negishi, J. Am. Chem. Soc., 89, 5477 (1967); H. C. Brown and W. C. Dickason, ibid, 91, 1226 (1969).

Chapter 7

HYDROBORATION OF ALKYNES

In general, alkynes readily undergo hydroboration to form either vinylboranes or diboro derivatives depending on the amounts of hydroborating agents used.

Vinylboranes are useful intermediates in the synthesis of alkenes, vinylhalides, cis, trans-dienes, and terminal allenes (Sec. 7.1.1), and are also easily oxidized to aldehydes and ketones (Sec. 7.1.2). 1,1-Diboro derivatives, derived from the dihydroboration of terminal alkynes, are particularly useful in the synthesis of carboxylic acids, malonic acid derivatives, cyclopropanols, and cyclobutanols (Sec. 7.2.2). The monohydroboration of cycloalkynes gives the corresponding cyclic vinylboranes which can be converted into cyclic alkenes (Sec. 7.3), while monohydroboration of conjugated diynes provides a useful route to cis-enynes and α,β-acetylenic ketones (Sec. 7.4.1). Cis,cis-dienes are synthesized via dihydroboration of diynes (Sec. 7.4.2), while the selective hydroboration of enynes provides another route to dienes as well as unsaturated carbonyl compounds (Sec. 7.5).

7.1 MONOHYDROBORATION OF ACYCLIC ALKYNES

The hydroboration of alkynes involves the initial formation of a vinylborane derivative; thereafter there exists a competition

227

between unreacted alkyne and the vinylborane for the residual
hydroborating reagent (<u>la</u>).

Reaction of <u>disubstituted alkynes</u> with sufficient borane to
achieve monohydroboration results in the predominant formation of
the cis trisvinylborane, indicating that the alkyne is more
reactive than the initially formed vinylborane (<u>la</u>).

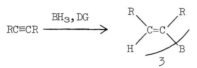

$$RC{\equiv}CR \xrightarrow{\quad BH_3, DG \quad} \underset{3}{\overset{\displaystyle R \quad\quad R}{\underset{\displaystyle H \quad\quad B}{C=C}}}$$

In the case of 3-hexyne, 16% of the initial alkyne remained after
completion of the reaction, indicating that approximately 16%
dihydroboration occurs (<u>la</u>). Use of disiamylborane, however,
leads to almost exclusive formation of the monovinylborane (<u>la</u>).

In the case of unsymmetrically disubstituted alkynes, a
higher preference for attack by boron at the less hindered position
of the triple bond is exhibited than is observed in the hydroboration of correspondingly substituted alkenes (Sec. 3.1.3) (<u>lb</u>).

$RC{\equiv}CCH_3$			$RCH=CHCH_3$ (trans)		
↑	↑		↑	↑	
18	53	($R=i-C_3H_7$)	39	51	($R=i-C_3H_7$)
17	63	($R=CMe_3$)	38	52	($R=CMe_3$)
55	19	($R=Ph$)	76	14	($R=Ph$)

Use of disiamylborane or dicyclohexylborane, however, results in
predominant attack at the less hindered position (<u>lb</u>).

R	R^1	$RC{\equiv}CR^1$		
		↑	↑	
$i-C_3H_7$	CH_3	(7)	(93)	
CMe_3	CH_3	(3)	(97)	
$i-C_3H_7$	$n-C_4H_9$	(13)	(87)	⎤
CMe_3	$n-C_4H_9$	(1)	(99)	⎬ (Sia_2BH)
Ph	CH_3	(19)	(81)	⎦

A similar regioselectivity is observed when benzo-1,3-dioxa-2-
borole [(3); Sec. 2.2.4] is used as the hydroborating reagent (1c).
The higher degree of selectivity exhibited in the hydroboration
of unsymmetrically disubstituted alkynes, compared to the
correspondingly substituted alkenes, is probably due to the
greater influence of steric interactions in the transition state
involving the linear alkynes (1b).

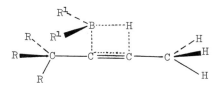

The above transition state will be markedly destabilized relative
to the other possible transition state, thereby favoring attack at
the less hindered position.

With 1-alkynes the initially formed vinylborane appears to
be more reactive than the residual alkyne, and preferential
dihydroboration occurs (1a).

$$RC\equiv CH \xrightarrow{\text{BH}_3,\text{DG}} RCH_2CH\text{-B} \quad + \text{ residual alkyne } (40\text{-}50\%)$$

This problem is overcome by use of selective hydroborating
reagents, such as disiamylborane (1). Thus, reaction of the
1-alkyne with an equimolar quantity of disiamylborane results in
almost exclusive monohydroboration to give the corresponding
trans-vinylborane (1a).

$$RC\equiv CH \xrightarrow[\text{THF}]{\text{Sia}_2\text{BH}} \underset{H}{\overset{R}{\diagdown}} C=C \underset{\text{BSia}_2}{\overset{H}{\diagup}}$$

The uses of vinylboranes as intermediates in organic
synthesis are discussed below.

7.1.1 Synthesis of Alkenes, Dienes, and Their Derivatives

The protonolysis of vinylboranes using glacial acetic acid proceeds readily at room temperature (1a). The hydroboration-protonolysis sequence thus provides a useful method for the conversion of disubstituted alkynes to cis-alkenes of high stereochemical purity (>98%). The yields obtained using borane as hydroborating reagent are usually 60-70% but are increased to 70-90% by use of disiamylborane (1a).

$$RC{\equiv}CR \xrightarrow[\text{2)CH}_3\text{CO}_2\text{H}]{\text{1)Sia}_2\text{BH,DG}} \begin{array}{c} R \\ \diagdown \\ H \end{array} C{=}C \begin{array}{c} R \\ \diagup \\ H \end{array} \quad 70\text{-}90\%$$

In the case of 1-alkynes hydroboration using disiamylborane, followed by protonolysis, gives terminal alkenes in yields of approximately 70% (1a).

The conversion of disubstituted alkynes to cis-alkenes has also been achieved by heating alkynes with triisobutylborane at 170°, followed by protonolysis of the intermediate trialkenylborane with acetic acid (2).

$$RC{\equiv}CR \xrightarrow[170°]{(i\text{-}C_4H_9)_3B} \begin{array}{c} R \\ \diagdown \\ H \end{array} C{=}C \begin{array}{c} R \\ \diagup \\ B \end{array}_3 \xrightarrow{CH_3CO_2H} \begin{array}{c} R \\ \diagdown \\ H \end{array} C{=}C \begin{array}{c} R \\ \diagup \\ H \end{array}$$

$$+ \; CH_2{=}C(CH_3)_2 \qquad 88\% \; (R{=}n\text{-}C_4H_9)$$

The method is not suitable for 1-alkynes due to polymer formation but does have the advantage of avoiding over-reduction in the case of disubstituted alkynes (2).

Trans addition of bromine to trans-vinylboranes, formed by reaction of 1-alkynes with disiamylborane (Sec. 7.1), gives dibromo-intermediates which, on alkaline hydrolysis, undergo trans elimination of Sia_2BBr to give the corresponding, cis-vinyl bromides (3); thermal decomposition in refluxing carbon tetrachloride results in cis elimination of Sia_2BBr to give the corresponding <u>trans-vinyl bromides</u> (3).

$$(40-75\%)(>85\% \text{ trans}) \quad (65-76\%)(>95\% \text{ cis})$$

In the case of phenylethyne (R =Ph) the stereochemistry of the vinyl bromides formed is the reverse of that obtained using l-alkynes (3). This reversal is possibly the result of cis addition of bromine to the phenyl-substituted vinylborane.

Treatment of vinylboranes with iodine does not give the corresponding vinyl iodides, but results in the transfer of an alkyl group from boron to the adjacent carbon atom to give good yields of the substituted alkenes (4a). Thus, reaction of l-alkynes with dicyclohexylborane, followed by iodination of an alkaline solution of the vinylborane, gives the corresponding <u>cis-alkenes</u> (4a).

$$75-83\%$$
$$(>90\% \text{ purity})$$

Disubstituted alkynes may be similarly converted to trisubstituted
alkenes (4a). Use of bis(trans-2-methylcyclohexyl)borane or
diisopinocampheylborane, in place of dicyclohexylborane, in the
above reactions results in the formation of the corresponding
alkenes, in which the cyclic group has migrated with retention
of configuration (4b).

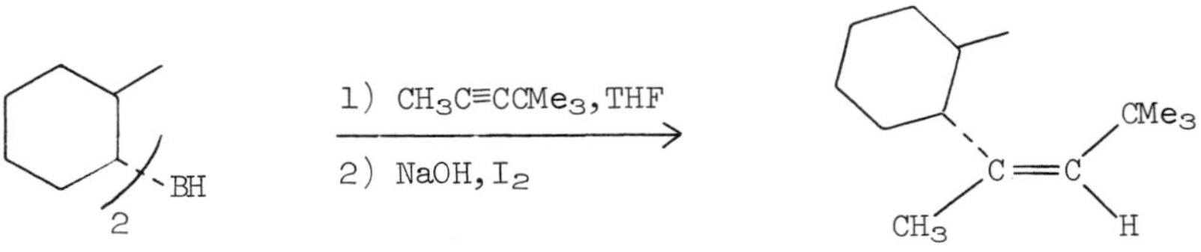

$$1)\ CH_3C\equiv CCMe_3, THF$$
$$2)\ NaOH, I_2$$

70%

The above iodination reaction has also been applied to the
synthesis of cis,trans-dienes (5). Thus, reaction of disubstituted
alkynes with borane in a 3:1 ratio gives trivinylboranes, which
are converted to the corresponding cis,trans-dienes of >99%
purity.

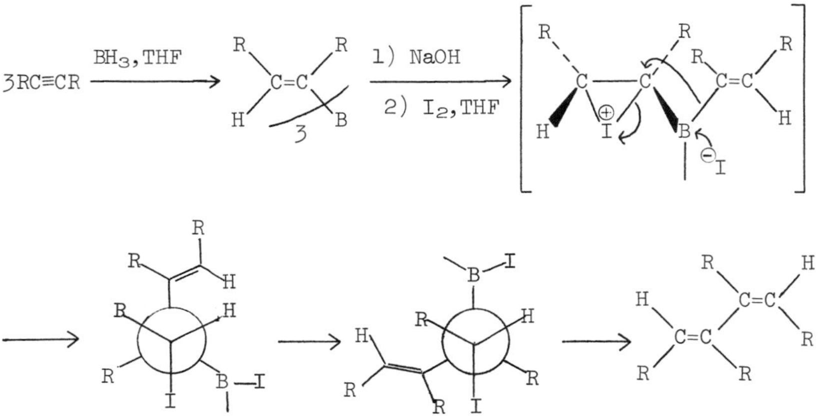

$$3RC\equiv CR \xrightarrow{BH_3, THF}$$

$$\begin{array}{c} 1)\ NaOH \\ \xrightarrow{\hspace{1cm}} \\ 2)\ I_2, THF \end{array}$$

Since 1-alkynes undergo preferential hydroboration to give
1,1-diboro derivatives (Sec. 7.1), use is made of thexylborane as
hydroborating reagent in these cases (5). In the iodination of

the divinylthexylborane, migration of the thexyl group competes
with that of the vinyl group; however, such competition is avoided
by prior selective oxidation of the thexyl moiety to the
corresponding divinylthexyloxyborane (Sec. 4.4), which is then
rearranged to give the corresponding cis,trans-diene (5).

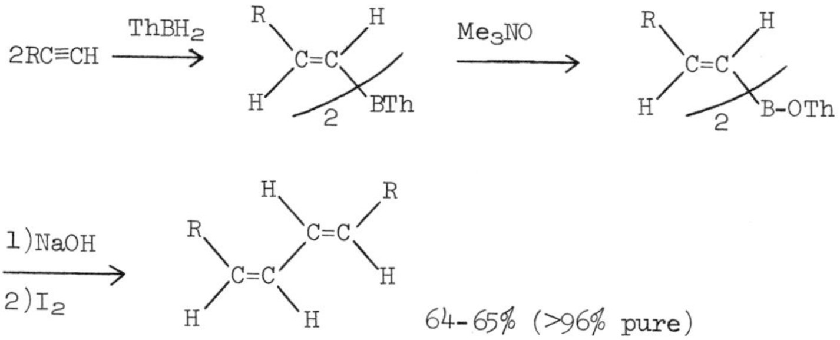

64-65% (>96% pure)

In the above reaction sequence, all of the alkyne is utilized
in diene formation. Thus, thexylborane is also used as
hydroborating reagent in the conversion of disubstituted alkynes
to dienes (5).

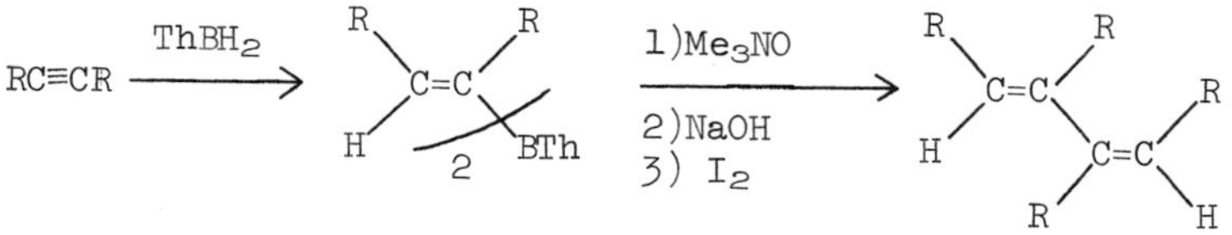

(R=CH$_3$, C$_2$H$_5$)

49-69% (100% pure)

Cis,cis-dienes are synthesized by dihydroboration-protonolysis
of diynes (Sec. 7.4.2).

In the reaction of l-bromo- or l-iodo-l-alkynes with
dicyclohexylborane stable trans-α-halovinylboranes are formed (6a).
Protonolysis of these α-halovinylboranes gives the corresponding
cis-vinyl bromides or iodides, while treatment with sodium
methoxide promotes migration of a cyclohexyl group from boron
to carbon. This migration occurs with inversion at the

halogen-bearing carbon atom due to expulsion of the halide by rearside attack of the alkyl group. Treatment of the intermediate methoxyvinylborane with acetic acid gives the corresponding, trans-alkenes (6a).

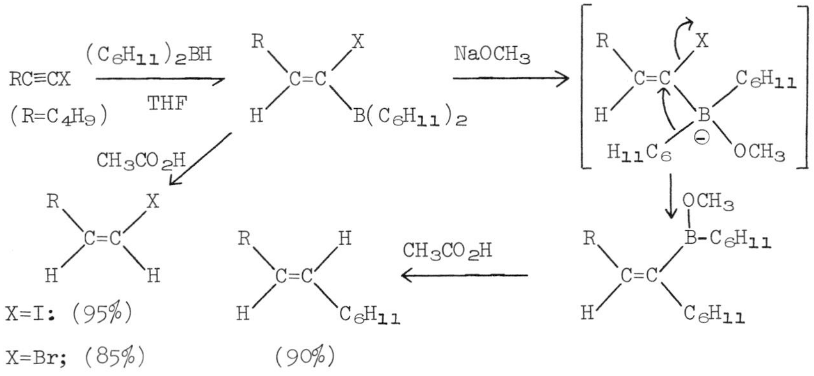
RC≡CX
(R=C4H9)
(C6H11)2BH
THF
NaOCH3
CH3CO2H
X=I: (95%)
X=Br; (85%)
(90%)

If the above reaction sequence is carried out with bis(trans-2-methylcyclohexyl)borane in place of dicyclohexylborane, migration of the cyclic group occurs with retention of configuration (4b).

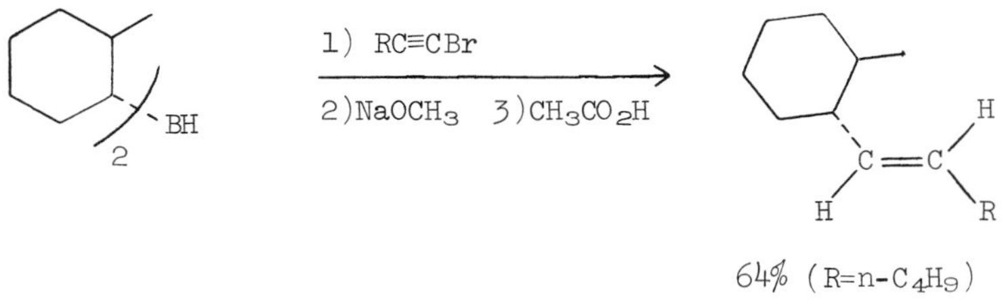
1) RC≡CBr
2)NaOCH3 3)CH3CO2H
64% (R=n-C4H9)

Use of cyclohexylthexylborane in the above reaction sequence results in the exclusive migration of the cyclohexyl group from boron to carbon (6b).

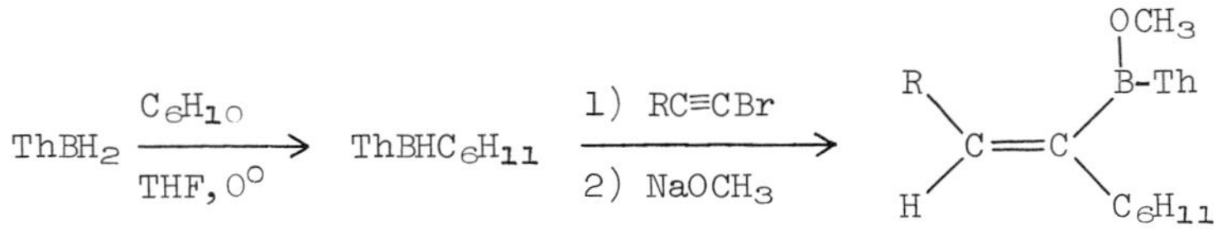
ThBH2
C6H10
THF, 0°
ThBHC6H11
1) RC≡CBr
2) NaOCH3

1) HOAc
————————————————→
2) Ag(NH$_3$)$_2$NO$_3$, H$_2$O
   75-80°

$$\begin{array}{c} R \\ \diagdown \\ C=C \\ \diagup \quad \diagdown \\ H \qquad C_6H_{11} \end{array} \quad H$$

$$\left[R= \begin{array}{c} -CH-n-C_5H_{11} \\ | \\ OTHP \end{array} \right]$$

65%

The use of aqueous silver ammonium nitrate complex in the final
step provides a mild method of cleavage of the vinylborane in the
presence of functional groups easily destroyed by refluxing acetic
acid.   This method has been adapted to the synthesis of
prostaglandin intermediates (6b).

1)ThBH$_2$,THF
————————————————→
2)RC≡CBr
3)NaOCH$_3$ 4) HOAc
5)Ag(NH$_3$)$_2$NO$_3$

$$\left[\begin{array}{c} R= -CH-n-C_5H_{11} \\ \quad | \\ \quad OTHP \\ \\ R^1=Me_2SiCMe_3 \end{array} \right]$$

Hydroboration of 1-chloro-2-alkynes with disiamylborane
gives the corresponding (β-chlorovinyl)borane which, on
protonolysis, gives the corresponding cis-1-chloro-2-alkene.
Treatment of the intermediate borane with sodium hydroxide,
however, results in facile elimination to give the corresponding
terminal allene (7).

RC≡CCH$_2$Cl

Sia$_2$BH
————————→

NaOH
————————→

CH$_3$CO$_2$H

RCH=C=CH$_2$

64-73%

(R=C$_4$H$_9$, C$_6$H$_{11}$, tert-C$_4$H$_9$, Ph)

83% (R=C$_4$H$_9$)

Allene has been prepared in a similar manner using tetraethyldiborane as hydroborating reagent (8). However, attempted
preparation of terminal allenes from 3-chloro-1-alkynes using a
similar reaction sequence gives mainly the product arising from
migration of one of the siamyl groups from boron to the adjacent
carbon atom (7). Oxidation of the allylic borane gives the
corresponding allylic alcohol.

$$
\underset{\underset{\text{Cl}}{|}}{\text{RCHC}\equiv\text{CH}} \xrightarrow{\text{Sia}_2\text{BH}} \underset{\underset{\text{Cl}}{|}}{\text{RCH-CH=CH-B}}\overset{\text{Sia}}{\underset{\text{Sia}}{\big\langle}} \xrightarrow{\text{NaOH}} \begin{array}{l}\text{RCH=CH-CH-Sia}\\[2pt]\text{HO-BSia}\end{array}
$$

(R=C$_4$H$_9$)

$$
\xrightarrow{\text{[o]}} \underset{\underset{\text{OH}}{|}}{\text{RCH=CH-CH-Sia}}
$$

68% (trans)

7.1.2 Synthesis of Aldehydes and Ketones

Monohydroboration of 1-alkynes using disiamylborane, followed
by oxidation, gives the corresponding aldehydes (1a).

$$
\text{RC}\equiv\text{CH} \xrightarrow{\text{Sia}_2\text{BH}} \underset{\text{H}}{\overset{\text{R}}{\diagdown}}\text{C=C}\underset{\text{BSia}_2}{\overset{\text{H}}{\diagup}} \xrightarrow{\text{[o]}} [\text{RCH=CHOH}] \longrightarrow \text{RCH}_2\text{CHO}
$$

70-88%

(R=C$_4$H$_9$, C$_6$H$_{13}$)

The reaction is further illustrated by the synthesis of
tricycloekasantalol (9).

In the case of symmetrically disubstituted alkynes,
monohydroboration using either disiamylborane (1a) or

triisobutylborane ($\underline{2}$), followed by oxidation, gives the corresponding ketones.

$$RC{\equiv}CR \xrightarrow[\substack{(i-C_4H_9)_3B, \ 170^{\circ} \\ 2) \ [O]}]{\substack{1)Sia_2BH,THF \quad \underline{or}}} RCOCH_2R$$

$$50-76\% \qquad (R=C_2H_5, \ C_4H_9 \ \ Ph)$$

The high degree of selectivity exhibited in the monohydroboration of unsymmetrically disubstituted alkynes (Sec. 7.1) permits the synthesis of a large variety of highly substituted ketones ($\underline{1b}$).

In the case of trans-α-halovinylboranes, formed by hydroboration of 1-halo-1-alkynes with dicyclohexylborane, treatment with sodium hydroxide followed by oxidation, gives cyclohexyl ketones ($\underline{6a}$).

$RC{\equiv}CX \xrightarrow{(C_6H_{11})_2BH}$

$(R=C_6H_{11})$
$(X=Br,I)$

$RCH_2COC_6H_{11}$
87%

7.2   DIHYDROBORATION OF ACYCLIC ALKYNES

7.2.1   Dihydroboration of 1-Alkynes: Synthesis of 1,1-Diboro Derivatives

The hydroboration of 1-hexyne with diborane in a 3:1 ratio gives a polymeric dihydroboration product, which, on oxidation, gives 80% 1-hexanol and 10% 1,2-hexanediol ($\underline{10}$). The 1-hexanol is the result of the rapid hydrolysis of the 1,1-diboro derivative prior to oxidation, while the 1,2-diol is formed by oxidation of the 1,2-diboro derivative. The facile hydrolysis of the

1,1-diboro derivative is rationalized in terms of nucleophilic attack by the base on boron with the formation of an intermediate carbanion which is stabilized by interaction with the vacant p orbital of the adjacent boron atom (10).

Use of thexylborane or dicyclohexylborane as hydroborating reagent gives the 1,1-diboro derivatives in 90-96% yields (10). 1,1-Diboro derivatives have been shown to be useful synthetic intermediates; their uses are discussed below.

7.2.2 Synthetic Uses of 1,1-Diboro Derivatives

As discussed in Sec. 7.2.1, dihydroboration-oxidation of 1-alkynes gives the corresponding primary alcohols (10). In addition, oxidation of the 1,1-diboro derivatives with m-chloroperbenzoic acid under anhydrous conditions gives the corresponding carboxylic acids (10).

Treatment of 1,1-diboro derivatives with 2 mole equivalents of methyl- or butyllithium gives 1-boro-1-lithio derivatives which possibly exist as the lithium tetraalkylboron derivatives shown below (11, 12).

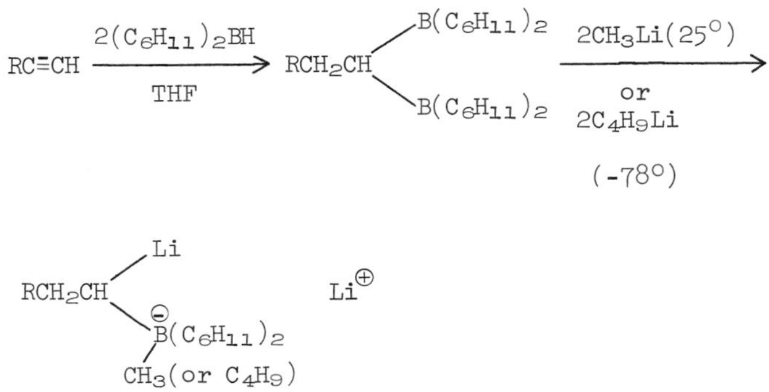

$$RC\equiv CH \xrightarrow[\text{THF}]{2(C_6H_{11})_2BH} RCH_2CH \Big\langle{}^{B(C_6H_{11})_2}_{B(C_6H_{11})_2} \xrightarrow[\substack{\text{or} \\ 2C_4H_9Li \\ (-78^{\circ})}]{2CH_3Li(25^{\circ})}$$

$$RCH_2CH \Big\langle{}^{Li}_{\substack{\overset{\ominus}{B}(C_6H_{11})_2 \\ CH_3(\text{or } C_4H_9)}} \quad Li^{\oplus}$$

These 1-boro-1-lithio derivatives react with carbon dioxide to give malonic acid derivatives (12). Reaction with aldehydes or ketones gives intermediate adducts which undergo elimination to give alkenes (13a), while treatment with ethyl bromide, followed by oxidation, gives alcohols as shown below (11).

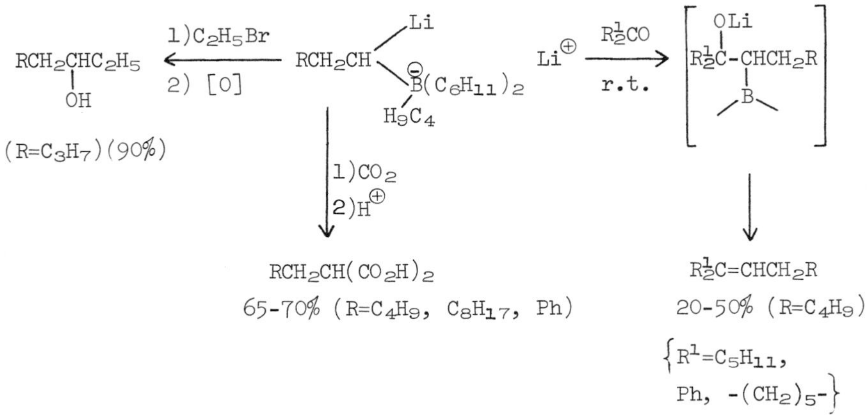

$$RCH_2\underset{\underset{OH}{|}}{CH}C_2H_5 \xleftarrow[\text{2) [O]}]{\text{1)}C_2H_5Br} RCH_2CH \Big\langle{}^{Li}_{\substack{\overset{\ominus}{B}(C_6H_{11})_2 \\ H_9C_4}} \quad Li^{\oplus} \xrightarrow[\text{r.t.}]{R_2^1CO} \left[{}^{OLi}_{\substack{R_2^1C-CHCH_2R \\ B}} \right]$$

$$(R=C_3H_7)(90\%)$$

$$\Big\downarrow {}^{1)CO_2}_{2)H^{\oplus}} \qquad\qquad \Big\downarrow$$

$$\begin{array}{cc} RCH_2CH(CO_2H)_2 & R_2^1C=CHCH_2R \\ 65\text{-}70\% \ (R=C_4H_9, \ C_8H_{17}, \ Ph) & 20\text{-}50\% \ (R=C_4H_9) \\ & \left\{{}^{R^1=C_5H_{11},}_{Ph, \ -(CH_2)_5-}\right\} \end{array}$$

In the reaction with benzaldehyde the cis-trans ratio of the alkene formed is dependent on the temperature of reaction; thus, at 40°, trans-alkene of 96% purity is formed, while at -78° the cis-alkene predominates to the extent of 70 to 30 (13a). A similar dependence of stereoselectivity on temperature has been

observed in other syntheses of alkenes from carbonyl compounds
(13b).

1,1-Diboro derivatives have also been applied to the synthesis
of cyclopropanols and cyclobutanols. Thus, dihydroboration of
1-bromo-2-propyne with 9-BBN, followed by treatment with sodium
hydroxide or methyllithium, gives B-cyclopropyl-9-BBN which, on
oxidation, gives cyclopropanol (14).

$$BrCH_2C{\equiv}CH \xrightarrow[\text{THF}]{\text{2 9-BBN}} BrCH_2CH_2CH\langle\overset{B}{\underset{B}{}} \xrightarrow[\text{or} \atop CH_3Li]{\text{NaOH}} \triangleright{-}B \xrightarrow[\text{NaOAc}]{H_2O_2,} \triangleright{-}OH \quad 65\%$$

In the synthesis of cyclobutanol, the tosylate of 3-butyn-1-ol
is dihydroborated with 9-BBN, followed by cyclization with
methyllithium and oxidation (14).

$$TsOCH_2CH_2C{\equiv}CH \xrightarrow[\text{THF}]{\text{2 9-BBN}} TsOCH_2CH_2CH_2CH\langle\overset{B}{\underset{B}{}} \xrightarrow[\text{2) [o]}]{\text{1)}CH_3Li} \square{-}OH \quad 65\%$$

Cyclopropanes have been prepared by treatment of
γ-chloroboranes with sodium hydroxide (Sec. 5.3.1), but attempts
to prepare cyclobutanes by similar treatment of δ-chloro- or
tosyloxyboranes (15) failed. The successful preparation using
the above 1,1-diboro derivative is probably due to the stabilized
carbanion (resulting from initial hydrolysis) (Sec. 7.2.1), which
is sufficiently nucleophilic to achieve the relatively unfavorable
displacement to form the four-membered ring (14).

7.2.3 Dihydroboration of Disubstituted Alkynes

While it has been established that dihydroboration of
1-alkynes gives mainly the 1,1-diboro derivatives (Sec. 7.2.1)

(1a,10), the reaction with disubstituted alkynes appears to be more complex.

Dihydroboration-oxidation of diphenylethyne gives d-ℓ-1,2-diphenyl-1,2-ethanediol, 1,2-diphenylethanol, and benzylphenylketone (deoxybenzoin) as the major products, with minor amounts of stilbene and bibenzyl also being formed (16,17). In the case of 3-hexyne, the major products are 3-hexanone, 3-hexanol, and 3,4-hexanediol (1a,17), with the formation of minor amounts of 3-hexene and 1-propanol also being reported (17).

The formation of monoalcohols could reasonably be ascribed to ready hydrolysis of the gem-diboro derivative, followed by oxidation (Sec. 7.2.1). However, on the basis of deuterium labeling studies, it has been suggested that monoalcohols are also formed from the vicinal diboro derivatives via formation of a bridged boron species (17).

Such a bridged species is also suggested as an intermediate in the formation of the alkenes (17). The diols arise from oxidation of the vicinal diboro derivatives, while ketones can be formed by oxidation of vinylboranes and gem-diboro derivatives (1a,17).

7.3 HYDROBORATION OF CYCLOALKYNES

Conversion of many-membered ring cycloalkynes to cis-cycloalkenes has been achieved by monohydroboration-protonolysis.

89% (n=11) (Ref. 18)

43% (n=8) (Ref. 19)
(81%-d$_2$)

I : n=2   II : n=2 : 87% (Ref. 20)
  : n=3   : n=3 : no reaction (Ref. 21)

The failure of I (n=3) to react under the usual conditions is surprising in view of the successful reaction of I (n=2). Use of excess borane gave only a small amount of II (n=3) together with nonvolatile material, probably a result of dihydroboration (21).

7.4 HYDROBORATION OF DIYNES

7.4.1 Monohydroboration of Diynes

Monohydroboration of symmetrically substituted conjugated diynes with disiamylborane proceeds to place the boron atom preferentially at the internal position of the diyne systems due to the less hindered nature of this position (22). Treatment of the intermediate boranes with acetic acid at 55-60° gives the corresponding <u>cis-enynes</u>, while oxidation gives the corresponding α,β-acetylenic ketones (<u>22</u>).

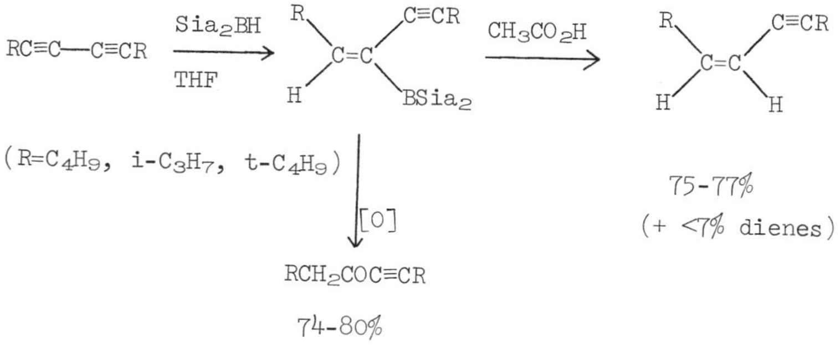

$$RC\equiv C-C\equiv CR \xrightarrow[THF]{Sia_2BH} \begin{array}{c} R \\ \diagdown \\ C=C \\ H \diagup \diagdown BSia_2 \end{array}\begin{array}{c} \diagup C\equiv CR \end{array} \xrightarrow{CH_3CO_2H} \begin{array}{c} R \\ \diagdown \\ C=C \\ H \diagup \diagdown H \end{array}\begin{array}{c} \diagup C\equiv CR \end{array}$$

$(R=C_4H_9, \ i-C_3H_7, \ t-C_4H_9)$

$\downarrow [O]$

75-77%
(+ <7% dienes)

$RCH_2COC\equiv CR$

74-80%

With unsymmetrically substituted conjugated diynes, approximately equal amounts of the two possible enynes are obtained, resulting from equal attack by boron at the two internal positions of the diyne system (<u>22</u>).

$$(CH_3)_3CC\equiv C-C\equiv C-C_4H_9 \xrightarrow[2)CH_3CO_2H]{1)Sia_2BH} (CH_3)_3CCH=CH-C\equiv CC_4H_9$$

$$+ \ (CH_3)_3CC\equiv C-CH=CH-C_4H_9$$

Selective monohydroboration of cyclic diynes has been achieved by reaction of a large excess of the diyne with triisobutylborane at 170° (<u>23</u>). Oxidation gives the corresponding cycloalkynones.

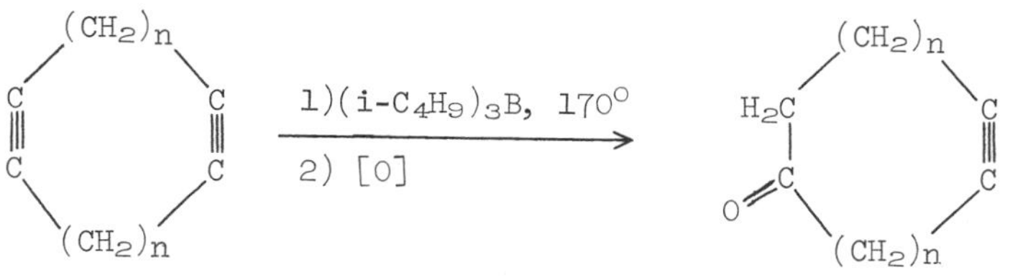

~50% (n=5, 6, 7)

7.4.2 Dihydroboration of Diynes

Attempted dihydroboration of conjugated diynes using disiamylborane proceeds very slowly beyond the monohydroboration stage. However, use of the sterically less demanding dicyclohexylborane achieves dihydroboration, except in the case of highly hindered diynes ($\underline{22}$). The second boron adds preferentially to the terminal carbon of the remaining triple bond of the monohydroboration product, and protonolysis gives the corresponding $\underline{cis, cis\text{-}diene}$.

In the case of hindered diynes, monohydroboration-protonolysis gives the cis-enyne, which is then treated with disiamylborane, followed by protonolysis, to give the $\underline{cis, cis\text{-}diene}$ ($\underline{22}$).

Obviously, the above methods can be used to convert unsymmetrical diynes to the corresponding dienes ($\underline{22}$).

Another method for the synthesis of cis,cis-dienes from diynes involves heating the diyne with triisobutylborane at 170° followed by protonolysis ($\underline{2}$).

$$RC\equiv C\text{-}C\equiv CR \xrightarrow[\substack{170^\circ \\ 2)CH_3CO_2H}]{1)(i\text{-}C_4H_9)_3B}$$

$$+ \; H_2C=C(CH_3)_2$$

36-64%

(R=CH_3, C_4H_9, C_6H_{13}, C_6H_{11}, Ph)

This method has also been used in the conversion of
1,3,10,12-cyclooctadecatetrayne to the all-cis-tetraene in
33% yield (2).

7.5 SELECTIVE HYDROBORATION OF ENYNES

Selective hydroboration of enynes has been achieved by
reaction of an excess of the enyne with diborane in diglyme (24).
However, rate studies with disiamylborane have shown that the
selective hydroboration of a triple bond in the presence of any
double bond, except an unhindered terminal double bond, should be
possible (25). The synthetic applicability of this proposal is
demonstrated by the following results.

$$RC\equiv CCH=CHR' \xrightarrow[\substack{2)CH_3CO_2H}]{1)Sia_2BH,DG}$$

Ref. 24

(R=H, CH_3, C_2H_5)   31-58%
(R'=H, CH_3)

$$HC\equiv C\overset{\overset{\displaystyle CH_3}{|}}{C}=CH_2 \xrightarrow[\substack{0\text{-}5^\circ \\ 2)CH_3CO_2H \; 3)[O]}]{1) \; (C_6H_{11})_2BH,THF} H_2C=CH\overset{\overset{\displaystyle CH_3}{|}}{C}=CH_2$$

Ref. 1b

$\xrightarrow[\substack{2)CH_3CO_2H \\ 3) \; [O] \;\; 4) \; NaOH}]{1)Sia_2BH,THF}$

Ref. 26

58%

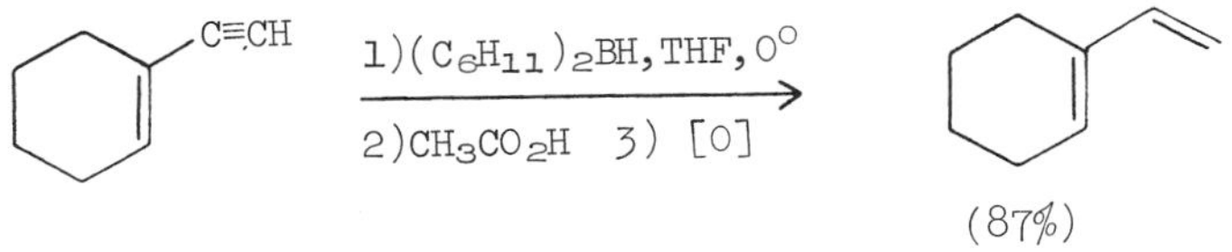

(87%)   Ref. 1b

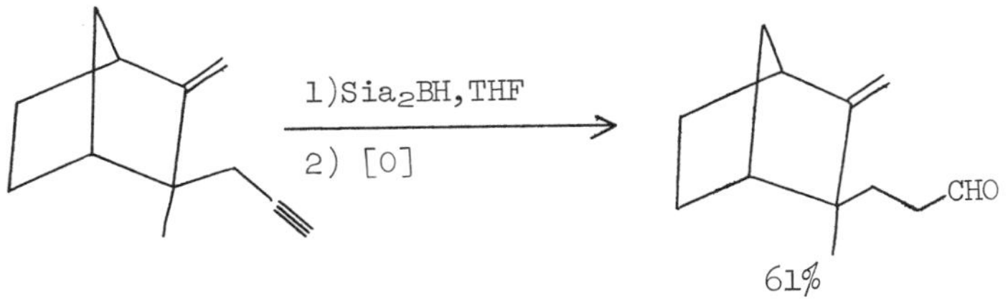

61%

REFERENCES

1. (a) H. C. Brown and G. Zweifel, J. Am. Chem. Soc., 83,
3834 (1961); (b) G. Zweifel, G. M. Clark, and N. L. Polston,
ibid, 93, 3395 (1971); (c) H. C. Brown and S. K. Gupta, ibid,
94, 4370 (1972).

2. A. J. Hubert, J. Chem. Soc., 1965, 6669.

3. H. C. Brown, D. H. Bowman, S. Misumi, and M. K. Unni, J. Am.
Chem. Soc., 89, 4531 (1967).

4. (a) G. Zweifel, H. Arzoumanian, and C. C. Whitney, J. Am.
Chem. Soc., 89, 3652 (1967); (b) G. Zweifel, R. P. Fisher,
J. T. Snow, and C. C. Whitney, ibid, 93, 6309 (1971).

5. G. Zweifel, N. L. Polston, and C. C. Whitney, J. Am. Chem.
Soc., 90, 6243 (1968).

6. (a) G. Zweifel and H. Arzounmanian, J. Am. Chem. Soc., 89,
5086 (1967); (b) E. J. Corey and T. Ravindranathan, ibid, 94,
4013 (1972).

7. G. Zweifel, A. Horng, and J. T. Snow, J. Am. Chem. Soc., 92,
1427 (1970).

8. P. Binger and R. Köster, Angew. Chem., 74, 652 (1962).

9.   R. G. Lewis, D. H. Gustafson, and W. F. Erman, Tetrahedron Letters, 1967, 401.

10.   G. Zweifel and H. Arzoumanian, J. Am. Chem. Soc., 89, 291 (1967).

11.   G. Zweifel and H. Arzoumanian, Tetrahedron Letters, 1966, 2535.

12.   G. Cainelli, G. Dal Bello, and G. Zubiani, Tetrahedron Letters, 1965, 3429.

13.   (a) G. Cainelli, G. Dal Bello and G. Zubiani, Tetrahedron, Letters, 1966, 4315;)(b) J. Reucroft and P. G. Sammes, Quart. Rev., 25, 135 (1971).

14.   H. C. Brown and S. P. Rhodes, J. Am. Chem. Soc., 91, 4306 (1969).

15.   H. C. Brown and K. A. Keblys, J. Am. Chem. Soc., 86, 1791 (1964).

16.   A. Hassner and B. H. Braun, J. Org. Chem., 28, 261 (1963).

17.   D. J. Pasto, J. Am. Chem. Soc., 86, 3039 (1964).

18.   D. S. Sethi, S. Vaidyanathaswamy, and D. Devaprabhakara, Org. Prep. Procedures, 2, 171 (1970).

19.   A. C. Cope, G. A. Berchtold, P. E. Peterson, and S. H. Sharman, J. Am. Chem. Soc., 82, 6370 (1960).

20.   J. Sicher, M. Svoboda, J. Závada, R. B. Turner, and P. Goebel, Tetrahedron, 22, 659 (1966).

21.   J. Sicher, M. Svoboda, B. J. Mallon, and R. B. Turner, J. Chem. Soc., B1968, 441.

22.   G. Zweifel and N. Polston, J. Am. Chem. Soc., 92, 4068 (1970).

23.   A. J. Hubert, J. Chem. Soc., 1965, 6679.

24.   V. V. Markova, V. A. Kormer, and A. A. Petrov. Zhur. Obshch, Khim, 35, 1669 (1965); Chem. Abstr., 63, 17870 (a)(1965).

25.   H. C. Brown and A. W. Moerikofer, J. Am. Chem. Soc., 85, 2063 (1963).

26.   E. J. Corey and D. K. Herron, Tetrahedron Letters, 1971, 1641.

27.   H. Kretschmar and W. F. Erman, Tetrahedron Letters, 1970, 41.

Chapter 8

SYNTHESIS OF CARBON CHAINS AND RINGS

The synthesis of a variety of organic compounds via the hydroboration of alkenes, dienes, and alkynes is discussed in earlier chapters and in Chap. 9, but most of the reactions discussed involve the introduction of a functional group into a molecule without the occurrence of significant change in the basic carbon structure of the molecule.

However, this chapter deals with the reaction of organoboranes with a variety of substrates in which alkyl or aryl groups are transferred from boron to carbon, thereby resulting in the synthesis of compounds containing extended carbon chains or in the synthesis of carbocyclic compounds. The chapter is divided into sections based on the type of functional group formed in the reactions. Thus, the synthesis of alcohols is dealt with in Sec. 8.3, aldehydes in Sec. 8.4, ketones in Sec. 8.5, esters in Sec. 8.6, nitriles in Sec. 8.7, and alkenes and dienes in Sec. 8.8. Section 8.9 deals with the synthesis of alkanes via coupling reactions. Of particular interest are those reactions which provide methods of achieving chain extension by a fixed number of carbon atoms. The relevant sections are listed below:

One carbon chain extension:

Synthesis of primary alcohols	(Sec. 8.3.1)
Synthesis of aldehydes	(Sec. 8.4)
Synthesis of carboxylic acids	(Sec. 8.5.1)

<u>Two carbon chain extension:</u>

Synthesis of aldehydes	(Sec. 8.4)
Synthesis of esters	(Sec. 8.6.1)
Synthesis of nitriles	(Sec. 8.7)

<u>Three carbon chain extension:</u>

Synthesis of aldehydes	(Sec. 8.4)
Synthesis of methyl ketones	(Sec. 8.5.3)

<u>Four carbon chain extension:</u>

Synthesis of primary alcohols	(Sec. 8.3.1)
Synthesis of methyl ketones	(Sec. 8.5.4)
Synthesis of β,γ-unsaturated esters	(Sec. 8.6.2)

<u>Four or more carbon chain extension:</u>

Synthesis of terminally functionalized ketones	(Sec. 8.5.1)

Since many of the reactions involve similar mechanisms these
are discussed in Sec. 8.1. In addition, some of the experimental
factors which must be considered in carrying out many of the
reactions are discussed in Sec. 8.2.

8.1 MECHANISTIC CONSIDERATIONS

The reactions in which organoboranes are used in the
construction of carbon chains or rings generally involve the
transfer of alkyl or aryl groups from boron to carbon. Many of
these reactions proceed either by coordination or radical chain
mechanisms depending on the nature of the substrate with which
the organoborane is reacted, and these mechanisms are considered
in the following section.

8.1.1 Coordination Mechanisms

The reactions which proceed by coordination mechanisms
usually involve initial Lewis acid-Lewis base interaction between
the organoborane and the substrate to form an adduct, which

subsequently undergoes rearrangement with migration of an alkyl
or aryl group from boron to carbon. The reactions proceed by
intramolecular transfer of the groups from boron to carbon, as
shown by experiments in which an equimolar mixture of two
different organoboranes is reacted with the relevant substrate.
In these experiments no trace of the mixed derivatives expected
from intermolecular transfer of groups has been found. Among the
reactions which proceed by coordination mechanisms are carbonylation
reactions (1), and reactions between organoboranes and α-halocarbanions (2), ylids (3), diazo compounds (4), and carbenes (5).

   Carbonylation reactions involve the reaction between carbon
monoxide and organoboranes. Early studies of this reaction
utilized very high pressures, of the order of 500-700 atm, and
temperatures varying from 50 to 150° (6). However, it was later
discovered that essentially quantitative reaction occurs in
diglyme at atmospheric pressure and temperatures of 100-125° (1).
In addition, the extent of intramolecular migration of groups from
boron to carbon, and hence the nature of the products formed, can
be readily controlled by choice of suitable reaction conditions (1).
Thus, when the reaction is carried out in diglyme at 100-125°,
migration of all three groups from boron to carbon occurs, while
in the presence of a small amount of water, migration of the third
group is inhibited due to cleavage of the intermediate boraepoxide
(3). In the presence of a complex metal hydride, however, migration
is limited to one group, probably as a result of the reduction of
the intermediate product (2). Hydrolysis or oxidation of the
intermediate products formed gives aldehydes, ketones or alcohols
as shown below (1).

$$R_3B + CO \; \rightleftharpoons \; R_2\overset{\ominus}{B}\text{-}\overset{\oplus}{C}O \longrightarrow$$

with the R group migrating:

$$(1)$$

$$\underset{(2)\ O}{RB\overset{R}{\underset{\|}{-C}}-R} \longrightarrow \underset{(3)\ O}{R-B\overset{R}{-C}-R} \longrightarrow O=B-CR_3$$

$$\downarrow MBH_4 \qquad\qquad \downarrow H_2O \qquad\qquad \downarrow [O]$$

$$R_2B-\overset{|}{C}HR \qquad\qquad R-B-CR_2 \qquad R_3C-OH$$

$$\underset{O\ominus}{|} \qquad\qquad HO\ \ OH$$

$$\qquad\qquad\qquad (Sec.\ 8.3.3)$$

$$[O]\diagdown_{OH\ominus} \qquad\qquad \downarrow [O]$$

RCHO   RCH$_2$OH   R$_2$CO

(Sec. 8.4) (Sec. 8.3.1)   (Sec. 8.5.1)

The reactions of organoboranes with α-halocarbanions (2), ylids (3), and diazo compounds (4) may be represented as proceeding by the following mechanisms:

$$BrCH_2COX \xrightarrow{Base} Br\overset{\ominus}{C}HCOX \xrightarrow{R_3B} R_2B\overset{R}{-}\overset{|}{C}HCOX \longrightarrow R_2B-\overset{R}{\overset{|}{C}}HCOX$$

(X=OC$_2$H$_5$, alkyl, or aryl) $\overset{\ominus}{}$ Br   (4) $\downarrow$ R'OH

$$RCH_2COX$$

(Secs. 8.5.3, 8.6.1)

$$G-CH_2^{\oplus\ominus} + R_3B \longrightarrow R_2B\overset{R}{-}\overset{\ominus}{C}H_2 \xrightarrow{[O]} R_2B-CH_2R \longrightarrow RCH_2OH$$

$$\underset{G}{\overset{\oplus}{|}}$$

(Sec. 8.3.1)

[G=(CH$_3$)$_2$S, (CH$_3$)$_3$N, Ph$_3$P]

$$N_2CHY + R_3B \longrightarrow R_2B\overset{R}{-}\overset{\ominus}{C}HY \longrightarrow R_2B-\overset{R}{\overset{|}{C}}HY \xrightarrow{R'OH} RCH_2Y$$

$$\underset{N_2}{\overset{\oplus}{|}}$$

(5)   (Secs. 8.5.3, 8.6.1, 8.7)

(Y=COR, CO$_2$C$_2$H$_5$, CN)

In the above mechanisms the β-ketoboranes (4) and (5; Y = COR or CO$_2$C$_2$H$_5$) are represented as undergoing direct protonolysis to give the final products. However, studies of the reaction of tri-n-propylborane with diazoacetophenone or ethyl diazoacetate have shown that vinyloxyboranes (6) are the intermediates which undergo protonolysis ($\underline{7}$). It is possible that the β-ketoborane is initially formed and rearranges rapidly to the vinyloxyborane as shown below.

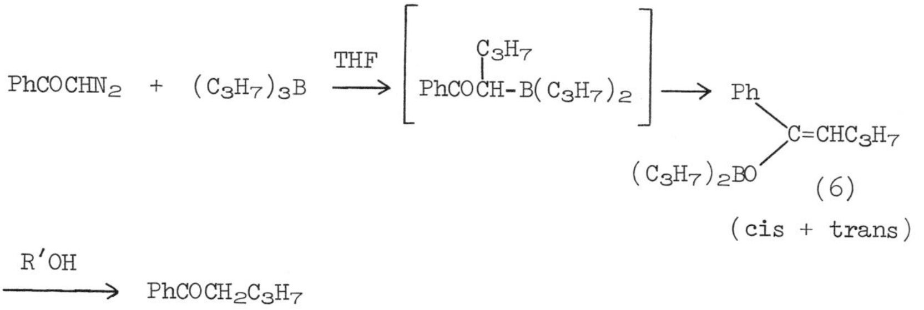

$$\text{PhCOCHN}_2 \; + \; (\text{C}_3\text{H}_7)_3\text{B} \; \xrightarrow{\text{THF}} \; \left[\begin{array}{c} \text{C}_3\text{H}_7 \\ | \\ \text{PhCOCH-B(C}_3\text{H}_7)_2 \end{array} \right] \; \longrightarrow \; \begin{array}{c} \text{Ph} \\ \\ (\text{C}_3\text{H}_7)_2\text{BO} \end{array} \!\!\! \begin{array}{c} \\ \diagdown \\ \diagup \end{array} \!\!\! \text{C=CHC}_3\text{H}_7$$

(6)

(cis + trans)

$$\xrightarrow{\text{R}'\text{OH}} \; \text{PhCOCH}_2\text{C}_3\text{H}_7$$

The reaction of organoboranes with <u>carbenes</u> ($\underline{5}$) also proceeds by intramolecular migration of groups from boron to carbon via the intermediate adduct (7).

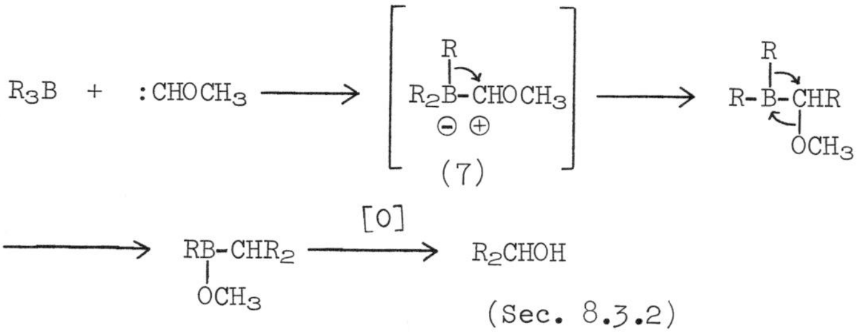

$$\text{R}_3\text{B} \; + \; :\text{CHOCH}_3 \; \longrightarrow \; \left[\begin{array}{c} \text{R} \\ | \\ \text{R}_2\overset{\ominus}{\text{B}}\text{-}\overset{\oplus}{\text{CHOCH}}_3 \end{array} \right] \; \longrightarrow \; \begin{array}{c} \text{R} \\ | \\ \text{R-B-CHR} \\ | \\ \text{OCH}_3 \end{array}$$

(7)

$$\longrightarrow \; \begin{array}{c} \text{RB-CHR}_2 \\ | \\ \text{OCH}_3 \end{array} \; \xrightarrow{[\text{O}]} \; \text{R}_2\text{CHOH}$$

(Sec. 8.3.2)

8.1.2 Radical Chain Mechanisms

Organoboranes undergo rapid 1,4-addition reactions with 3-buten-2-one (methyl vinyl ketone) or 2-propenal (acrolein) to give vinyloxyboranes, which undergo hydrolysis on treatment with water ($\underline{8},\underline{9}$).

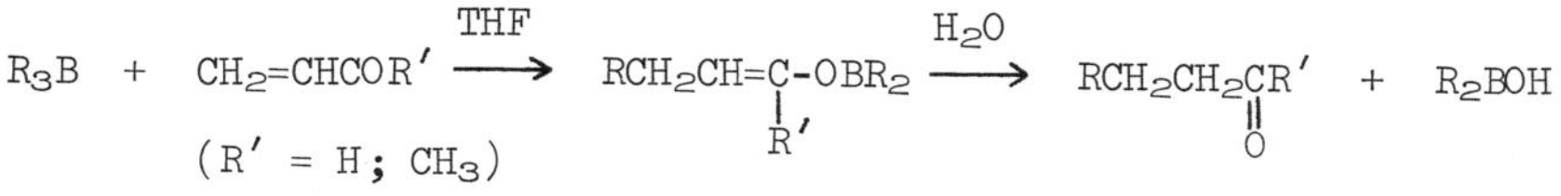

$$\text{R}_3\text{B} \; + \; \text{CH}_2\text{=CHCOR}' \; \xrightarrow{\text{THF}} \; \text{RCH}_2\text{CH=C-OBR}_2 \; \xrightarrow{\text{H}_2\text{O}} \; \text{RCH}_2\text{CH}_2\text{CR}' \; + \; \text{R}_2\text{BOH}$$

(R' = H; CH₃)

The reaction was initially thought to proceed via a cyclic transition state (8)

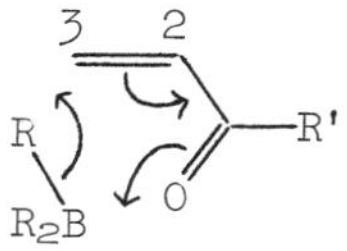

(8)

The reaction proceeds readily whan a substituent is present in the
2 position, such as in the case of 2-methylpropenal (Sec. 8.4),
but fails when an alkyl substituent is present in the 3 position
as in the cases of trans-3-penten-2-one or trans-2-butenal (10a).
Terminal methyl groups have been observed to have an inhibiting
effect on free radical polymerizations (10b). Furthermore, it was
discovered that galvinoxyl inhibits the reaction between
trialkylboranes and 2-propenal or 3-buten-2-one (10a), and that
reaction with 3-substituted enone systems is effectively promoted
by the addition of acyl peroxides (11) or air (12), or by photochemical means (11). These reactions therefore proceed by radical
chain mechanisms as shown below.

$R_3B + O_2 \longrightarrow R_2BO_2\cdot + R\cdot$ (Sec. 4.1)

$R\cdot + CH_2=CHCOR' \longrightarrow RCH_2\overset{\bullet}{C}HCOR' \longleftrightarrow RCH_2CH=\underset{R'}{\overset{|}{C}}-O^{\bullet} \overset{R_3B}{\longrightarrow}$

$RCH_2CH=\underset{R'}{\overset{|}{C}}-OBR_2 + R\cdot$

The nature of the chain propagation step may be associated with
the formation of the very strong boron-oxygen bond in the
vinyloxyborane, and is similar to the reaction of alkoxy radicals
with organoboranes (Sec. 4.1). The failure of the reaction of
3-substituted enone systems in the absence of air or radical
initiators may possibly be the result of the relatively short
chain length for radical addition to these systems compared to
the addition to unsubstituted systems.

8.1.3 Stereochemistry of Transfer Reactions

Many of the reactions involving the intramolecular transfer
of a group from boron to another atom, such as protonolysis
(Sec. 10.11), oxidation with alkaline hydrogen peroxide (Sec. 4.3),
or amine N-oxides (Sec. 4.4), and amination with O-hydroxylamine-
sulfonic acid (Sec. 9.2), proceed with retention of configuration
of the migrating group.

Likewise, the reactions involving intramolecular transfer of
a group from boron to carbon, such as those discussed in Sec. 8.1.1,
all proceed with retention of configuration of the migrating group.
This has been demonstrated by the conversion of the trans-2-methyl-
cyclopentylboryl moiety (9) into the derivatives shown below (13a).

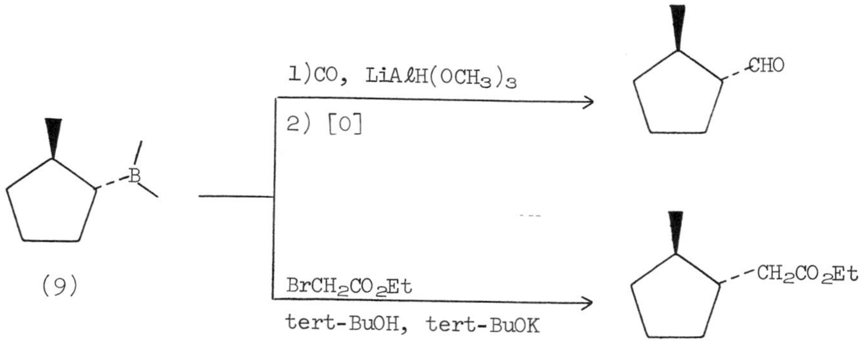

Intramolecular transfer reactions involving α,β-unsaturated enone
systems, which proceed by radical chain mechanisms (Sec. 8.1.2),
were also initially thought to proceed with retention of
configuration of the migrating group (13a). Later studies have,
however, shown that the nature of the products formed in such
reactions is determined by the relative thermodynamic stabilities
of the various possible product isomers (13b). Thus, in the
reaction of the trans-2-methylcyclopentylboryl moiety (9) with
2-propenal (13a), the trans product is thermodynamically favored
to the extent of >90%.

(9)   >90%   <10%

8.2 PRACTICAL SYNTHETIC CONSIDERATIONS

8.2.1 Reaction Conditions

While most of the reactions discussed in the following sections use temperatures not exceeding room temperature (~25°), certain of the carbonylation reactions use temperatures exceeding 100° (Sec. 8.3.3). Under such conditions, isomerization of organoboranes is possible (Sec. 2.3.2); however, where the possibility of isomerization of the organoborane exists, milder methods of synthesis are usually available (Sec. 8.3.3).

A number of the reactions involve the initial generation of a suitable carbanion by treatment of an active methylene compound with base (Secs. 8.5.3, 8.6, 8.7). In such reactions the <u>strength of the base</u> used can have a marked effect on the efficiency of the reaction, since many of the substrates are sensitive to strong bases such as potassium tert-butoxide. In addition, the base can coordinate the organoborane, thereby preventing effective reaction between the organoborane and the generated carbanion. After investigation of a large number of bases <u>potassium 2,6-di-tert-butylphenoxide</u> (10) has been found to be the most effective base in a variety of reactions involving organoboranes and carbanions (<u>14</u>). This base can be prepared by reaction of potassium and 2,6-di-tert-butylphenol or by reaction of the phenol with the calculated amount of potassium tert-butoxide (<u>14</u>).

(10)

This base is so mild that ketones, esters, and nitriles are quite stable to it in tetrahydrofuran solution at 25° (15). In addition, the bulky tert-butyl groups prevent its coordination with the organoborane. However, the hindered 2,6-di-tert-butylphenol is often slow to protonolyze the intermediate boranes formed after transfer of groups from boron to carbon (Sec. 8.1.1), and the addition of ethanol or tert-butanol might be necessary in order to liberate the final product.

In many of the reactions, oxidation with alkaline hydrogen peroxide is included as a final step, even though this is not necessary for the formation of the final product. Such an oxidation step, however, facilitates subsequent separation procedures by oxidizing any other boranes formed in the reactions to the corresponding alcohols. Where mild oxidation conditions are required sodium hydroxide may be replaced by either sodium hydrogen carbonate or sodium acetate in the standard procedure (Sec. 4.3).

8.2.2 Yields

While many of the reactions discussed in the following sections proceed in very high yields, these yields are often based on the production of one mole of the product from one mole of the organoborane. However, based on starting alkenes, these yields correspond to a maximum theoretical yield of 33%. In the case of valuable starting alkenes, this constitutes a serious disadvantage.

Fortunately, in many of the reactions this disadvantage can be overcome by use of B-alkyl-9-BBN-derivatives, formed by reaction of the selective hydroborating reagent, 9-BBN, with alkenes (Sec. 2.2.1c). B-alkyl- and B-aryl-9-BBN derivatives may also be prepared by the reaction of 9-BBN with the corresponding alkyl- or aryllithium reagents, thus extending the use of these reagents to alkyl groups not available via hydroboration (16a). In the

subsequent transfer reactions the B-alkyl bond usually migrates
in preference to the B-cyclooctyl bond. However, in certain
reactions, notably those involving the use of α-diazo compounds
(16b) (Secs. 8.4, 8.5.3, 8.6.1, and 8.7) and radical 1,4-additions
to conjugated enone systems (Sec. 8.5.4), migration of the
B-cyclooctyl bond predominates.

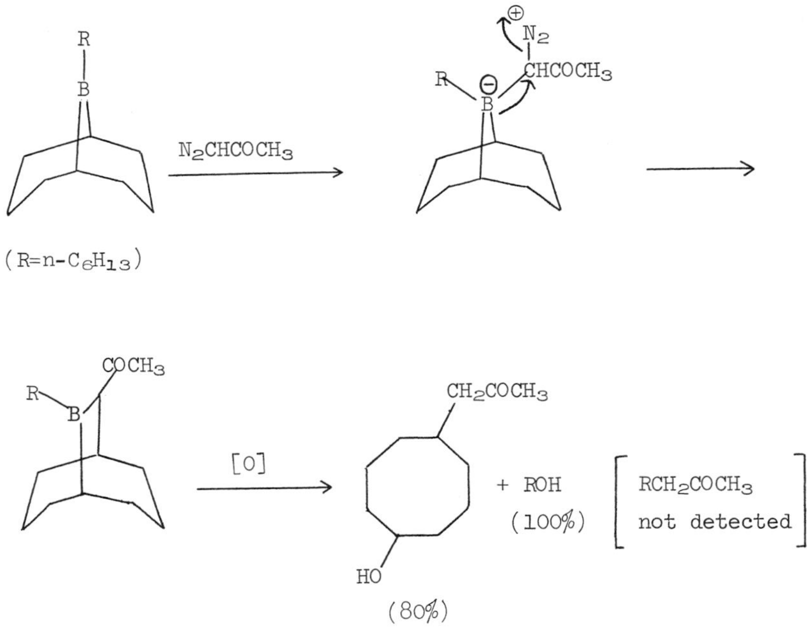

(R=n-C$_6$H$_{13}$)

(80%)

In the latter reactions B-alkylboracyclanes have been successfully
used (17). Derivatives, such as B-alkylboracyclohexane,
B-alkyl-3,5-dimethylboracyclohexane, or B-alkyl-3,6-dimethyl-
boracycloheptane are readily prepared by reaction of alkenes with
the corresponding hydroborating reagents (Sec. 2.2.1.c), or by
the reaction of B-methoxyboracyclanes with alkyllithium reagents
(17). Use of these reagents results in preferential migration

of the B-alkyl bond for a variety of alkyl groups (17). (Sec. 8.5.4).

$$\text{[cyclohexyl]}\!-\!\text{B}\!-\!\text{R} \ + \ CH_2{=}CHCOCH_3 \ \xrightarrow[\text{2) [O]}]{\text{1)THF,H}_2\text{O}} \ RCH_2CH_2COCH_3 \ + \ HO(CH_2)_5OH$$

$$65\text{-}80\%$$

8.2.3 Scope of the Reactions

In the development of many of the reactions discussed in the following sections, a number of organoboranes, containing a variety of alkyl groups, has been used. Some of the most frequently used organoboranes are listed below:

R_3B: R = C_2H_5, n=C_3H_7, n-C_4H_9, i-C_4H_9, s-C_4H_9, n-C_6H_{13}

cyclopentyl, cyclohexyl, exo-norbornyl

In discussing the reactions, they are considered as being applicable to the above, and similar, reagents, Where notable exceptions are observed, these are referred to specifically under the reaction concerned. In addition, the presence of various common functional groups, such as esters, halides and nitriles, can be accommodated in these reactions; this applies particularly to those reaction sequences using selective reagents, such as 9-BBN and thexylborane.

8.3 SYNTHESIS OF ALCOHOLS

8.3.1 Synthesis of Primary Alcohols

Reaction of trialkylboranes with carbon monoxide at atmospheric pressure and in the presence of lithium or sodium borohydride proceeds rapidly in tetrahydrofuran at 45°; hydrolysis of the intermediate organoboranes gives the corresponding homologated alcohols in 70-85% yields (based on R_3B) (18). Yields

of 50-85% (based on starting alkene) of the homologated alcohol
may be obtained by carrying out the reaction in the presence of
lithium trimethoxyaluminumhydride and using B-alkyl-9-BBN
derivatives (19).

$$R\text{-}B\text{-}9\text{-}BBN + CO + \text{LiA}\ell\text{H}(\text{OCH}_3)_3 \xrightarrow[0^\circ]{\text{THF}} \overset{\ominus}{O}\text{-}\underset{R}{\text{CH}}\text{-}B\text{-}9\text{-}BBN \xrightarrow[\text{EtOH}]{\text{KOH}} RCH_2OH +$$

$$(50\text{-}85\%)$$

HO-9-BBN

Homologated alcohols may also be synthesized by reaction of
trialkylboranes with a variety of ylids. Among the ylids used are
triphenylphosphonium methylide ($Ph_3\overset{\oplus}{P}CH_2^{\ominus}$) (20), trimethylammonium
methylide [$(CH_3)_3\overset{\oplus}{N}CH_2^{\ominus}$] (21), dimethyloxosulfonium methylide
[$(CH_3)_2\overset{\oplus}{SO}CH_2^{\ominus}$], (22) and dimethylsulfonium methylide
[$(CH_3)_2\overset{\oplus}{S}CH_2^{\ominus}$] (23). Of these, dimethylsulfonium methylide appears
to be the most useful reagent synthetically (23).

$$R_3B + (CH_3)_2\overset{\oplus}{S}CH_2^{\ominus} \xrightarrow[\substack{-10^\circ \\ 2)\ [o]}]{1)\text{DMSO-THF}} \underset{50\text{-}90\%}{RCH_2OH + 2ROH}$$

Trialkylboranes readily react with 1,3-butadiene monoxide in
the presence of catalytic amounts of oxygen to give trans-4-alkyl-
2-buten-1-ols of about 90% stereochemical purity (24).

$$R_3B + CH_2\text{=}CHCH\text{---}CH_2 \xrightarrow[2)\ [o]]{1)\text{PhH, Air, } 25^\circ} \underset{(44\text{-}75\%)}{RCH_2CH\text{=}CHCH_2OH + 2ROH}$$

This reaction thus provides a means of achieving <u>four carbon atom</u>
<u>chain extension.</u> Use of tetrahydrofuran as solvent results in
the formation of considerable quantities of 4-(2'-tetrahydrofuryl)-
2-buten-1-ol (11), providing supporting evidence for the radical
nature of the reaction.

$$\text{(structure)} \quad CH_2CH=CHCH_2OH$$

$$(11)$$

The mechanism of the reaction is proposed as follows (24):

$$R_3B + O_2 \longrightarrow R_2BO_2\cdot + R\cdot \text{ (Sec. 4.1)}$$

$$R\cdot + CH_2=CHCH\text{---}CH_2 \longrightarrow RCH_2\overset{\cdot}{C}HCH\text{---}CH_2 \longleftrightarrow RCH_2CH=CHCH_2O\cdot$$

$$\overset{R_3B}{\longrightarrow} RCH_2CH=CHCH_2OBR_2 + R\cdot$$

It is interesting that radical addition in this case occurs to a relatively "unactivated" double bond compared to radical additions to conjugated enone systems (Sec. 8.1.2).

8.3.2 Synthesis of Secondary Alcohols

Carbonylation of trialkylboranes in the presence of water, followed by hydrolysis of the reaction mixture with base, gives secondary alcohols (1,25).

$$R_3B + CO \xrightarrow[100°]{DG, H_2O} \underset{\underset{HO\ OH}{|\ \ |}}{RB\text{-}CR_2} \xrightarrow[\underset{\Delta}{H_2O}]{KOH} \underset{(80\text{-}90\%)}{R_2CHOH + RB(OH)_2}$$

Trialkylboranes readily react with methoxycarbene, derived from dichloromethyl methyl ether, to give the corresponding alcohols (12) and (13) (Sec. 8.1.1) (5).

$$(RCH_2)_3B + C\ell_2CHOCH_3 \xrightarrow[2)\ [O]]{1)CH_3Li,\ THF,\ r.t.} (RCH_2)_2CHOH + \underset{\underset{OH}{|}}{RCHCH_2R}$$

$$(12) \qquad\qquad (13)$$

The ratio of alcohols (12) and (13) formed depends on the relative amounts of methyllithium and trialkylborane used in the reaction.

Use of a ten-fold excess of methyllithium gives (12) and (13) in yields of 21-50% and 25-39%, respectively. If the trialkylborane is first treated with methyl lithium followed by dichloromethyl methyl ether, only alcohol (13) is obtained (5).

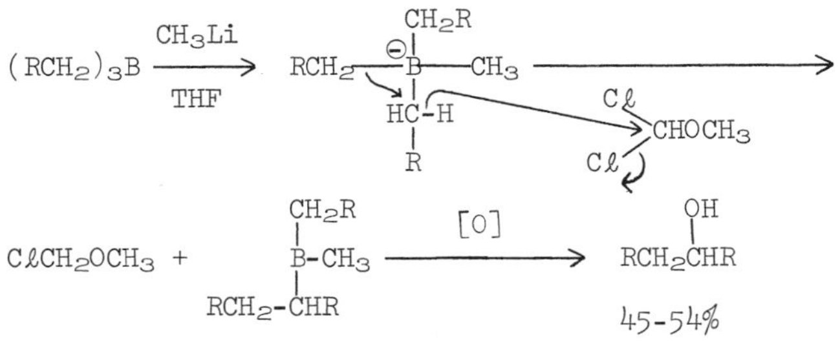

$$(RCH_2)_3B \xrightarrow[\text{THF}]{CH_3Li} RCH_2 \overset{\overset{CH_2R}{|}}{\underset{\underset{R}{|}}{\underset{HC-H}{B}}} -CH_3 \xrightarrow{}$$

$$C\ell CH_2OCH_3 + \overset{\overset{CH_2R}{|}}{\underset{\underset{RCH_2-CHR}{|}}{B-CH_3}} \xrightarrow{[O]} RCH_2\overset{\overset{OH}{|}}{CHR}$$

45-54%

Reaction of dichloromethyllithium and triphenylborane gives, after oxidation, diphenylmethanol as the major product (26). The reaction possibly proceeds via formation of chlorocarbene as shown below (Sec. 8.1.1).

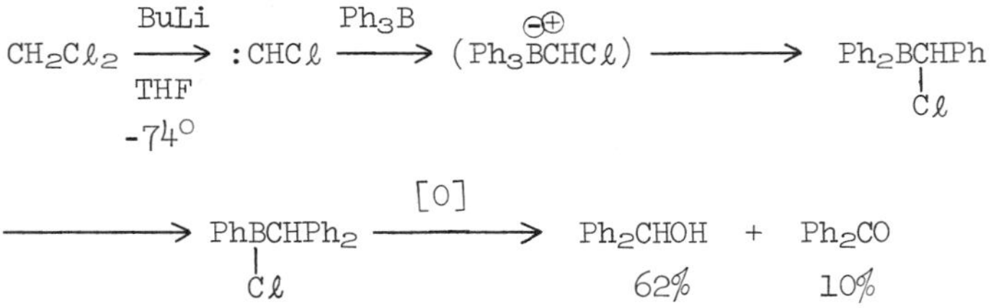

$$CH_2C\ell_2 \xrightarrow[\substack{THF \\ -74°}]{BuLi} :CHC\ell \xrightarrow{Ph_3B} (Ph_3\overset{\ominus\oplus}{B}CHC\ell) \xrightarrow{} Ph_2B\underset{\underset{C\ell}{|}}{CHPh}$$

$$\xrightarrow{} PhB\underset{\underset{C\ell}{|}}{CHPh_2} \xrightarrow{[O]} Ph_2CHOH + Ph_2CO$$

62%   10%

Photolytically induced bromination of trialkylboranes with 1 mole equivalent of bromine in inert solvents gives α-bromo-organoboranes (27a).

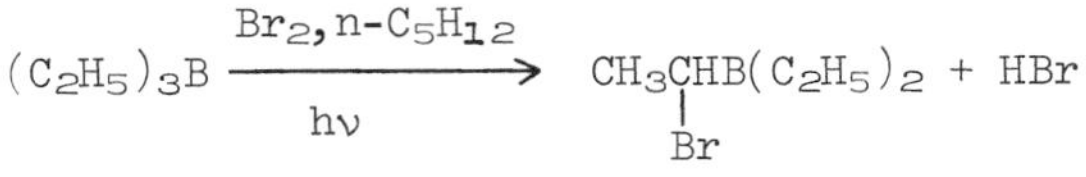

$$(C_2H_5)_3B \xrightarrow[h\nu]{Br_2, n-C_5H_{12}} CH_3\underset{\underset{Br}{|}}{CHB}(C_2H_5)_2 + HBr$$

The hydrogen bromide formed can have undesirable effects such as protonolysis of the product borane (Sec. 9.1.2) or reaction with functional groups present in the molecule. Such disadvantages may

be overcome by the use of bromotrichloromethane as the brominating
reagent (27b).

$$(C_2H_5)_3B + BrCCl_3 \xrightarrow{CH_2Cl_2, h\nu} CH_3\underset{\underset{Br}{|}}{C}HB(C_2H_5)_2 + CHCl_3$$

The α-bromoorganoboranes undergo ready rearrangment on treatment
with nucleophilic reagents such as water or pyridine to give, after
oxidation, the corresponding secondary alcohols (27a).

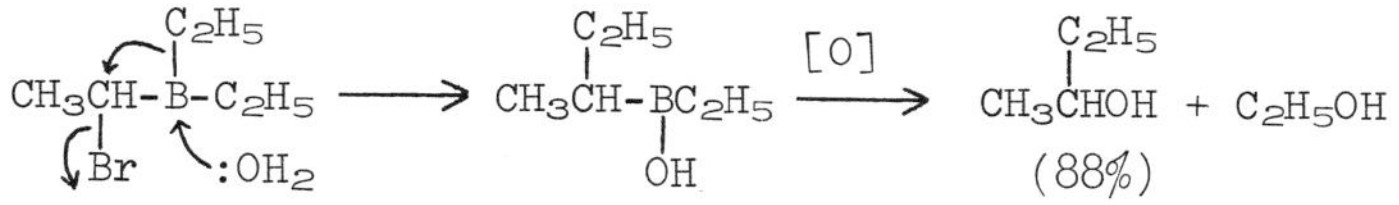

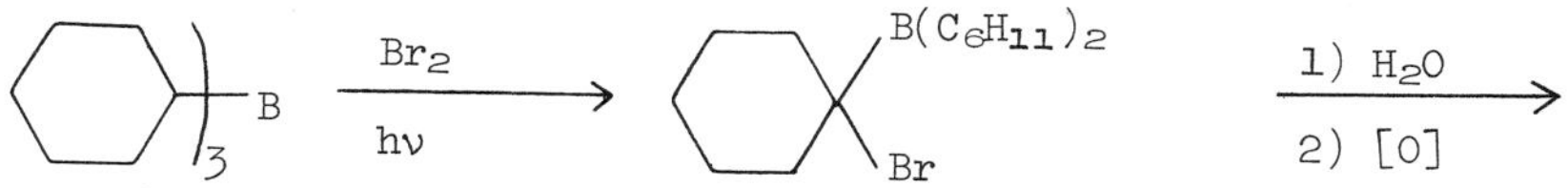

Ref. 35

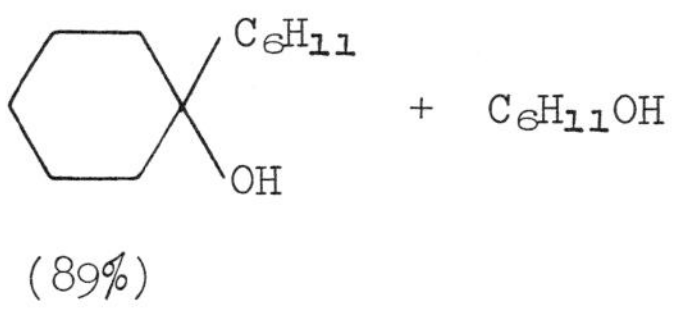

(89%)

Similar rearrangements can also be induced by electrophilic
reagents such as aluminum tribromide or aluminum trichloride (27c).

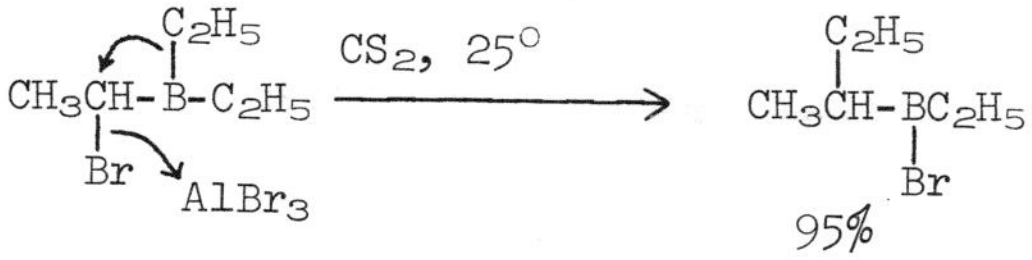

Cyclohexene undergoes photochemically induced reaction with
unhindered trialkylboranes to give, after oxidation, cis-2-alkyl-
cyclohexanols (27d). Similar reaction occurs with certain

1-alkyl-cyclohexenes but fails with other cycloalkenes such as
cyclopentene, cyclooctene, and cyclodecene (27d). The reaction
probably occurs via electrophilic attack by the trialkylborane
on the highly strained trans-cyclohexene, followed by migration
of an alkyl group from boron to carbon.

$70-86\%$ ($R=C_2H_5, n-C_4H_9; R^1=H, C_2H_5$)

Similar reaction of 2,7-cyclooctadienone with triethylborane
gives 3-ethyl-7-cyclooctenone in 90% yield (27d).

8.3.3 Synthesis of Tertiary Alcohols

Trialkylcarbinols are readily synthesized by carbonylation
of trialkylboranes in diglyme at 100-125°, followed by oxidation
of the intermediate oxyborane, $(R_3CBO)_x$ ($\underline{1},\underline{28}$). Oxidation is
difficult in certain cases due to polymerization of the oxyborane;
however, this difficulty can be overcome by carrying out the
carbonylation in the presence of 1,2-ethanediol ($\underline{28}$).

$80-90\%$

This method has been applied to the conversion of organoboranes containing a tertiary alkyl group to the corresponding highly branched trialkylcarbinols without isomerization of the tertiary alkyl group occurring during the reaction (28b).

$$R^1-\underset{\underset{R^3}{|}}{\overset{\overset{R^2}{|}}{C}}-BR^4R^5 \quad \xrightarrow[\substack{100-125^\circ \\ 2)\ [o]}]{1)CO,(CH_2OH)_2} \quad R^1-\underset{\underset{R^3}{|}}{\overset{\overset{R^2}{|}}{C}}-\underset{\underset{R^5}{|}}{\overset{\overset{R^4}{|}}{C}}-OH$$

43-81% ($R^1=R^2=CH_3$; $R^3=CH_3$, i-C_3H_7)

The above method suffers from the disadvantage of the high temperatures used during the carbonylation reaction, and the resultant danger of isomerization of the organoboranes (Sec. 2.3.2). A more convenient route to trialkylcarbinols involves the reaction of trialkylboranes with trihalosubstituted methanes, such as chlorodifluoromethane or chloroform, under the influence of lithium triethylcarboxide (29).

$$(n-C_4H_9)_3B + HCClF_2 \quad \xrightarrow[\substack{THF-Hexane \\ 25-60^\circ \\ 2)\ [o]}]{1)\ LiOCEt_3} \quad (n-C_4H_9)_3COH$$

(98%)

With chloroform a yield of 85% of the alcohol is obtained (29). The yields obtained using lithium triethylcarboxide are considerably higher thanthose obtained using potassium tert-butoxide, probably due to the lower degree of coordination of the organoborane by the more hindered alkoxide.

Another convenient synthesis of trialkylcarbinols utilizes trialkylcyanoborates (14) which are readily formed by treatment of trialkylboranes with a suspension of sodium cyanide in tetrahydrofuran (30).

$$R_3B + NaCN \xrightarrow[22^\circ]{THF} R_3\overset{\ominus}{B}-CN \underset{Na^\oplus}{} \xrightarrow[\substack{45^\circ \\ \\ 2)\ NaOH\ \ 3)\ [O]}]{1)\ (CF_3CO)_2O} R_3COH \quad 70\text{-}80\%$$

$$(14)$$

The following mechanism of reaction has been postulated (30).

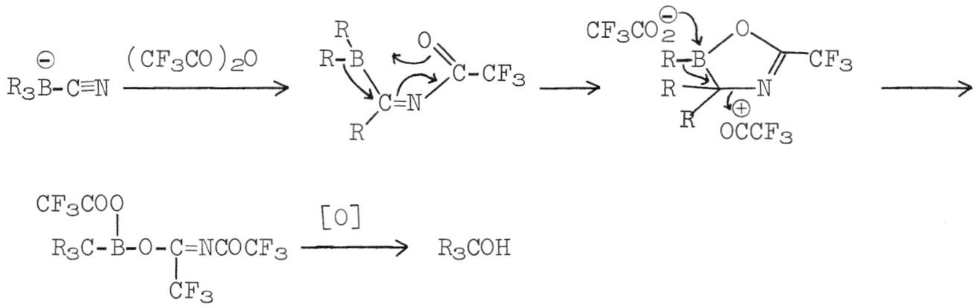

$$\underset{\substack{| \\ CF_3}}{\overset{\substack{CF_3COO \\ |}}{R_3C-B-O-C}}=NCOCF_3 \xrightarrow{[O]} R_3COH$$

The development of convenient syntheses of mixed trialkylboranes (Sec. 2.2.4) widens the scope of the above reactions
considerably.

The ready synthesis of a wide variety of cyclic organoboranes
(Sec. 2.2.3), combined with the carbonylation-oxidation reaction
sequence, provides a useful approach to many cyclic alcohols.
However, in these cases it is usually necessary to carry out the
carbonylation reactions under pressure.

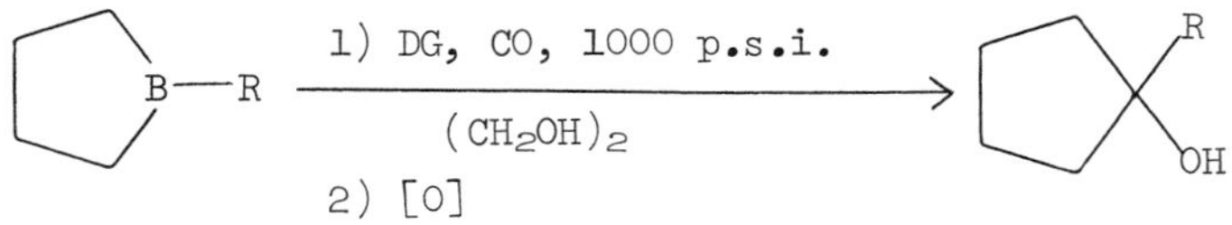

Ref. 31

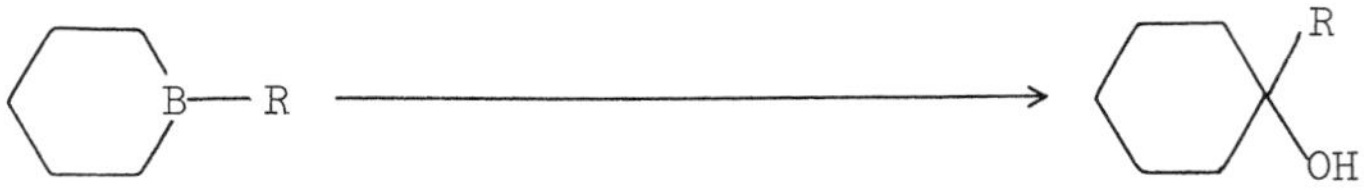

Ref. 32

Ref. 33

89% (cis:trans = 4:1)

Ref. 34

Bromination of trialkylboranes with 2 mole equivalents of bromine in the presence of water, followed by oxidation of the product, gives tertiary alcohols in yields generally exceeding 70% (35a). The reaction proceeds via a series of α-bromination and rearrangement steps (see Sec. 8.3.2).

In the case of bulky alkyl groups, the final migration is sluggish and results in lower yields (<50%) of the tertiary alcohols (35a). In a similar manner dialkylhydroxyboranes containing at least one alkyl group with a tertiary hydrogen atom α to the boron atom are readily converted to the corresponding tertiary alcohols in 30-90% yields (35b).

The above bromination reaction sequences utilize symmetrical trialkylboranes and dialkylhydroxyboranes. Use of thexyldialkylboranes, however, permits the combination of two different alkyl groups, and such groups may contain functional substituents, such as chloro or ester groups (35c).

(89%)

Migration of the thexyl group in the above rearrangement is very slow.

8.3.4 Synthesis of Diols

α-Lithium furan reacts with organoboranes via initial coordination, followed by transfer of two groups from boron to carbon, to give the cyclic borate (15). Oxidation of the borate gives 4,4-dialkyl-cis-2-buten-1,4-diols in good yields (36a).

(15) 60-90%   80-100%
                              (R=n-Alkyl)

1,3-Diols may be prepared by the hydroboration-oxidation of the allyllithium derivatives of 3-arylpropenes as shown below (36b).

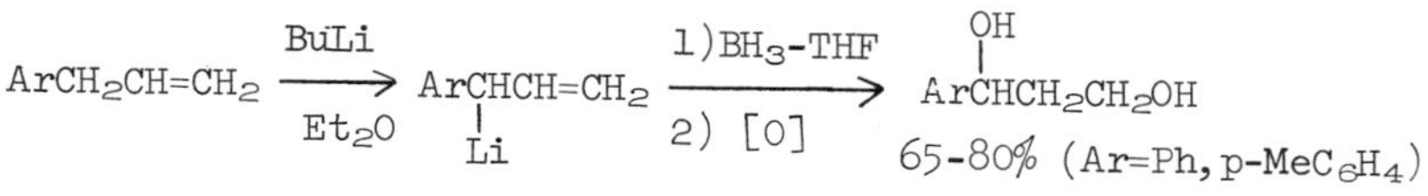

$$ArCH_2CH=CH_2 \xrightarrow[Et_2O]{BuLi} ArCHCH=CH_2 \xrightarrow[2) \ [o]]{1)BH_3-THF} ArCHCH_2CH_2OH$$

with Li on the ArCHCH=CH_2, OH on the product, and 65-80% (Ar=Ph, p-MeC_6H_4)

8.3.5 Synthesis of Quinols

Trialkylboranes react with 1,4-benzoquinone to give the corresponding alkylquinols in high yields (37).

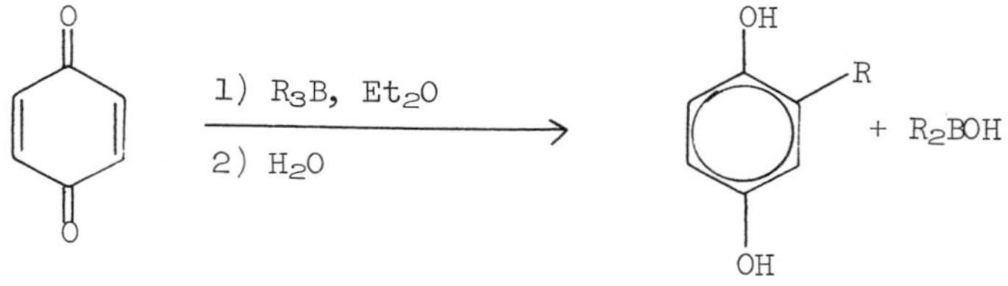

Reagents: 1) R_3B, Et_2O 2) H_2O, giving product + R_2BOH

Similar reaction occurs with triallylborane, but in this case, the major product is reported to be the 2,5-diallyl derivative (38a). Reaction of trialkylboranes with 1,4-naphthaquinone gives the corresponding 2-alkyl-1,4-naphthalenediols (38b).

8.4 SYNTHESIS OF ALDEHYDES

Carbonylation of trialkylboranes in tetrahydrofuran at 0-25° in the presence of lithium trimethoxyaluminumhydride, followed by oxidation, gives the corresponding aldehydes in 87-98% yields (based on R_3B) (39). Use of B-alkyl-9-BBN derivatives gives yields of 50-80% based on starting alkenes (19) while use of a milder reducing agent, such as lithium-tri-tert-butoxyaluminum-hydride, in place of lithium trimethoxyaluminumhydride permits the presence of reducible goups in the alkyl moiety (40). Thus,

good yields of aldehydes containing cyano and ester groups may be
obtained.

$$R\text{-}B\text{-}9\text{-}BBN + CO + LiAlH(\text{tert-BuO})_3 \xrightarrow[\substack{-20 \text{ to} -35°}]{1)\ THF} RCHO$$

$$2)\ [O] \qquad (70\text{-}87\%)$$

(R can contain OCOR
and CN groups)

It is interesting that lithium tri-tert-butoxyaluminumhydride
fails to react with the intermediates formed using normal
trialkylboranes in place of R-B-9-BBN (Sec. 8.1.1), thus reflecting
the less hindered environment at the bridgehead position of
B-R-9-BBN derivatives (40).

The above reaction provide methods for achieving one carbon
atom chain extension. Two carbon atom chain extension may be
achieved by reaction of trialkylboranes with diazo acetaldehyde
(41). In general, the reaction proceeds in yields exceeding 70%
for organoboranes derived from 1-alkenes and for triphenylborane,
but yields are lower for more hindered trialkylboranes, such as
tris(2-butyl)borane.

$$R_3B + N_2CHCHO \xrightarrow[25°]{THF,\ H_2O} RCH_2CHO + R_2BOH$$

$$(33\text{-}98\%;\ \text{based on } R_3B)$$

Unfortunately B-R-9-BBN derivatives cannot be used in the above
reaction due to migration of the B-cyclooctyl bond (Sec. 8.2.2).

Trialkylboranes undergo fast 1,4-addition to propenal in the
presence of water to give the corresponding 3-alkylpropanals (9).

$$R_3B + CH_2=CHCHO \xrightarrow[25°]{THF,\ H_2O} RCH_2CH_2CHO + R_2BOH$$

$$(77\text{-}96\%)$$

This reaction thus provides a method for achieving three carbon
atom chain extension. Unfortunately, B-alkyl-9-BBN derivatives

cannot be used in this reaction, but B-alkylboracyclanes have
been successfully used in reactions with 3-buten-2-one (Sec. 8.5.4)
(17) and should be applicable to the above reaction.

Substituents in the 2-position of propenal do not affect the
reaction adversely, and both 2-methyl- and 2-bromopropenal react
readily (42).

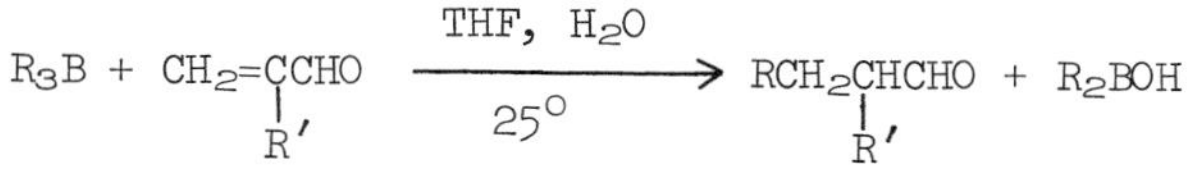

$$R_3B + CH_2=\underset{\underset{R'}{|}}{C}CHO \xrightarrow[25°]{\text{THF, } H_2O} RCH_2\underset{\underset{R'}{|}}{C}HCHO + R_2BOH$$

$$R' = CH_3 \qquad\qquad (> 90\%)$$
$$R' = Br \qquad\qquad (65\text{-}85\%)$$

However, the above reaction fails when a substituent is present
in the 3-position, as in the case of trans-2-butenal (crotonalde-
hyde) (10a), but in the presence of air the reaction proceeds
readily to give the corresponding aldehydes (Sec. 8.1.2) (12).

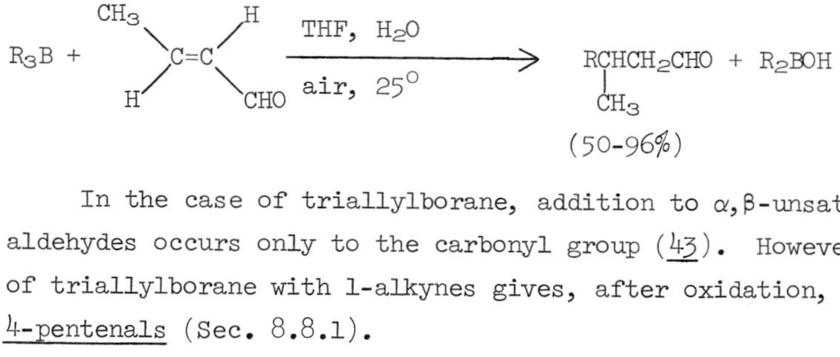

$$R_3B + \quad \underset{H}{\overset{CH_3}{>}}C=C\underset{CHO}{\overset{H}{<}} \xrightarrow[\text{air, } 25°]{\text{THF, } H_2O} R\underset{\underset{CH_3}{|}}{C}HCH_2CHO + R_2BOH$$

$$(50\text{-}96\%)$$

In the case of triallylborane, addition to α,β-unsaturated
aldehydes occurs only to the carbonyl group (43). However, reaction
of triallylborane with 1-alkynes gives, after oxidation, 2-alkyl-
4-pentenals (Sec. 8.8.1).

8.5 SYNTHESIS OF KETONES

A number of very useful methods for the synthesis of ketones
have been developed based on the reaction of trialkylboranes with
a variety of substrates. In the following discussion these
methods have been divided into sections based on the types of
reactions used.

8.5.1 Carbonylation Reactions

Reaction of trialkylboranes with carbon monoxide in diglyme in the presence of a small amount of water, followed by oxidation, gives the corresponding symmetrical ketones (Sec. 8.1.1) (25).

$$R_3B + CO \xrightarrow[\text{2) [o]}]{\text{1) DG, H}_2\text{O, 100}^\circ} R_2CO + ROH$$
$$(80\text{-}90\%)$$

Application of the reaction to mixed trialkylboranes can give a mixture of ketones with the relative yields being dependent on the relative migratory aptitudes of the alkyl groups.

$$R_2R'B \longrightarrow \underset{\underset{HO\ \ OH}{|\ \ \ |}}{RB\text{-}CRR'} + \underset{\underset{HO\ \ OH}{|\ \ \ |}}{R'B\text{-}CR_2} \xrightarrow{[o]} RCOR' + R_2CO$$

In the case of n-alkyldicyclohexylboranes, however, carbonylation in tetrahydrofuran at 45° results in a marked preference for the migration of the alkyl group relative to the cyclohexyl groups, and high yields ($\sim$80%) of the corresponding alkylcyclohexyl ketones are obtained (44). By use of terminal alkenes containing functional groups, a variety of functionally substituted cyclohexyl ketones may be prepared (44).

$$X(CH_2)_nCH=CH_2 \xrightarrow[\text{THF}]{(C_6H_{11})_2BH} X(CH_2)_nCH_2CH_2B(C_6H_{11})_2 \xrightarrow[\text{45}^\circ]{\text{1) CO, THF}}$$

$$(X = OAc,\ CO_2R,\ CN) \qquad\qquad\qquad \text{2) [o]}$$
$$X(CH_2)_nCH_2CH_2COC_6H_{11}$$
$$43\text{-}61\%$$

Baeyer-Villager oxidation of the alkyl cyclohexyl ketones usually proceeds with preferred migration of the cyclohexyl group to form the corresponding cyclohexyl carboxylates which, on hydrolysis, give the functionally substituted alkane carboxylic acids (44).

$$CH_2=CH(CH_2)_8CH_2OH \xrightarrow[\text{2) CO \quad 3) [O]}]{\text{1)}(C_6H_{11})_2BH} C_6H_{11}CO(CH_2)_{10}CH_2OH \xrightarrow{\text{1) } RCO_3H}$$

$$C_6H_{11}O_2C(CH_2)_{10}CH_2OH \xrightarrow{KOH} HO_2C(CH_2)_{10}CH_2OH$$

The use of thexylborane as hydroborating agent permits the synthesis of a variety of <u>unsymmetrical ketones</u> (45a). Thus, reaction of thexylborane with 1 mole equivalent of an alkene gives the monoalkylthexylborane, which can be reacted with a second mole equivalent of another alkene to give the corresponding dialkyl-thexylborane. The hydroboration reactions proceed readily provided the alkenes are not too hindered. However, high pressures are required in the carbonylation of the dialkylthexylboranes due to their hindered nature, but the thexyl group has a low migratory aptitude, thus permitting conversion into the corresponding dialkylketones in yields of 45-90% (45a). As before, a variety of functional substituents are tolerated.

$$\text{Alkene A} \xrightarrow[\text{THF,0}^\circ]{ThBH_2} \underset{\overset{|}{Th}}{R_A}BH \xrightarrow[\text{THF, 0}^\circ]{\text{Alkene B}} R_A R_B BTh \xrightarrow[\substack{70 \text{ atm} \\ 50^\circ}]{CO,H_2O} ThB-\underset{\overset{|}{HO}\ \overset{|}{OH}}{C}R_A R_B$$

$$\xrightarrow{[O]} R_A R_B CO \quad (45\text{-}90\%)$$

The above sequence of reactions is adaptable to the homologation of a wide range of alkenes by four or more carbon atoms with the incorporation of a terminal functional group. This is illustrated by the use of ω-alkenyl acetates as the homologating moieties (45b).

$$\text{Alkene} \xrightarrow[\text{THF}]{ThBH_2} \underset{\overset{|}{Th}}{R}BH \xrightarrow[25^\circ]{H_2C=CH(CH_2)n\text{-}OAc} R\text{-}\underset{\overset{|}{Th}}{B}\text{-}(CH_2)_{n+2}\text{-}OAc$$

$$\xrightarrow[\text{2)}H_2O_2,\ NaOAc]{\text{1) }CO,H_2O,70 \text{ atm.}} \begin{array}{l} RCO\text{-}(CH_2)_{n+2}\text{-}OAc \\ > 60\% \ (n=1\text{-}4) \end{array}$$

It is important to add the more hindered alkene first, followed
by the less hindered one.

The use of thexylborane in the synthesis of cyclic organoboranes from suitable dienes has been discussed in Sec. 6.2.
Carbonylation of the B-thexylboracyclanes provides a convenient
synthesis of cyclic ketones.

(62%)   (17%)

66% (100% trans)

8.5.2 Syntheses using Trialkylcyanoborates

A number of the carbonylation reactions discussed in Sec.
8.5.1 require the use of high reaction temperatures (> 100°)
or high pressures (70 atm). These practical disadvantages have
been overcome by use of trialkylcyanoborates. Reaction of
trialkylcyanoborates with trifluoroacetic anhydride at room
temperature, followed by oxidation, gives the corresponding

symmetrical dialkylketones in high yields (48). The mechanism
of reaction is similar to that proposed for the synthesis of
trialkylcarbinols using trialkylcyanoborates (Sec. 8.3.3).

$$R_3B \xrightarrow[\text{DG, r.t.}]{\text{NaCN}} R_3\overset{\ominus}{B}\text{-CN} \xrightarrow[\text{2) [O]}]{\text{1) } (CF_3CO)_2O, \text{ r.t.}} R_2CO$$

$$84\text{-}100\%$$

Use of benzoyl chloride in place of trifluoroacetic anhydride
gives the ketones in yields of 65-89% (48).

As in the case of carbonylation reactions (Sec. 8.5.1),
thexylborane has been applied to the synthesis of unsymmetrical
ketones and cyclic ketones (49).

$$ThBH_2 \xrightarrow[\text{2) Alkene B}]{\text{1) Alkene A}} R_A R_B BTh \xrightarrow[\substack{\text{2) } (CF_3CO)_2O, \text{ r.t.} \\ \text{3) [O]}}]{\text{1) NaCN}} R_A R_B CO$$

$$76\text{-}95\%$$

1) ThBH$_2$ 2) NaCN

3) (CF$_3$CO)$_2$O, r.t.

4) [O]

95% (1:1 isomer ratio)

The reaction conditions are sufficiently mild to tolerate
functional groups such as the iodo-group (49).

8.5.3 α-Alkylation and Arylation of Ketones

α-Bromo ketones, readily synthesized by the reaction of
ketones with cupric bromide in ethyl acetate-chloroform (50),
react with tri-n-alkylboranes in tetrahydrofuran in the presence
of potassium tert-butoxide to give the corresponding mono-
alkylated ketones in yields of 60-100% (based on R$_3$B) (51).

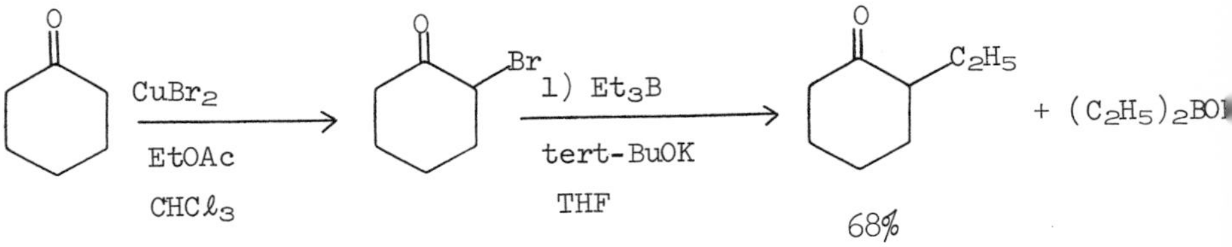

Branched chain organoboranes, such as tri-isobutylborane, fail to react under the above conditions, indicating that the steric environment of the boron atom is an important limiting factor in the reaction. Use of B-alkyl-9-BBN derivatives overcomes this difficulty, once more illustrating the relatively unhindered environment of the bridgehead position in these derivatives (see carbonylation in presence of LiAlH (o-tert-BuO)$_3$; Sec. 8.4)(52).

$$R-B-9-BBN + R'COCH_2Br \xrightarrow[\text{THF, }0^\circ]{\text{tert-BuOK}} R'COCH_2R$$

  (R'=Ph, tert-C$_4$H$_9$)   (30-90%)

Use of B-aryl-9-BBN derivatives (Sec. 8.2.2) gives the α-arylated ketones in yields exceeding 90% (16a)

Attempts to extend the above reactions to α-bromo acetone failed, probably due to the sensitivity of the compound to potassium tert-butoxide. However, alkylation of α-bromo acetone has been achieved using potassium 2,6-di-tert-butylphenoxide (10) (Sec. 8.2.1) as base (2), thus providing a convenient synthesis of a variety of methyl ketones and a method of achieving three carbon atom chain extension.

$$CH_3COCH_2Br + R-B-9-BBN + (10) \xrightarrow[\text{2) EtOH}]{\text{1) THF}} CH_3COCH_2R + EtO-B-9-BBN$$

                                                  (62-80%)

Another useful method for the synthesis of methyl ketones involves the reaction of diazo acetone with trialkylboranes, followed by hydrolysis of the intermediates (53). The reaction proceeds readily at room temperature using tri-n-alkylboranes,

but with more hindered organoboranes, refluxing is required and
yields of the corresponding ketones are lower.

$$R_3B + CH_3COCHN_2 \xrightarrow[\text{2) KOH}]{\text{1) THF}} CH_3COCH_2R + R_2BOH$$
$$36\text{-}98\%$$

The reaction has been applied to the synthesis of α-alkylcyclo-
alkanones in yields exceeding 80% (54), and to the synthesis of
diketones from bisdiazo ketones (55).

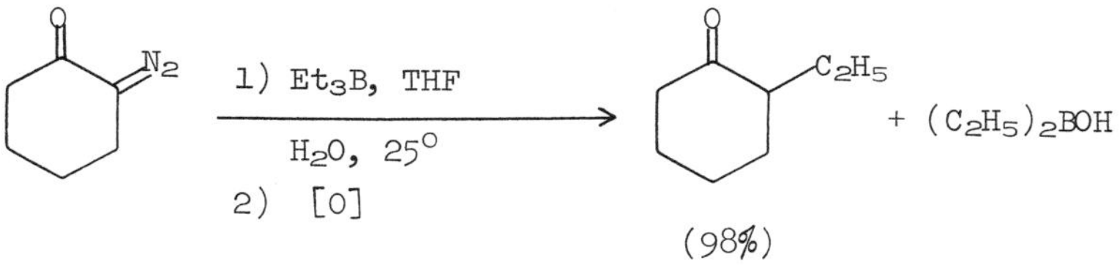

$$N_2CHCO(CH_2)nCOCHN_2 \xrightarrow[\text{H}_2\text{O}]{\text{1) R}_3\text{B,THF,25}^\circ} RCH_2CO(CH_2)nCOCH_2R$$
$$(n=1,2) \qquad\qquad\qquad\qquad (52\text{-}92\%)$$

If the above reactions are carried out in the presence of
deuterium oxide, high yields of the α-monodeuterated ketones are
obtained (4), while similar reactions starting with α-deuterated
diazo ketones give the α,α-dideuterated ketones (4).

$$CH_3COCHN_2 \xrightarrow[\text{D}_2\text{O}]{\text{R}_3\text{B, THF}} CH_3COCHR$$
$$\underset{D}{|} \quad (>90\%)$$

$$CH_3COCDN_2 \xrightarrow[\text{D}_2\text{O}]{\text{R}_3\text{B, THF}} CH_3COCD_2R$$
$$(100\%; \; R=C_6H_{13})$$

Unfortunately, B-R-9-BBN derivatives cannot be used in the
above reactions involving diazo ketones due to predominant
migration of the B-cyclooctyl bond in the transfer step (Sec.
8.2.2) (16b).

   Reaction of the intermediate vinyloxyboranes, formed in the
reaction of trialkylboranes with diazo ketones (Sec. 8.1.1), with
2 mole equivalents of alkyllithium reagents gives the corresponding
lithium enolates, which undergo facile alkylation with alkyl
halides to give α,α-dialkyl ketones ($\underline{56}$).

   $(n-C_4H_9)_3B + N_2CHCOPh \xrightarrow{\text{THF}} \begin{array}{c} (n-C_4H_9)_2B\text{O} \\ \diagdown \\ \text{Ph} \end{array} C=CHC_4H_9$

$\xrightarrow{n-C_4H_9Li} \begin{array}{c} (n-C_4H_9)_3\overset{\ominus}{B}\text{O} \quad Li^{\oplus} \\ \diagdown \\ Ph \end{array} C=CHC_4H_9 \rightleftharpoons \begin{array}{c} LiO \\ \diagdown \\ Ph \end{array} C=CHC_4H_9 + (C_4H_9)_3B$

(RX = CH$_3$I : 72%)

(RX = PhCH$_2$Cl : 47%)

with RX arrow: PhCO-$\underset{R}{\overset{|}{C}}HC_4H_9$

with C$_4$H$_9$Li arrow: $(C_4H_9)_4\overset{\ominus}{B}Li^{\oplus}$

1)distil.
2)2BuLi

CH$_3$I

+ $(C_4H_9)_4\overset{\ominus}{B}Li^{\oplus}$

(61%)

8.5.4 1,4-Addition to α,β-Unsaturated Ketones

   Trialkylboranes undergo fast 1,4-addition to 3-buten-2-one
(methyl vinyl ketone) to give, after hydrolysis, the corresponding
methyl ketones in yields of 50-86% ($\underline{8}$); this reaction thus provides
a method for achieving four carbon atom chain extension.

$$R_3B + CH_2=CHCOCH_3 \xrightarrow[25°]{1)\ THF,\ H_2O} RCH_2CH_2COCH_3 + R_2BOH$$
$$(50-86\%)$$

Attempted reaction with β-substituted enones, such as trans-3-penten-2-one and cyclic enones, under similar reaction conditions, fails (10a), but in the presence of air or radical chain initiators, the reaction proceeds readily to give the corresponding ketones (11,12).

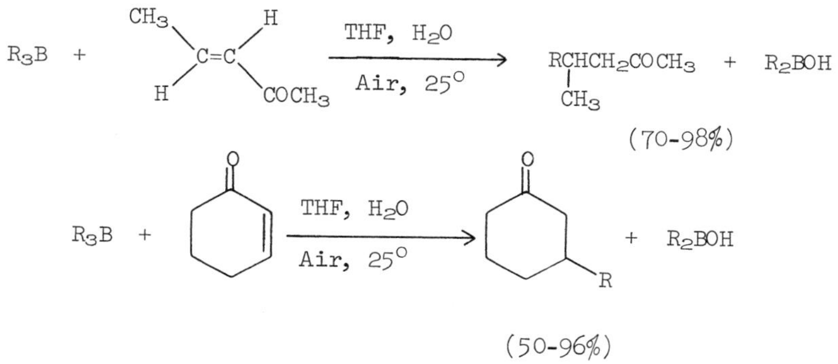

$$R_3B + \underset{H}{\overset{CH_3}{\diagdown}}C=C\underset{COCH_3}{\overset{H}{\diagup}} \xrightarrow[Air,\ 25°]{THF,\ H_2O} RCHCH_2COCH_3 + R_2BOH$$
$$\underset{CH_3}{|}$$
$$(70-98\%)$$

$$R_3B + \text{(cyclohexenone)} \xrightarrow[Air,\ 25°]{THF,\ H_2O} \text{(cyclohexanone-R)} + R_2BOH$$
$$(50-96\%)$$

As in the case of 1,4-addition reactions to propenal, B-alkyl-9-BBN derivatives cannot be used in the above reactions due to preferential migration of the boron-cyclooctyl bond (Sec. 8.2.2). However, reaction with B-alkylboracyclanes, such as B-alkylboracyclohexane, proceeds with migration of the boron-alkyl bond to give the corresponding ketones in yields generally exceeding 70% (Sec. 8.2.2) (17).

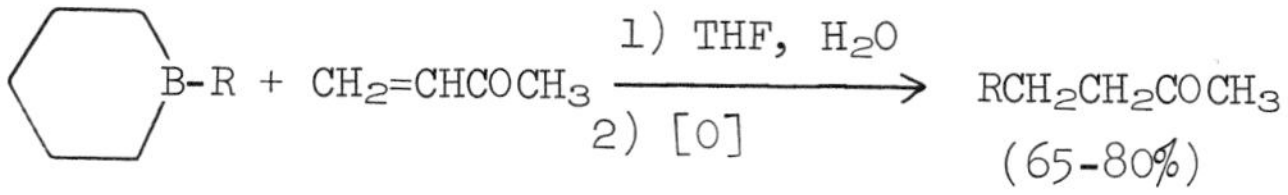

$$\text{(boracyclohexane)}B-R + CH_2=CHCOCH_3 \xrightarrow[2)\ [O]]{1)\ THF,\ H_2O} RCH_2CH_2COCH_3$$
$$(65-80\%)$$

B-tert-Alkylboracyclanes, which are prepared by the reaction of B-methoxyboracyclanes with tert-alkyllithium reagents, also react readily to give the corresponding tert-alkyl substituted ketones (17).

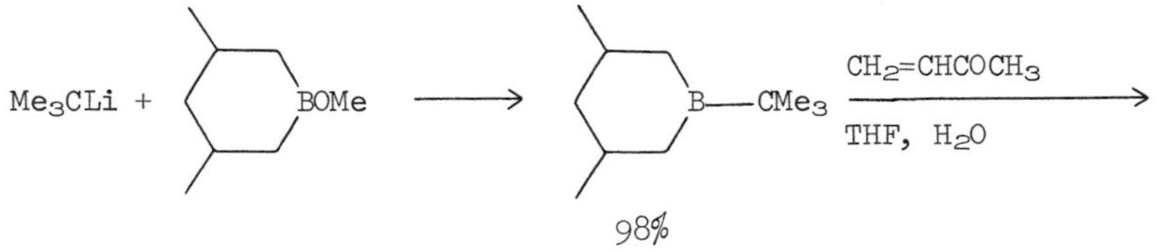

Me_3CLi + [structure: BOMe] $\longrightarrow$ [structure: B—CMe_3] $\xrightarrow{\begin{array}{c}CH_2=CHCOCH_3\\ \hline THF,\ H_2O\end{array}}$

98%

$Me_3CCH_2CH_2COCH_3$

76%

Cyclic organoboranes, derived from the hydroboration of dienes (Sec. 6.2), also react with 3-buten-2-one to give, after oxidation, ω-hydroxy ketones (57).

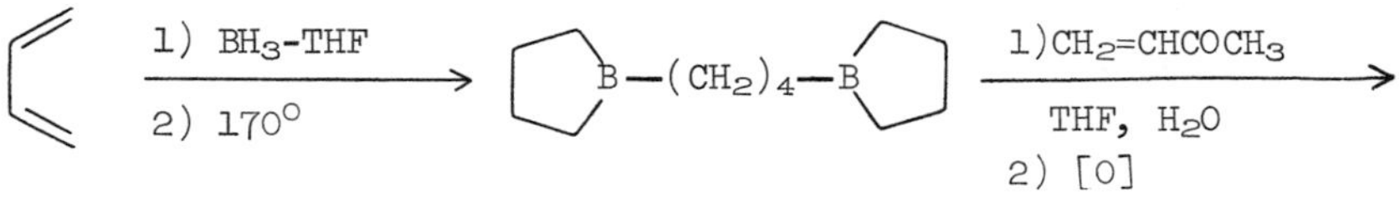

[structure: diene] $\xrightarrow[\text{2) } 170°]{\text{1) } BH_3\text{-THF}}$ [structure: B—(CH_2)_4—B] $\xrightarrow[\begin{array}{c}THF,\ H_2O\\ \text{2) [o]}\end{array}]{\text{1)}CH_2=CHCOCH_3}$

$HO(CH_2)_6COCH_3$

If the above 1,4-addition reactions are carried out in the presence of deuterium oxide in place of water, deuterolysis of the intermediate vinyloxyboranes (Sec. 8.1.2) occurs to give α-deuterated carbonyl compounds (4).

$(C_2H_5)_3B$ + $CH_2=CHCOCH_3$ $\xrightarrow{THF,\ D_2O}$ $C_2H_5CH_2\underset{D}{\overset{|}{C}HCOCH_3}$

(93% d_1)

Quaternization of Mannich bases derived from a variety of cyclic ketones results in in situ formation of the unstable α,β-unsaturated ketones, which react readily with organoboranes to give the corresponding α-alkylated ketones (58).

$$(60\text{-}90\%)$$

Treatment of the intermediate vinyloxyborane, formed by 1,4-addition of tri-n-butylborane to 3-buten-2-one (Sec. 8.1.2), with butyllithium and methyl iodide, gives the α-methylated ketone as the major product. This reaction thus provides a method of synthesis of α,β-dialkylated ketones from α,β-unsaturated ketones (56).

$$(n\text{-}C_4H_9)_3B + CH_2=CHCOCH_3 \xrightarrow[\text{air}]{\text{THF, } H_2O} C_4H_9CH_2CH=\underset{\underset{CH_3}{|}}{C}\text{-}OB(C_4H_9)_2$$

$$\xrightarrow[\text{2) BuLi}]{\text{1) Distil}} C_4H_9CH_2CH=\underset{\underset{CH_3}{|}}{C}\text{-}OLi \xrightarrow{CH_3I} C_4H_9CH_2\underset{\underset{CH_3}{|}}{C}HCOCH_3 + C_6H_{13}COC_2H_5$$

$$59\% \qquad 11\%$$

α-β-Unsaturated ketones may be prepared by the reaction of acetylethyne with trialkylboranes in the presence of catalytic amounts of oxygen (59). In each case a mixture of cis- and trans-isomers is obtained with the cis-isomer predominating. Further reaction of the resultant α,β-unsaturated ketone with a different organoborane provides a convenient synthetic route to a wide variety of ketones (59).

$$HC\equiv CCOCH_3 \xrightarrow[\text{H}_2O, \text{ air}]{R_3B, \text{ THF}} RCH=CHCOCH_3 \xrightarrow[\text{H}_2O, \text{ air}]{R'_3B, \text{ THF}} RR'CHCH_2COCH_3$$

$$(34\text{-}77\%)$$
$$(\text{Mainly cis})$$

In the case of triallylborane, addition to α,β-unsaturated ketones occurs only to the carbonyl group (60a). Reaction of triallylborane with alkoxyalkynes, however, gives 4-penten-2-ones (Sec. 8.8.1).

Enolizable β-diketones react with trialkylboranes under photochemical conditions to give mixtures of β-alkylated-β-hydroxy ketones and β-alkylated-α,β-unsaturated ketones (60b). The reaction probably proceeds via conjugate addition of the trialkylborane to the enol form of the diketone.

15%   35% (cis+trans)

Similar reaction has been observed with enolizable β-keto esters to give β-alkylated-β-hydroxy esters (60b). The reaction is not promoted by the presence of oxygen, and fails to give the desired products with non-enolizable β-keto esters.

8.5.5 Miscellaneous Methods of Ketone Synthesis

α-Bromination-rearrangement reactions have been applied to the synthesis of secondary (Sec. 8.3.2) and tertiary alcohols (Sec. 8.3.3) from trialkylboranes. Bromination of n-butoxy-n-hexylmethylborane with 2 mole equivalents of bromine proceeds in a similar manner to give, after oxidation, 2-heptanone (61).

$$C_5H_{11}CH_2\overset{|}{\underset{CH_3}{B}}OC_4H_9 \xrightarrow[CCl_4,\ h\nu]{Br_2} C_5H_{11}\overset{Br}{\underset{CH_3}{CH}}\overset{|}{B}OC_4H_9 \xrightarrow{Br_2} C_5H_{11}\overset{Br}{\underset{CH_3}{C}}\overset{Br}{-B}OC_4H_9$$

$$\xrightarrow{[O]} C_5H_{11}COCH_3$$

70%

Low yields of phenyl ketones have been obtained by reaction of the diarylchlorovinyllithium compounds (16) with triphenyl-borane, followed by oxidation of the intermediate vinylboranes (17) (26).

$$Ar_2C=CHCl \xrightarrow[\substack{THF \\ -90°}]{BuLi} Ar_2C=C\overset{Li}{\underset{Cl}{\diagdown}} \xrightarrow{Ph_3B} Ar_2C=\overset{|}{\underset{BPh_2}{C}}Ph \xrightarrow{[O]}$$

$$(16) \qquad\qquad (17)$$

$Ar_2CHCOPh + ArC\equiv CAr + Ar_2CO$

~20% (Ar = Ph, p-CH$_3$C$_6$H$_4$)

Substantial yields of diarylalkynes and diaryl ketones are also isolated.

8.6 SYNTHESIS OF ESTERS

8.6.1 α-Alkylation and Arylation of Esters

Organoboranes react readily with ethyl bromoacetate in tetrahydrofuran at 0° under the influence of potassium tert-butoxide to give the corresponding esters in yields exceeding 80% (based on R$_3$B) (62). Reaction with ethyl dihaloacetates gives the corresponding α-halocarboxylic acid esters, which can be reacted further with a different organoborane to give dialkylacetate derivatives (63).

$$R_3B + Br_2CHCO_2C_2H_5 \xrightarrow[\substack{tert\text{-}BuOK \\ tert\text{-}BuOH}]{THF} \underset{\substack{| \\ Br \\ (80\text{-}98\%)}}{RCHCO_2C_2H_5} \xrightarrow[\substack{tert\text{-}BuOK \\ tert\text{-}BuOH}]{R_3B} RR'CHCO_2C_2H_5$$

(47-90%)

The use of hindered trialkylboranes, such as tricyclohexylborane, in the second stage of the above reaction gives low yields of the dialkylated esters (63). Use of B-alkyl or B-aryl-9-BBN derivatives in the above reactions gives good yields (50-90%) of the corresponding esters (16a,64).

In all the above reactions use of excess base must be carefully avoided due to the extreme sensitivity of esters to potassium tert-butoxide. However, esters are stable to potassium 2,6-di-tert-butylphenoxide (10) in tetrahydrofuran (Sec. 8.2.1), and hence the use of this base in the above reactions is preferable (15). tert-Butanol must be present in the reaction mixture to ensure rapid protonolysis of the intermediate borane (Sec. 8.2.1).

$$R\text{-}B\text{-}9\text{-}BBN + BrCH_2CO_2C_2H_5 + (10) \xrightarrow[\substack{tert\text{-}BuOH \\ 25^\circ}]{THF,} RCH_2CO_2C_2H_5$$
$$(50\text{-}83\%)$$

Reaction of organoboranes with ethyl bromoacetate provides a means of achieving two carbon atom chain extension. This may also be achieved by reaction of trialkylboranes with ethyl diazo acetate (65a). With hindered organoboranes, refluxing of the reaction mixture is required, and yields of the corresponding esters are generally lower (65a).

$$R_3B + N_2CHCO_2C_2H_5 \xrightarrow[\substack{2)\ H_2O}]{1)\ THF} RCH_2CO_2C_2H_5 + R_2BOH$$
$$(40\text{-}83\%)$$

Use of dialkylchloroboranes in place of trialkylboranes results in ready reaction at -78° even when bulky alkyl groups are present (65b).

$$R_2BCl + N_2CHCO_2C_2H_5 \xrightarrow[\substack{2)CH_3OH,\ H_2O}]{1)Et_2O,\ -78^\circ} RCH_2CO_2C_2H_5 + RB(OCH_3)_2$$
$$(89\text{-}98\%)$$

As in the case of the reactions of organoboranes with other α-diazo compounds (Sec. 8.2.2), use of B-alkyl-9-BBN derivatives results in predominant B-cyclooctyl bond migration (16b). Hydrolysis of the intermediates with deuterium oxide gives the corresponding α-deuterated esters, while use of the deuterio diazo ester gives the α,α-dideuterated esters (4).

$$R_3B + N_2CDCO_2C_2H_5 \xrightarrow[\text{2) D}_2\text{O}]{\text{1) THF}} RCD_2CO_2C_2H_5$$

α-Alkylated acetate esters may also be prepared by the reaction of ethyl (dimethylsulfuranylidene)acetate (18) with trialkylboranes (3).

$$R_3B + (CH_3)_2\overset{\oplus\ominus}{S}CHCO_2C_2H_5 \xrightarrow[\text{2) [O]}]{\text{1) THF}} RCH_2CO_2C_2H_5 + 2ROH$$
$$\quad\quad\quad (18) \quad\quad\quad\quad\quad\quad\quad\quad 31\text{-}52\%$$

8.6.2 Alkylation of Ethyl 4-Bromo-2-Butenoate

Trialkylboranes react readily with ethyl 4-bromo-2-butenoate in tetrahydrofuran under the influence of potassium 2,6-di-tert-butylphenoxide (10) to give the corresponding unsaturated esters in which the double bond is in the 3-position and is trans (66).

R₃B + BrCH₂CH=CHCO₂C₂H₅ + (10) $\xrightarrow[0°]{\text{THF}}$

(70-90%)

This reaction thus provides a method of achieving <u>four carbon atom chain extension</u>.

8.7 SYNTHESIS OF NITRILES

Reaction of chloroacetonitrile with trialkylboranes in tetrahydrofuran under the influence of potassium 2,6-di-tert-butylphenoxide (10) gives the corresponding nitriles (2). The reaction proceeds readily with B-alkyl-9-BBN derivatives (2).

$$R\text{-}B\text{-}9\text{-}BBN + C\ell CH_2CN + (10) \xrightarrow[0^\circ]{THF} RCH_2CN \atop (57\text{-}77\%)$$

Use of dichloroacetonitrile permits the introduction of two different alkyl groups as shown below (67).

$$R\text{-}B\text{-}9\text{-}BBN + Cl_2CHCN \xrightarrow[25^\circ]{THF,(10)} \underset{(70\text{-}90\%)}{\underset{Cl}{\overset{|}{R}CHCN}} \xrightarrow[THF, tert\text{-}BuOH]{R'\text{-}B\text{-}9\text{-}BBN,\,(10)} RR'CHCN \atop (46\text{-}97\%)$$

The dialkylation proceeds readily for two primary alkyl groups, but introduction of more hindered groups requires refluxing in tetrahydrofuran for up to 24 hr (67). The above reactions fail in the presence of potassium tert-butoxide as base (2).

Reaction of trialkylboranes with diazo acetonitrile gives the corresponding nitriles, with the reaction proceeding slowly and in relatively low yield with hindered trialkylboranes (65a).

$$R_3B + N_2CHCN \xrightarrow[2)\,KOH]{1)\,THF} RCH_2CN + R_2BOH \atop (50\text{-}100\%)$$

Trialkylboranes also react with ethyl bromocyanoacetate and bromodicyanomethane (bromomalononitrile) under the influence of potassium 2,6-di-tert-butylphenoxide (10) to give α-cyano esters and 1,1-dicyano compounds respectively (68).

$$R\text{-}B\text{-}9\text{-}BBN + \underset{\underset{CN}{|}}{Br}CHCO_2C_2H_5 \xrightarrow[\text{tert-BuOH}]{THF, (10)} \underset{\underset{CN}{|}}{R}CHCO_2C_2H_5$$

$$94\% \ (R=C_2H_5); \ 48\% \ (R=i\text{-}C_4H_9)$$

$$R\text{-}B\text{-}9\text{-}BBN + BrCH(CN)_2 \xrightarrow[\text{tert-BuOH}]{THF, (10)} RCH(CN)_2$$

$$> 80\%$$

8.8 SYNTHESIS OF ALKENES, DIENES AND DERIVATIVES

The synthesis of alkenes and dienes via the hydroboration of alkynes has been discussed in Chap. 7.

8.8.1 Addition Reactions of Allylboranes

Triallylborane reacts with alkynes and alkenes via addition accompanied by allylic rearrangement to give, after protonolysis or oxidation, 1,4-pentadienes or unsaturated carbonyl compounds.

The reaction of triallylborane with 1-alkynes proceeds slowly at 20° to give mono-addition products (19), which slowly cyclize to give 3-alkyl-1,5-diallyl-1-bora-2-cyclohexenes (20) (69-71), Faster reaction occurs at 70° to give (20) (69). while at 140° 3-borabicyclo[3.3.1]non-6-ene derivatives (21) are the major products (72a).

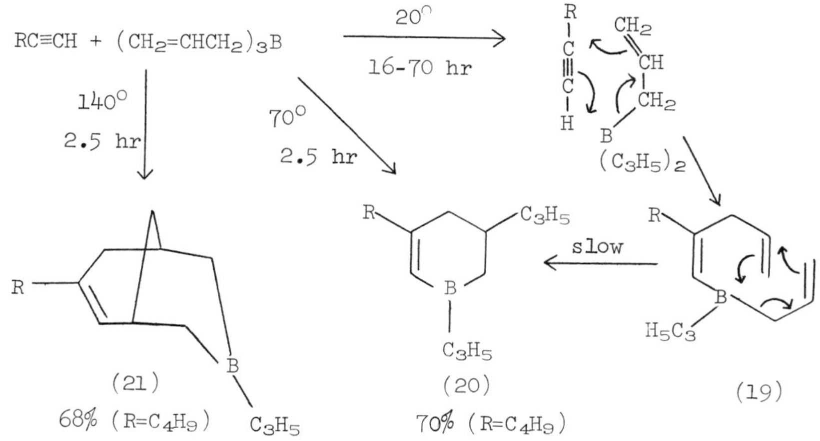

$$RC\equiv CH + (CH_2=CHCH_2)_3B \xrightarrow[\text{16-70 hr}]{20^\circ}$$

140°
2.5 hr

70°
2.5 hr

slow

(21)
68% (R=C₄H₉)

(20)
70% (R=C₄H₉)

(19)

Adducts (19) give <u>2-alkyl-1,4-pentadienes</u> on treatment with
methanol and acetic acid (<u>70</u>), while oxidation gives 2-alkyl-4-
pentenals (<u>70</u>).

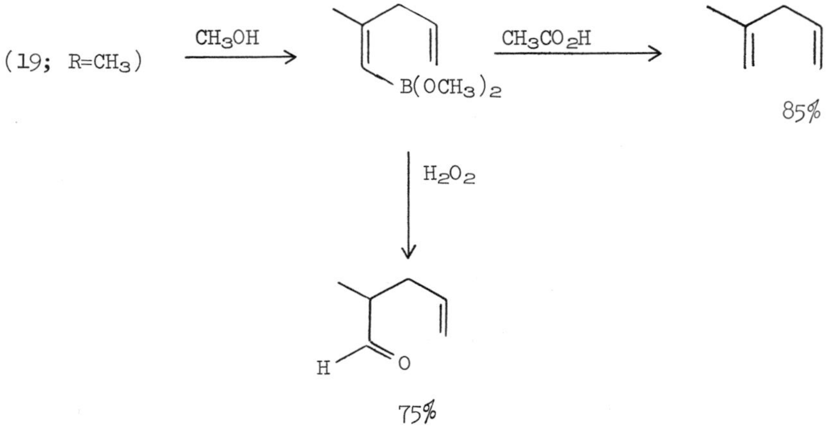

(19; R=CH₃) $\xrightarrow{\text{CH}_3\text{OH}}$ [structure with B(OCH₃)₂] $\xrightarrow{\text{CH}_3\text{CO}_2\text{H}}$ [structure]

85%

$\downarrow$ H₂O₂

[aldehyde structure]
H   O

75%

The 1,5-dimethyl derivative of (21), which is formed by the
reaction of ethyne with tris(methallyl)borane at 140°, on
treatment with methanol followed by acetic acid, gives 3,3,5,5-
tetramethylcyclohexene (<u>72b</u>).

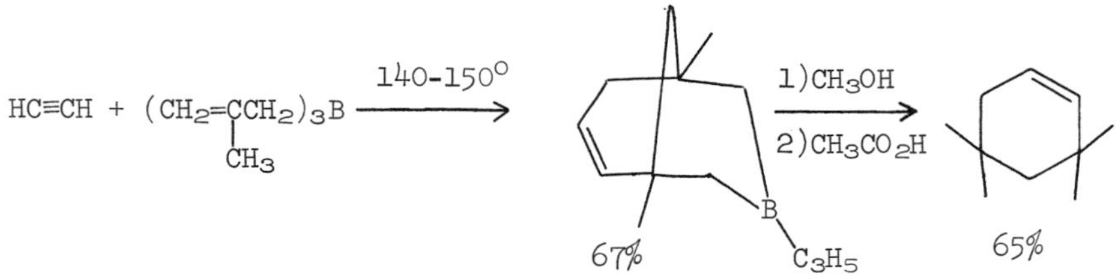

HC≡CH + (CH₂=CCH₂)₃B $\xrightarrow{\text{140-150}°}$ [bicyclic structure] $\xrightarrow[\text{2)CH}_3\text{CO}_2\text{H}]{\text{1)CH}_3\text{OH}}$ [cyclohexene structure]
           |
          CH₃

67%   B   65%
       C₃H₅

Reaction of triallylborane with <u>alkoxyalkynes</u> proceeds
readily at room or lower temperatures, with the nature of the
products depending on the relative amounts of reactants used (<u>70</u>).

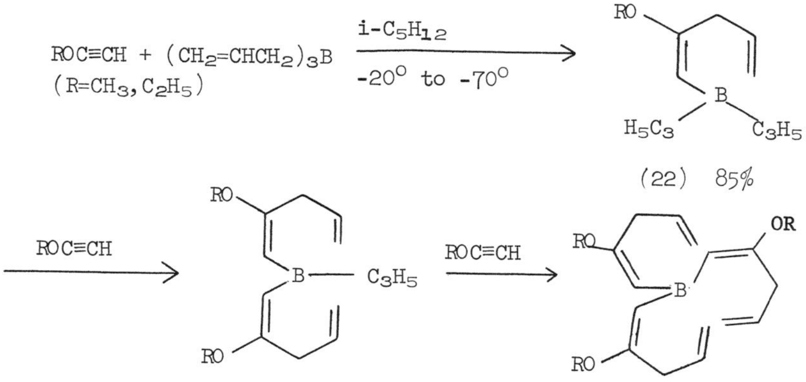

$$ROC\equiv CH + (CH_2=CHCH_2)_3B \xrightarrow[\text{-20}^\circ \text{ to -70}^\circ]{\text{i-C}_5\text{H}_{12}}$$
$$(R=CH_3, C_2H_5)$$

(22) 85%

$$ROC\equiv CH \longrightarrow$$

(23) 80%

$$ROC\equiv CH \longrightarrow$$

(24) 65%

Treatment of adducts (22) or (23) with water or alcohols gives the corresponding 2-alkoxy-1,4-pentadienes, while acid hydrolysis gives methyl allyl ketone (70).

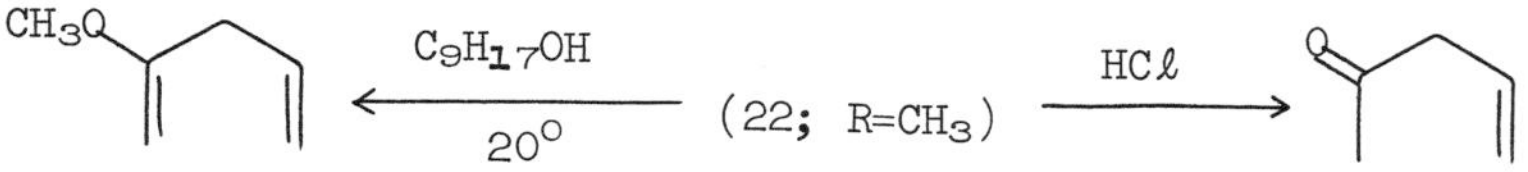

$$\xleftarrow[\text{20}^\circ]{\text{C}_9\text{H}_{17}\text{OH}} (22;\ R=CH_3) \xrightarrow{\text{HC}\ell}$$

92%

~50%

Similar reactions occur using 3-substituted allylic boranes such as tris (2-butenyl)borane to give 2-alkoxy-3-alkyl-1,4-pentadienes (70).

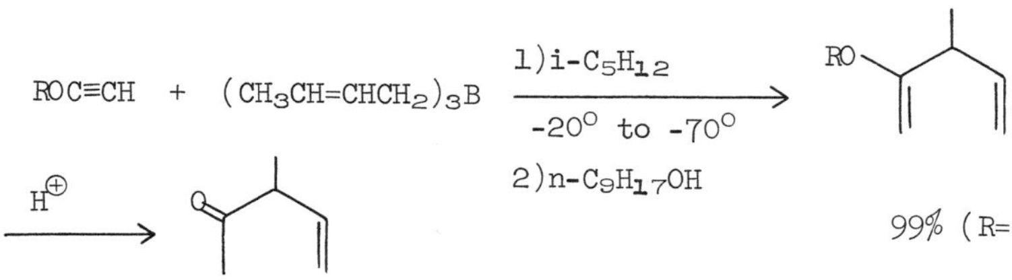

$$ROC\equiv CH + (CH_3CH=CHCH_2)_3B \xrightarrow[\text{-20}^\circ \text{ to -70}^\circ]{\begin{array}{l}1)\text{i-C}_5\text{H}_{12}\\ 2)\text{n-C}_9\text{H}_{17}\text{OH}\end{array}}$$

$$\xrightarrow{\text{H}^\oplus}$$

99% (R=C$_2$H$_5$)

Allylic boranes also react with vinyl ethers at temperatures in the region of 100° to give penta-1,4-dienes (73). The diene formation involves a β-elimination process (Sec. 5.1).

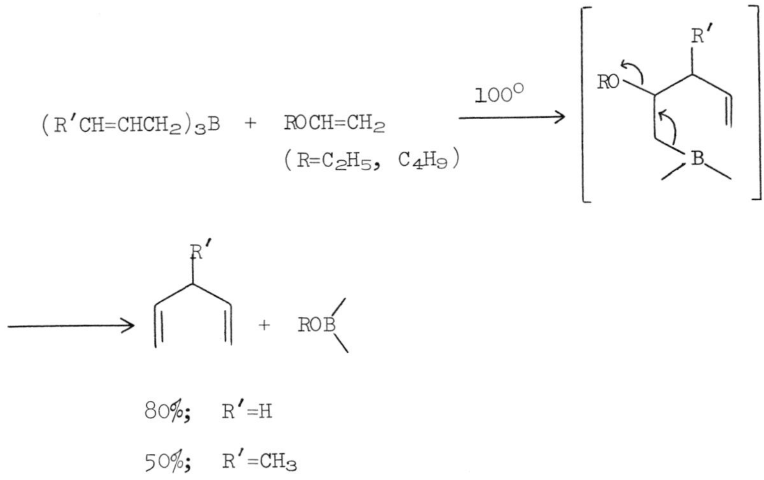

$$(R'CH=CHCH_2)_3B \ + \ ROCH=CH_2 \xrightarrow{\ 100° \ }$$

$$(R=C_2H_5, \ C_4H_9)$$

$$\longrightarrow \quad + \ ROB$$

80%; R'=H

50%; R'=CH₃

With tris(methallyl)borane, 2-methyl-1,4-pentadiene is obtained (73).

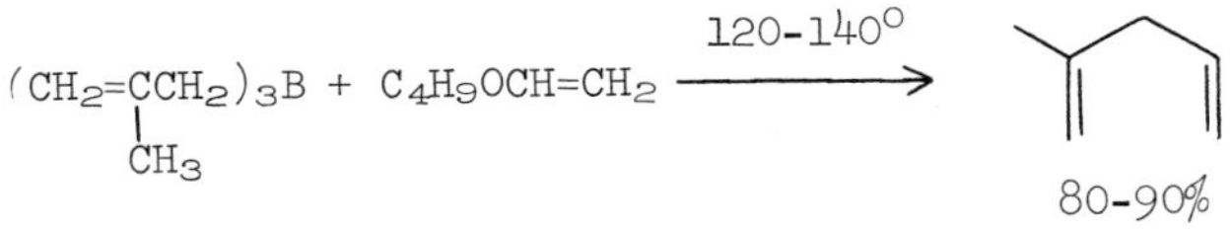

$$(CH_2=CCH_2)_3B + C_4H_9OCH=CH_2 \xrightarrow{\ 120-140° \ }$$
$$\quad\quad\quad\ |$$
$$\quad\quad\ CH_3$$

80-90%

The reaction has also been applied to dihydrofurans as shown below (73).

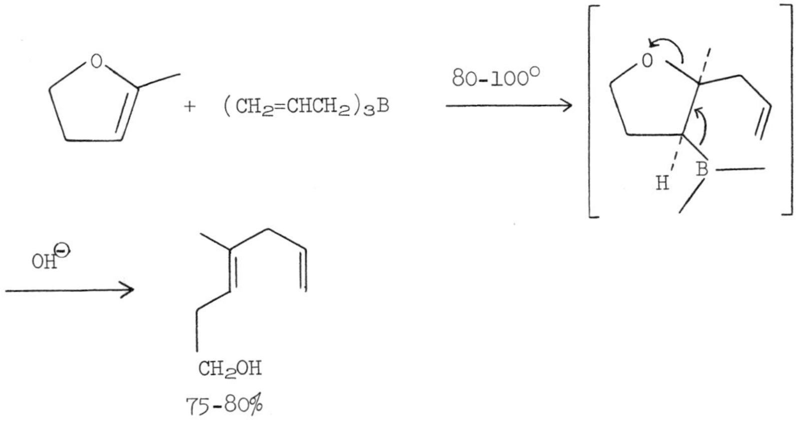

$$+ \ (CH_2=CHCH_2)_3B \xrightarrow{\ 80-100° \ }$$

$$\xrightarrow{\ OH^{\ominus} \ }$$

CH₂OH
75-80%

Reaction of allylic boranes with 1-methylcyclopropene gives allyl-substituted cyclopropane derivatives (74).

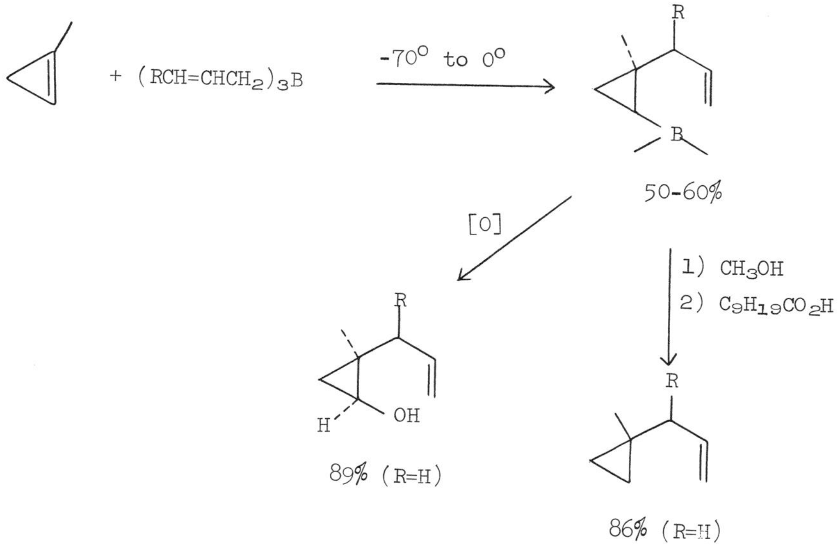

$$+ (RCH=CHCH_2)_3B \xrightarrow{-70^\circ \text{ to } 0^\circ}$$

50-60%

[O]

1) CH_3OH
2) $C_9H_{19}CO_2H$

89% (R=H)

86% (R=H)

8.8.2 Reaction of Trialkylboranes with Dichlorocarbene

Reaction of equimolar quantities of phenyl(bromodichloromethyl) mercury with trialkylboranes derived from C_n-terminal alkenes proceeds with precipitation of phenylmercuric bromide, and gives, after hydrolysis, the corresponding C_{2n+1} internal alkenes in yields of 58-68% (based on the mercury compound used) (75).

$$(RCH_2CH_2)_3B + PhHgCCl_2Br \xrightarrow[\substack{60-70^\circ \\ 2) \ H_2O}]{1) \ PhH} RCH_2CH_2CH=CHCH_2R + PhHgBr$$

$(R=C_3H_7, \ C_4H_9, \ C_6H_{13})$

58-68%

The ratio of cis- and trans-isomers varies depending on the nature of the trialkylborane used in the reaction. The following mechanism has been proposed (75).

$$(RCH_2CH_2)_3B + PhHgCC\ell_2Br \longrightarrow (RCH_2CH_2)_2\overset{\ominus}{B}-\overset{\oplus}{C}C\ell_2$$
$$RCH_2CH_2$$

$$\longrightarrow RCH_2CH_2\overset{\text{Cl}}{\underset{\underset{RCH_2CH_2}{|}}{B}}{-}\overset{\text{Cl}}{\underset{\text{Cl}}{C}}{-}CH_2CH_2R \longrightarrow RCH_2CH_2\overset{\text{Cl}}{\underset{\text{Cl}}{B}}{-}\overset{}{C}(CH_2CH_2R)_2$$

$$\longrightarrow RCH_2CH_2BC\ell_2 + RCH_2CH_2\overset{..}{C}CH_2CH_2R \longrightarrow RCH_2CH_2CH=CHCH_2R$$

8.8.3 Contrathermodynamic Isomerization of Alkenes

The synthetic usefulness of the hydroboration-isomerization-displacement sequence of reactions as a means of shifting a double bond from a thermodynamically stable internal position to the less stable terminal position has already been mentioned (Sec. 2.3.2 and 2.3.3). While this process cannot be regarded as a direct means of synthesis of terminal alkenes, it does provide a convenient synthetic route to such alkenes. The process has been applied to a variety of acyclic alkenes (76), and to the conversion of endocyclic to exocyclic alkenes (77).

$$CH_3CH=C-C_2H_5 \quad \xrightarrow[\text{3) 1-Decene, }\Delta]{\text{1)BH}_3\text{DG 2) }\Delta} \quad CH_2=CHCH(C_2H_5)_2$$
$$C_2H_5 \qquad\qquad\qquad\qquad\qquad 82\% \quad (98\% \text{ pure})$$

62% + endocyclic isomers

~30%

54% + endocyclic isomers

8.9   COUPLING REACTIONS OF ORGANOBORANES

Hydroboration of alkenes followed by in situ treatment with potassium hydroxide and silver nitrate gives coupled products ($\underline{78}$). With terminal alkenes the yields are 60-80%, while with more hindered internal alkenes yields of 35-50% are obtained.

$$C_4H_9CH=CH_2 \quad \xrightarrow[\substack{2)\ KOH,\ AgNO_3,\ H_2O \\ 0^\circ}]{1)\ BH_3, \qquad DG} \quad \substack{C_{12}H_{26} \\ 71\%}$$

$$\substack{C_3H_7C=CH_2 \\ |\ \ \\ CH_3} \quad \xrightarrow{\hspace{3cm}} \quad \substack{C_3H_7CHCH_2CH_2CHC_3H_7 \\ |\qquad\quad\ | \\ CH_3\qquad\ CH_3}$$

$$61\%$$

When applied to the coupling of two different alkyl groups, statistical yields of the three possible products are obtained using terminal alkenes, while, in the case of alkenes forming secondary alkylboranes, the yield of the desired coupled product drops ($\underline{79}$). However, use of a large excess of the more readily available alkene ensures a more complete conversion of the less readily available alkene into the desired product ($\underline{79}$).

$$\underset{\substack{3\ mole.}}{C_4H_9CH=CH_2} + \underset{\substack{| \\ CH_3 \\ 1\ mole}}{CH_3CH_2C=CH_2} \quad \xrightarrow[\substack{2)KOH,AgNO_3 \\ H_2O,\ CH_3OH}]{1)BH_3,DG} \quad \left[\substack{C_2H_5CCH_2 \\ |\ \ \ \\ CH_3}\right]_2 \ +$$

$$4\%$$

$$\substack{C_2H_5CHCH_2C_6H_{13} \\ |\ \ \ \\ CH_3} \quad + \quad C_{12}H_{26}$$

$$22\% \qquad\qquad\qquad 36\%$$

Similar reaction of the organoborane derived from longifolene (25) proceeds via autoxidation of the organoborane (Sec. 4.1) followed by transannular radical transfer, and results in the formation of the epimeric 3-alcohols, together with the 15-alcohol (80).

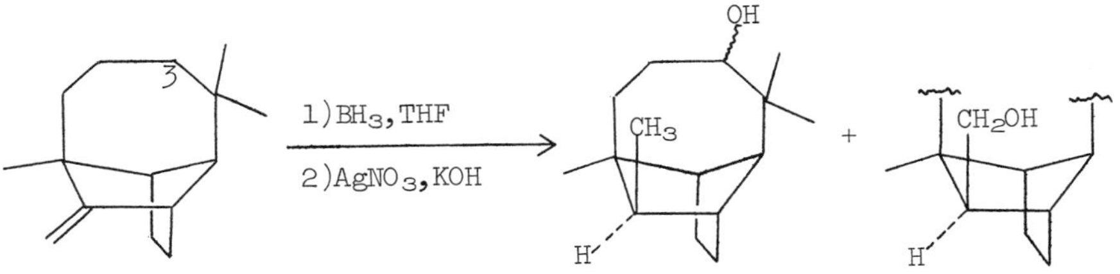

(25)   20-30%   30-35%

Reaction of benzylic or allylic iodides with a simple trialkylborane, such as triethylborane, in the presence of air, gives coupled products (81). These products are formed by a transfer mechanism as shown below.

$$(C_2H_5)_3B + O_2 \longrightarrow (C_2H_5)_2BO_2\cdot + C_2H_5\cdot \qquad (1)$$

$$C_2H_5\cdot + R\text{-}I \longrightarrow C_2H_5I + R\cdot \qquad (2)$$

$$R\cdot + R\cdot \longrightarrow R\text{-}R \qquad (3)$$

$$R\cdot + O_2 \longrightarrow RO_2\cdot \qquad (4)$$

$$RO_2\cdot + (C_2H_5)_3B \longrightarrow RO_2B(C_2H_5)_2 + C_2H_5\cdot \qquad (5)$$

$$C_2H_5\cdot + O_2 \longrightarrow C_2H_5O_2\cdot \qquad (6)$$

Control of the amount of oxygen promotes Reaction (2) thus providing a convenient method of synthesis of alkyl iodides (Sec. 9.1.3). However, in order to obtain high yields of the coupled product, R-R [Reaction (3)], the reaction of R· with oxygen [Reaction (4)] must be minimized. This is achieved by using a 100% excess of triethylborane, thereby ensuring complete utilization of the oxygen by this compound [Reaction (6)] (81).

$$RI + (C_2H_5)_3B \xrightarrow[\text{air (50 ml/min)}]{\text{THF, 45 min.}} R\text{-}R$$

10 mmole   20 mmole   87%;   R=PhCH$_2$

                                           97%;   R=CH$_2$-CH=CH$_2$

Mixed coupling products may be obtained using a mixture of allylic and benzylic iodides. As in the case of silver nitrate induced coupling, a statistical distribution of the three possible coupled products is obtained, but use of a large excess of the more readily available iodide increases the yield of the desired product (81).

REFERENCES

1. H. C. Brown, Acc. Chem. Res., 2, 65 (1969).

2. H. C. Brown, H. Nambu, and M. M. Rogić, J. Am. Chem. Soc., 91, 6854 (1969); and references cited therein.

3. J. J. Tufariello, L. T. C. Lee, and P. Wojtkowski, J. Am. Chem. Soc., 89, 6804 (1967); and references cited therein.

4. J. Hooz and D. M. Gunn, J. Am. Chem. Soc., 91, 6195 (1969); and references cited therein.

5. A. Suzuki, S. Nozawa, N. Miyaura, M. Itoh, and H. C. Brown, Tetrahedron Letters, 1969, 2955.

6. M. E. D. Hillmann, J. Am. Chem. Soc., 84, 4715 (1962); 85, 982, 1626 (1963).

7. D. J. Pasto and P. W. Wojtkowski, Tetrahedron Letters, 1970, 215.

8. A. Suzuki, A. Arase, H. Matsumoto, M. Itoh, H. C. Brown, M. M. Rogić, and M. W. Rathke, J. Am. Chem. Soc., 89, 5708 (1967).

9. H. C. Brown, M. M. Rogić, M. W. Rathke, and G. W. Kabalka, J. Am. Chem. Soc., 89, 5709 (1967).

10. (a) G. W. Kabalka, H. C. Brown, A. Suzuki, S. Honma, A. Arase, and M. Itoh, J. Am. Chem. Soc., 92, 710 (1970); (b) R. E. Lutz and W. G. Revely, J. Am. Chem. Soc., 63, 3184 (1941).

11. H. C. Brown and G. W. Kabalka, J. Am. Chem. Soc., 92, 712 (1970).

12. H. C. Brown and G. W. Kabalka, J. Am. Chem. Soc., 92, 714 (1970).

13. (a) H. C. Brown, M. M. Rogić, M. W. Rathke, and G. W. Kabalka, J. Am. Chem. Soc., 91, 2150 (1969); (b) H. C. Brown and E. Negishi, personal communication.

14. H. C. Brown, H. Nambu, and M. M. Rogić, J. Am. Chem. Soc., 91, 6852 (1969).

15. H. C. Brown, H. Nambu, and M. M. Rogić, J. Am. Chem. Soc., 91, 6855 (1969).

16. (a) H. C. Brown and M. M. Rogić, J. Am. Chem. Soc., 91, 4304 (1969); (b) J. Hooz and D. M. Gunn, Tetrahedron Letters, 1969, 3455.

17. H. C. Brown and E. Negishi, J. Am. Chem. Soc., 93, 3777 (1971).

18. M. W. Rathke and H. C. Brown, J. Am. Chem. Soc., 89, 2740 (1967).

19. H. C. Brown, E. F. Knights, and R. A. Coleman, J. Am. Chem. Soc., 91, 2144 (1969).

20. R. Köster and B. Rickborn, J. Am. Chem. Soc., 89, 2782 (1967).

21. W. K. Musker, Fortschr. Chem. Forsch., 14, 295 (1970); pp. 311-314.

22. J. J. Tufariello and L. T. C. Lee, J. Am. Chem. Soc., 88, 4757 (1966).

23. J. J. Tufariello, P. Wojtkowski, and L. T. C. Lee, Chem. Commun., 1967, 505.

24. A. Suzuki, N. Miyaura, M. Itoh, H. C. Brown, G. W. Holland, and E. Negishi, J. Am. Chem. Soc., 93, 2792 (1971).

25. H. C. Brown and M. W. Rathke, J. Am. Chem. Soc., 89, 2738 (1967).

26. G. Köbrich and H. Merkle, Chem. Ber., 100, 3371 (1967).

27. (a) H. C. Brown and Y. Yamamoto, J. Am. Chem. Soc., 93, 2796 (1971); (b) H. C. Brown and Y. Yamamoto, Chem. Commun., 1971, 1535; (c) H. C. Brown and Y. Yamamoto, ibid., 1972, 71; (d) N. Miyamoto, S. Isiyama, K. Utimoto, and H. Nozaki, Tetrahedron Letters, 1971, 4597.

28. (a) H. C. Brown and M. W. Rathke, J. Am. Chem. Soc., 89, 2737 (1967); (b) E. Negishi and H. C. Brown, Synthesis, 1972, 197.

29. H. C. Brown, B. A. Carlson, and R. H. Prager, J. Am. Chem. Soc., 93, 2070 (1971).

30.   A. Pelter, M. G. Hutchings, and K. Smith, Chem. Commun., 1971, 1048.

31.   H. C. Brown, E. Negishi, and S. K. Gupta, J. Am. Chem. Soc., 92, 2460 (1970).

32.   H. C. Brown and E. Negishi, J. Organometal. Chem., 26, C67 (1971).

33.   H. C. Brown and E. Negishi, J. Am. Chem. Soc., 91, 1224 (1969).

34.   H. C. Brown and W. C. Dickason, J. Am. Chem. Soc., 91, 1226 (1969).

35.   (a) C. F. Lane and H. C. Brown, J. Am. Chem. Soc., 93, 1025 (1971); (b) H. C. Brown and C. F. Lane, Synthesis, 1972, 303; (c) H. C. Brown, Y. Yamamoto, and C. F. Lane, Synthesis, 1972, 304.

36.   (a) A. Suzuki, N. Miyaura, and M. Itoh, Tetrahedron, 27, 2775 (1971); (b) J. Klein, and A. Medlick, J. Am. Chem. Soc., 93, 6313 (1971).

37.   M. F. Hawthorne and M. Reintjes, J. Am. Chem. Soc., 87, 4585 (1965).

38.   (a) B. M. Mikhailov and G. S. Ter-Sarkisyan, Izv. Akad. Nauk. SSSR, Ser. Khim, 1966, 380; Chem. Abstr., 64, 15907h (1966); (b) G. W. Kabalka, J. Organometal. Chem., 33, C25 (1971); B. M. Mikhailov, G. S. Ter-Sarkisyan, and N. A. Nikolaeva, Zh. Obshch. Khim., 41, 1721 (1971); Chem. Abstr., 76, 3934n (1972).

39.   H. C. Brown, R. A. Coleman, and M. W. Rathke, J. Am. Chem. Soc., 90, 499 (1968).

40.   H. C. Brown and R. A. Coleman, J. Am. Chem. Soc., 91, 4606 (1969).

41.   J. Hooz and G. F. Morrison, Can. J. Chem., 48, 868 (1970).

42.   H. C. Brown, G. W. Kabalka, M. W. Rathke, and M. M. Rogić, J. Am. Chem. Soc., 90, 4165 (1968).

43.   G. Ter-Sarkisyan, N. A. Nikolaeva, and B. M. Mikhailov, Izv. Akad. Nauk. SSSR, Ser. Khim., 1970, 876; Chem. Abstr., 73, 44822(f) (1971).

44. H. C. Brown, G. W. Kabalka, and M. W. Rathke, J. Am. Chem. Soc., 89, 4530 (1967).

45. (a) H. C. Brown and E. Negishi, J. Am. Chem. Soc., 89, 5285 (1967); (b) E. Negishi and H. C. Brown, Synthesis, 1972, 196.

46. H. C. Brown and E. Negishi, J. Am. Chem. Soc., 89, 5477 (1967).

47. H. C. Brown and E. Negishi, Chem. Commun., 1968, 594.

48. A. Pelter, M. G. Hutchings, and K. Smith, Chem. Commun., 1970, 1529.

49. A. Pelter, M. G. Hutchings, and K. Smith, Chem. Commun., 1971, 1048.

50. L. C. King and G. K. Ostrum, J. Org. Chem., 29, 3459 (1964).

51. H. C. Brown, M. M. Rogić, and M. W. Rathke, J. Am. Chem. Soc., 90, 6218 (1968).

52. H. C. Brown, M. M. Rogić, H. Nambu, and M. W. Rathke, J. Am. Chem. Soc., 91, 2147 (1969).

53. J. Hooz and S. Linke, J. Am. Chem. Soc., 90, 5936 (1968).

54. J. Hooz, D. M. Gunn, and H. Kono, Can. J. Chem., 49, 2371 (1971).

55. J. Hooz and D. M. Gunn, Chem. Commun., 1969, 139.

56. D. J. Pasto and P. W. Wojtkowski, J. Org. Chem., 36, 1790 (1971).

57. A. Suzuki, S. Nozawa, M. Itoh, H. C. Brown, E. Negishi, and S. K. Gupta, Chem. Commun., 1969, 1009.

58. H. C. Brown, M. W. Rathke, G. W. Kabalka, and M. M. Rogić, J. Am. Chem. Soc., 90, 4166 (1968).

59. A. Suzuki, S. Nozawa, M. Itoh, H. C. Brown, G. W. Kabalka, and G. W. Holland, J. Am. Chem. Soc., 92, 3503 (1970).

60. (a) G. S. Ter-Sarkisyan, N. A. Nikolaeva, and B. M. Mikhailov, Izv. Akad. Nauk. SSSR, Ser. Khim., 1968; 2516; Chem. Abstr., 70, 67808(h) (1969); (b) K. Utimoto, T. Tanaka, and H. Nozaki, Tetrahedron Letters, 1972, 1167.

61. D. J. Pasto and K. McReynolds, Tetrahedron Letters, 1971, 801.

62.   H. C. Brown, M. M. Rogić, M. W. Rathke, and G. W. Kabalka, J. Am. Chem. Soc., 90, 818 (1968).

63.   H. C. Brown, M. M. Rogić, M. W. Rathke, and G. W. Kabalka, J. Am. Chem. Soc., 90, 1911 (1968).

64.   H. C. Brown and M. M. Rogić, J. Am. Chem. Soc., 91, 2146 (1969).

65.   (a) J. Hooz and S. Linke, J. Am. Chem. Soc., 90, 6891 (1968); (b) H. C. Brown, M. M. Midland, and A. B. Levy, ibid. 94, 3662 (1972).

66.   H. C. Brown and H. Nambu, J. Am. Chem. Soc., 92, 1761 (1970).

67.   H. Nambu and H. C. Brown, J. Am. Chem. Soc., 92, 5790 (1970).

68.   H. Nambu and H. C. Brown, Organometal. Chem. Synth., 1, 95 (1970).

69.   N. Yu Bubnov, S. I. Frolov, V. G. Kiselev, V. S. Bogdanov, and B. M. Mikhailov, Organometal. Chem. Synth., 1, 37 (1970); Zh Obshch. Khim., 40, 1131 (1970); Chem. Abstr., 74, 53885(b) (1971).

70.   B. M. Mikhailov, Yu. N. Bubnov, S. A. Korobeinikova, and S. I. Frolov, J. Organometal. Chem., 27, 165 (1971); and references cited therein.

71.   B. M. Mikhailov and K. L. Cherkasova, Izv. Akad. Dauk. SSSR, Ser. Khim, 1971, 1244; Chem. Abstr., 75, 76883(j) (1971).

72.   (a) Yu. N. Bubnov, S. I. Frolov, V. G. Kiselev, and B. M. Mikhailov, Zh. Obshch. Khim., 40, 1316 (1970); Chem. Abstr., 74, 53884(a) (1971); (b) B. M. Mikhailov, Yu. N. Bubnov, and M. S. Grigoryan, Izv. Akad. Nauk. SSSR. Ser. Khim., 1971, 1842; Chem. Abstr., 75, 151861F (1971).

73.   B. M. Mikhailov and Yu. N. Bubnov, Tetrahedron Letters, 1971, 2127.

74.   Yu. N. Bubnov, O. A. Nesmeyanova, T. Yu Rudashevskaya, B. M. Mikhailov, and B. A. Kazansky, Tetrahedron Letters, 1971, 2153.

75.   D. Seyferth and B. Prokai, J. Am. Chem. Soc., 88, 1834 (1966).

76.   H. C. Brown and M. V. Bhatt, J. Am. Chem. Soc., 88, 1440 (1966).

77.   H. C. Brown, M. V. Bhatt, T. Munekata, and G. Zweifel, J. Am. Chem. Soc., 89, 567 (1967).

78.   H. C. Brown and C. H. Snyder, J. Am. Chem. Soc., 83, 1002 (1961).

79.   H. C. Brown, C. Verbrugge, and C. H. Snyder, J. Am. Chem. Soc., 83, 1001 (1961).

80.   Y. Tanahashi, J. Lhomme, and G. Ourisson, Tetrahedron, 28, 2655 (1972).

81.   A. Suzuki, S. Nozawa, M. Harada, M. Itoh, H. C. Brown, and M. M. Midland, J. Am. Chem. Soc., 93, 1508 (1971).

Chapter 9

SYNTHESIS OF FUNCTIONAL DERIVATIVES VIA ORGANOBORANES

The anti-Markownikoff hydration of alkenes via hydroboration has been discussed in Sec. 3.1, while the use of hydroboration in the synthesis of a variety of β-substituted alcohols and diols is discussed in Chapts. 5 and 6, respectively. The use of vinylboranes, formed by monohydroboration of alkynes, in the synthesis of vinyl halides and aldehydes and ketones has been discussed in Chap. 7; 1,1-diboro derivatives, formed from the dihydroboration of 1-alkynes, are also useful intermediates in the synthesis of carboxylic acids (Chap. 7).

In this capter the synthesis of a variety of other functional derivatives from organoboranes is discussed. Thus, organoboranes may be converted into alkyl halides (Sec. 9.1), amines (Sec. 9.2.1), sulfides (Sec. 9.3.1), and organomercury compounds (Sec. 9.4). In addition, use of labeled diborane or other labeled hydroborating reagents provides synthetic routes to a variety of labeled organic compounds (Sec. 9.5).

9.1 SYNTHESIS OF ALKYL HALIDES

9.1.1 Synthesis of Alkyl Chlorides

Reaction of trialkylboranes with N-chlorodialkylamines, such as N-chlorodiethylamine or N-chloropiperidine, gives the

corresponding alkyl chlorides in yields of 30-53% (based on R_3B)
(1).

$$R_3B \; + \; \text{[cyclohexyl]}NCl \xrightarrow[\text{NaOH}]{\text{THF, } H_2O} RCl$$

$$38\%; \quad R = n\text{-}C_8H_{17}$$
$$53\%; \quad R = C_6H_{11}$$

Studies of the reaction of tri-n-butylborane with N-chlorodimethyl-
amine have shown that concurrent polar and radical processes occur
to give amine and chloride, respectively (2).

$$(n\text{-}C_4H_9)_3B \; + \; (CH_3)_2NCl \xrightarrow[35°]{PhCl} \begin{cases} \xrightarrow{\text{Polar}} C_4H_9N(CH_3)_2 + (C_4H_9)_2BCl \\ \xrightarrow{\text{Radical}} C_4H_9Cl + (C_4H_9)_2BN(CH_3)_2 \end{cases}$$

The four products are formed in approximately equal yields.
Addition of galvinoxyl inhibits the radical process, but the
polar process cannot be suppressed; thus the reaction is
unsuitable for the efficient preparation of alkyl chlorides (2).

Reaction of trialkylboranes with cupric chloride in
tetrahydrofuran-water gives the corresponding alkyl chlorides in
yields of 43-77% (based on R_3B) (3).

$$R_3B \; + \; 2CuCl_2 \xrightarrow[20-25°, \; 24 \; hr]{THF\text{-}H_2O} RCl + R_2BOH + Cu_2Cl_2 + HCl$$
$$(43\text{-}77\%)$$

$(R=C_6H_{13}, \; C_8H_{17}, \; C_6H_{11}, \text{norbornyl})$

Use of B-alkyl-9-BBN derivatives results in selective attack of
the boron-cyclooctyl bond (3).

9.1.2 Synthesis of Alkyl Bromides

Primary trialkylboranes, derived from the hydroboration of
1-alkenes, react with 3 mole equivalents of mercuric acetate to
form the corresponding alkyl mercuric acetates. These derivatives

undergo in situ bromination to give high yields of the primary alkyl bromides ($\underline{4}$).

$$(RCH_2)_3B + 3\ Hg(OAc)_2 \xrightarrow[\Delta]{THF} 3RCH_2HgOAc \xrightarrow[CCl_4]{Br_2} 3RCH_2Br \\ (71\text{-}86\%)$$

Secondary trialkylboranes such as tricyclohexylborane, however, give considerably lower yields ($<30\%$) of the corresponding secondary bromides ($\underline{4}$).

Reaction of trialkylboranes with bromine in dichloromethane gives the corresponding alkyl bromides in yields exceeding 80% (based on R_3B) ($\underline{5}$). The reaction proceeds with complete retention of configuration at reacting carbon atom.

$$R_3B + Br_2 \xrightarrow[25^\circ,\ \sim24\ hr]{CH_2Cl_2,\ dark} RBr + R_2BBr \\ (80\text{-}99\%)$$

$[R=n\text{-},\ i\text{-},\ s\text{-}C_4H_9;\ -(CH_2)_{5,6},$

norbornyl$]$

The reaction proceeds via radical bromination of the α-position of one of the alkyl groups. The resultant α-bromoorganoborane then undergoes protonolysis by the hydrogen bromide produced to give the alkyl bromide ($\underline{5}$).

$$R_2B\text{-}\overset{|}{\underset{|}{C}}\text{-}H + Br\cdot \xrightarrow{-HBr} R_2B\text{-}\overset{|}{\underset{|}{C}}\cdot \xrightarrow{Br_2} R_2B\text{-}\overset{|}{\underset{|}{C}}\text{-}Br + Br\cdot$$

$$\downarrow HBr$$

$$R_2BBr + \quad \rangle CHBr$$

Use of B-alkyl-9-BBN derivatives results in selective bromination of the alkyl group to give high yields of the alkyl bromides ($\underline{6}$).

$$R\text{-}B\text{-}9\text{-}BBN + Br_2 \xrightarrow[0\text{-}25^\circ]{CH_2Cl_2} RBr + Br\text{-}B\text{-}9\text{-}BBN \\ (85\text{-}90\%)$$

Selective bromination of the alkyl group occurs since bridgehead radicals, generated by radical abstraction of the bridgehead hydrogen atoms of the 9-BBN structure, are unable to interact with the vacant p-orbital of the boron atom, and hence are not stabilized (6). α-Bromoorganoboranes have also been used in the synthesis of secondary and tertiary alcohols (Secs. 8.3.2 and 8.3.3).

Reaction of organoboranes with bromine in the presence of sodium methoxide gives the corresponding alkyl bromides in yields of 60-99% (7).

$$R_3B \ + \ Br_2 \ \xrightarrow[\text{CH}_3\text{OH, THF}]{\text{CH}_3\text{ONa, } < 5^\circ} \ 3RBr$$

(R=primary and secondary   (60-99%)
         alkyl)

Application of the reaction to tri-exo-norbornylborane gives mainly endo-2-bromonorbornane indicating that the reaction proceeds with inversion of configuration (8). This result is in contrast to that obtained using radical bromination (see above) and has been explained in terms of the formation of an intermediate "ate" complex.

$$\longrightarrow \quad Br-R$$

In the case of trinorbornylborane only one of the norbornyl groups is converted into the bromide.

Alkyl bromides are also formed by the reaction of organoboranes with cupric bromide in tetrahydrofuran-water, though only one of the alkyl groups is utilized in bromide formation (3).

$$R_3B + 2CuBr_2 \xrightarrow[\text{20-25}^\circ, \text{ 24 hr}]{\text{THF-H}_2\text{O}} RBr + R_2BOH + Cu_2Br_2 + HBr$$

$$(75\text{-}92\%)$$

(R=n-alkyl, C_6H_{11}, norbornyl)

As in the case of the synthesis of alkyl chlorides (Sec. 9.1.1), use of B-alkyl-9-BBN derivatives results in selective attack of the boron-cyclooctyl bond (3).

9.1.3 Synthesis of Alkyl Iodides

Organoboranes, derived from terminal alkenes, react rapidly with iodine in the presence of sodium hydroxide to give the corresponding primary iodides (9).

$$(RCH_2)_3B + 2I_2 + NaOH \xrightarrow[\substack{CH_3OH \\ 25^\circ}]{\text{THF}} 2RCH_2I + RCH_2B(OH)_2$$

$$(60\text{-}65\%)$$

The reaction with secondary alkylboranes is sluggish and gives low yields (30-40%) of the corresponding iodides (9). Use of disiamylborane as hydroborating reagent in the hydroboration of terminal alkenes ensures full utilization of the primary alkyl groups in the above reaction.

$$RCH{=}CH_2 \xrightarrow{\text{Sia}_2BH} RCH_2CH_2BSia_2 \xrightarrow[\text{THF, CH}_3\text{OH}]{I_2, \text{ NaOH}} RCH_2CH_2I + Sia_2BOH$$

$$(90\%)$$

Alkyl iodides are also synthesized by the reaction of equimolar amounts of trialkylboranes and allyl iodide in tetrahydrofuran in the presence of air (10).

$$R_3B + CH_2{=}CHCH_2I \xrightarrow[\text{THF}]{\text{Air}} RI$$

$$(75\text{-}100\%; \text{ based on } CH_2{=}CHCH_2I)$$

(R=n-alkyl, C_6H_{11})

The reaction proceeds by a radical chain mechanism similar to
that discussed in the coupling of allylic iodides (Sec. 8.9).

9.2 SYNTHESIS OF NITROGEN DERIVATIVES

9.2.1 Synthesis of Amines

Organoboranes, derived from terminal and relatively
unhindered internal and cyclic alkenes, react with hydroxylamine-
O-sulfonic acid in refluxing tetrahydrofuran to give the
corresponding primary amines in yields of 50-60% (11).
Organoboranes, derived from hindered alkenes, such as
1-methylcyclohexene, only react on heating in diglyme at 100° (12).

$$R_3B + H_2NOSO_3H \xrightarrow[\text{2) HCl 3) OH}^{\ominus}]{\text{1) DG, 100}^\circ} 2RNH_2 + RB(OH)_2$$
$$(40\text{-}50\%)$$

The reaction is highly stereospecific proceeding with complete
retention of configuration at the reacting carbon atom (12).
Asymmetric hydroboration of cis-2-butene with (-)-diisopino-
campheylborane (Sec. 3.6.1), followed by treatment with
hydroxylamine-O-sulfonic acid, gives (R)-(-)-2-butylamine of
75% optical purity (13).

Application of the above reaction to the organoborane mixture
derived from 2,4,5-trimethoxy-propenylbenzene proceeds with some
carbon-carbon bond cleavage to give 20% 2,4,5-trimethoxyaniline (14).
This product has been shown to be formed from the α-arylorgano-
borane on treatment with hydroxylamine-O-sulfonic acid via
intramolecular electrophilic amination (14).

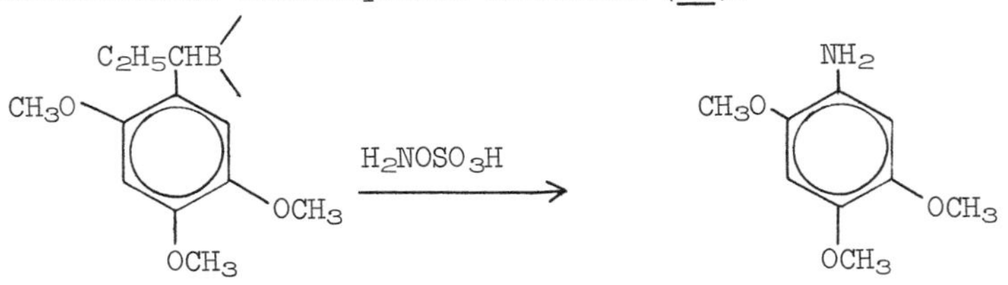

Primary amines are also formed in the reaction of organoboranes with chloramine in tetrahydrofuran (11). The reaction proceeds stereospecifically.

$$R_3B \; + \; H_2NCl \; \xrightarrow[\text{1 hr, r.t.}]{\text{THF, NaOH}} \; RNH_2$$

(R=Primary, secondary   28-50%
        alkyl)

Reaction of organic azides with triethylborane in refluxing xylene, followed by treatment with methanol, gives secondary amines in yields exceeding 70% (15a). The reaction proceeds via initial coordination between the azide and triethylborane, followed by transfer of an ethyl group from boron to nitrogen.

$$[\text{R=n-, i-, s-}C_4H_9;$$
$$-(CH_2)_{\underline{5},\,6},\; Ph]$$

The reaction is slow with sterically hindered azides and fails when both sterically hindred azides and organoboranes are used (15a). Use of B-alkyl-9-BBN derivatives results in exclusive migration of the B-cyclooctyl bond (Sec. 8.2.2). However, when dialkyl-chloroboranes (Sec. 2.2.5) are used in place of trialkylboranes, fairly rapid reaction occurs even in the case of relatively hindered organoboranes and azides (15b).

9.2.2 Synthesis of Miscellaneous Nitrogen Derivatives

Reaction of carboxylic acids with tris(dialkylamino)boranes in
benzene gives the corresponding amides in yields of 62-87% (16).

$$RCO_2H \ + \ B(NR_2')_3 \ \xrightarrow[\text{2) HCl}]{\text{1) PhH, r.t., } >40 \text{ hr}} \ \begin{array}{l} RCONR_2' \\ 62\text{-}87\% \end{array}$$

(R=n-C_5H_{11}, tert-C_4H_9,   (R_2'=Pyrrolidyl)
   Ph, $PhCH_2$)

Complete reaction requires the use of 1 mole equivalent of the
aminoborane (16), and has been shown to proceed via formation of
the triacyloxyborane-amine adduct, $(RCO_2)_3B{:}NR_2'$ (17). The
synthesis of amides by use of 1 mole equivalent of amine and
trimethoxyborane or dimethoxychloroborane only proceeds
satisfactorily at high temperature; hence the reactions are of
little synthetic value in peptide synthesis since racemization
occurs on warming the reaction mixtures (18).

Tris(dialkylamino)boranes have also been used in the synthesis
of enamines, β-enaminoketones, and enamine amides from ketones,
β-diketones, and β-keto esters, respectively (16).

9.3 SYNTHESIS OF SULFUR DERIVATIVES

9.3.1 Synthesis of Sulfides

Trialkylboranes react readily with disulfides in the presence
of air to give the corresponding sulfides (19).

$$R_3B \ + \ 2CH_3SSCH_3 \ \xrightarrow[\substack{\text{or} \\ h\nu, \text{ hexane}}]{\text{THF, air, r.t.}} \ \begin{array}{l} 2RSCH_3 \ + \ (CH_3S)_2BR \\ (68\text{-}95\%) \end{array}$$

(R=n, i, s-C_4H_9;
   -$(CH_2)_{\overline{5}, 6}$, norbornyl)

The reaction generally proceeds in better yields under the
influence of light when hexane is the solvent (19). Use of
B-alkyl-3,5-dimethylboracyclohexane derivatives gives high yields
of the corresponding alkylmethyl sulfides (19).

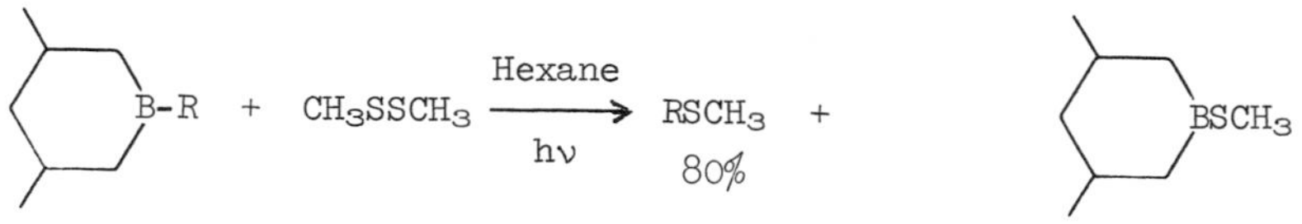

$$\text{B-R} \quad + \quad CH_3SSCH_3 \xrightarrow[h\nu]{\text{Hexane}} \quad RSCH_3 \quad + \quad \text{BSCH}_3$$
$$80\%$$

The reaction also proceeds using phenyl disulfide, and involves
a radical chain mechanism as shown below (19).

$$R_3B + O_2 \longrightarrow R_2BO_2\cdot + R\cdot$$
$$R\cdot + R'SSR' \longrightarrow RSR' + R'S\cdot$$
$$R'S\cdot + R_3B \longrightarrow R'SBR_2 + R\cdot$$
$$R'S\cdot + R'SBR_2 \longrightarrow (R'S)_2BR + R\cdot$$

Sulfides are also formed in low yields by the reaction of
sulfenyl chlorides with organoboranes (20), while addition of
borane to thiols in ether solvents results in alkyl transfer from
the solvent to sulfur to give sulfides (21).

9.3.2   Synthesis of Miscellaneous Sulfur Derivatives

Heating of organoboranes with sulfur at 130-200° gives
symmetrical disulfides in variable yields (22).

$$R_3B + S \xrightarrow[\text{2) Base}]{\text{1) 130°, N}_2\text{, 10 hr}} RSSR$$
$$50\% \ (R=n\text{-}C_8H_{17}, \ C_6H_{11})$$

Reaction of thioboranes with sulfenic esters gives
unsymmetrical disulfides (23).

$$(RS)_3B + 3PhSOCH_3 \longrightarrow 3RSSPh + B(OCH_3)_3$$
$$75\text{-}82\%$$

The synthesis and chemistry of thioboranes has been reviewed (24). Thioboranes have been used in the synthesis of thioacetals, thio-enol ethers, and thioesters (25).

9.4 METALLATION REACTIONS

Organoboranes have been particularly useful in the synthesis of organomercury compounds. Thus, organoboranes derived from terminal alkenes react readily with 3 equivalents of mercuric acetate in tetrahydrofuran at room temperature to give the corresponding alkylmercuric acetates (26). These are converted to alkylmercuric halides on treatment with the appropriate sodium halide (26), or into alkyl bromides on treatment with bromine (Sec. 9.1.2) (4).

$$(RCH_2)_3B \xrightarrow[\text{THF, r.t.}]{3Hg(OAc)_2} 3RCH_2HgOAc \xrightarrow[\text{H}_2\text{O}]{\text{NaX}} \underset{>90\%}{3RCH_2HgX}$$

The mercuration reaction is very sensitive to steric hindrance about the boron atom, and hence secondary alkyl groups fail to react under the above conditions. However, reaction of secondary benzylic groups competes with that of primary alkyl groups (26). Thus, hydroboration of styrene using borane-tetrahydrofuran, followed by reaction with mercuric acetate, gives mixtures of the primary and secondary alkylmercuric derivatives. Use of dicyclohexylborane as hydroborating reagent overcomes this problem since the primary organoborane is formed almost exclusively with this reagent (26).

$$PhCH=CH_2 \xrightarrow[\text{THF}]{(C_6H_{11})_2BH} (PhCH_2CH_2)_3B(C_6H_{11})_2 \xrightarrow[\text{2)NaX}]{1)Hg(OAc)_2} \underset{99\%;\ X=Cl}{PhCH_2CH_2HgX}$$

The reaction of secondary trialkylboranes with a variety of mercuric salts has been investigated (27a). Mercuric benzoate is the most effective reagent, but the reaction only proceeds satisfactorily for tricyclopentyl- and tricyclohexylborane.

$$R_3B + 2Hg(OCOPh)_2 \xrightarrow[\Delta]{THF} RB(OCOPh)_2 + 2RHgOCOPh \xrightarrow[H_2O]{NaCl} 2RHgCl$$

$$93\text{-}98\%;$$

$$[R= -(CH_2)_{5,6}]$$

More hindered trialkylboranes, such as tris(2-butyl)borane, give negligible yields of the alkylmercuric salts (27a). In addition, only two of the three alkyl groups react, and attempts to overcome this limitation by use of B-alkyl-9-BBN derivatives failed due to preferential cleavage of the boron-cyclooctyl bond (27a).

Dicyclohexylvinylboranes, formed by hydroboration of terminal and internal alkynes with dicyclohexylborane, react with mercuric acetate to give the corresponding vinylmercuric acetates (27b); these derivatives are readily converted to vinylmercuric chlorides on treatment with aqueous sodium chloride.

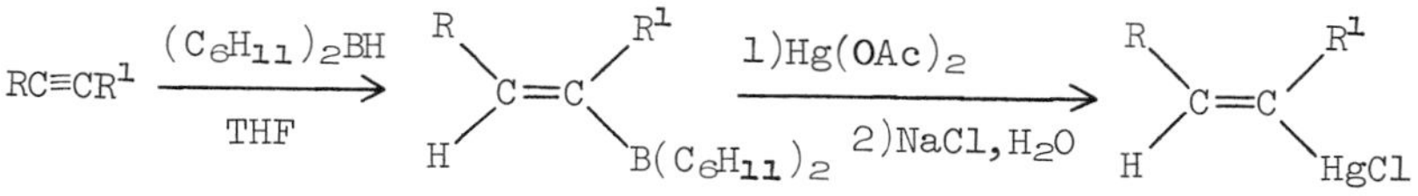

$$RC\equiv CR^1 \xrightarrow[THF]{(C_6H_{11})_2BH} \underset{H}{\overset{R}{\diagdown}}C=C\underset{B(C_6H_{11})_2}{\overset{R^1}{\diagup}} \xrightarrow[2)NaCl, H_2O]{1)Hg(OAc)_2} \underset{H}{\overset{R}{\diagdown}}C=C\underset{HgCl}{\overset{R^1}{\diagup}}$$

The reaction proceeds in high yield for terminal alkynes (>80%; R^1=H), but yields are lower for internal alkynes (59%; R=R^1=C_2H_5) (27b). Use of benzo-1,3-dioxa-2-borole [(3); Sec. 2.2.4] as hydroborating reagent in the above reaction sequence, however, results in the conversion of both terminal and internal alkynes to the corresponding vinylmercuric halides in yields exceeding 95% (27c).

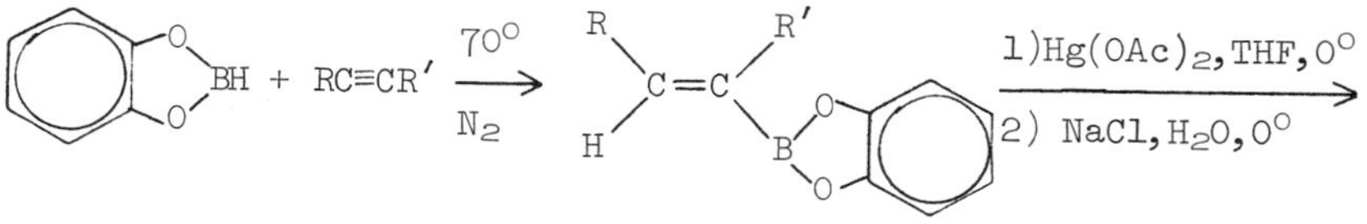

$$\underset{H}{\overset{R}{\diagdown}}C=C\underset{HgCl}{\overset{R'}{\diagup}}$$

97-99%

Organoboranes derived from terminal alkenes may be converted into dialkylmercuric compounds by treatment with 1.5 equivalents of mercuric acetate ($\underline{27d}$).

$$R_3B + 1.5\ Hg(OAc)_2 \xrightarrow[\Delta]{THF} 1.5\ R_2Hg + B(OAc)_3$$

70-90%

As in the formation of alkylmercuric acetates, secondary alkyl groups react very slowly. Alkyldicyclohexylboranes thus provide a convenient route to the corresponding dialkylmercuric compounds ($\underline{27d}$).

$$2(C_6H_{11})_2BH \xrightarrow[THF]{2RCH=CH_2} 2(C_6H_{11})_2BCH_2CH_2R \xrightarrow[r.t.]{Hg(OAc)_2,THF}$$

$(RCH_2CH_2)_2Hg + 2(C_6H_{11})_2BOAc$
40-94%

An alternative route to dialkylmercuric compounds involves the symmetrization of alkylmercuric acetates using zinc dust ($\underline{27d}$).

$$R_3B \xrightarrow[r.t.,\ THF]{3\ Hg(OAc)_2} 3\ RHgOAc + B(OAc)_3 \xrightarrow[r.t.]{Zn} R_2Hg + Zn(OAc)_2$$

65-90%

The above mercuration reactions all proceed with retention of configuration at the reacting carbon atom and exhibit all the characteristics of electrophilic substitution reactions ($\underline{27a}$).

Geminal diboro derivatives, such as compound (1), react with mercuric chloride to give the geminal dimercuric chlorides ($\underline{28}$).

$$CH_3CH[B(OC_4H_9)_2]_2 \xrightarrow[\text{NaOH}]{2HgCl_2} CH_3CH(HgCl_2)_2$$

(1)

The synthesis of a variety of organometallic compounds containing metals such as lead, aluminum, zinc, and magnesium from organoboranes using transmetallation reactions has been reviewed (29).

9.5 SYNTHESIS OF LABELED COMPOUNDS

Deuterated or tritiated diborane may be prepared from complex metal deuterides or tritiides using the procedures discussed in Sec. 2.2.1a. Likewise, labeled selective reagents may be prepared according to procedures discussed in Secs. 2.2.1c and d. Use of these labeled reagents in the various hydroboration and reduction reactions discussed elsewhere in this book provides convenient synthetic routes to a great variety of labeled compounds.

β-Labeled functional derivatives may be prepared from alkenes by reaction of the intermediate β-labeled organoboranes with suitable reagents.

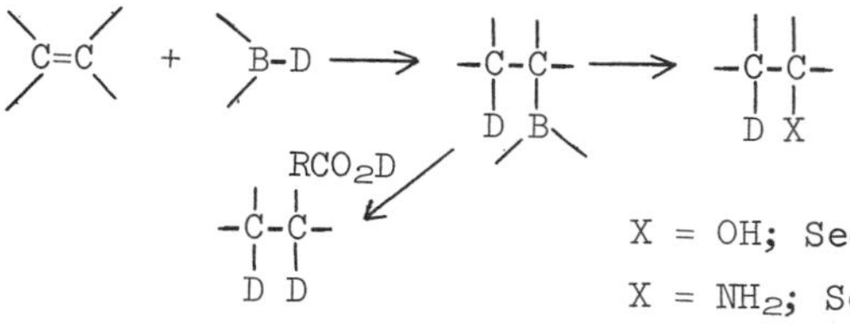

X = OH; Secs. 3.1 and 4.3; Ref.30
X = NH₂; Sec. 9.2.1; Ref. 31

Treatment of the intermediate β-labeled organoborane with a labeled carboxylic acid gives the corresponding doubly labeled compound (Sec. 10.11).

Labeled terminal alkenes may be prepared from 1-alkynes as shown below (Sec. 7.1.1) (32).

<u>Labeled internal alkenes</u> may be prepared in a similar manner from disubstituted alkynes (Secs. 7.1.1 and 7.3).

The synthesis of <u>α-deuterated ketones</u> from the reaction of organoboranes with diazo acetone and 3-buten-2-one has been discussed in Secs. 8.5.3 and 8.5.4, respectively ($\underline{33}$).

$$CH_3COCHN_2 \xrightarrow{R_3B, \ THF, \ D_2O} CH_3COCHDR$$

$$CH_2=CHCOCH_3 \xrightarrow{R_3B, \ THF, \ D_2O} RCH_2CHDCOCH_3$$

α-Labeled esters may be prepared in a similar manner using ethyl diazoacetate (Sec. 8.6.1) ($\underline{33}$).

$$N_2CHCO_2Et \xrightarrow{R_3B, \ THF, \ D_2O} RCHDCO_2Et$$

Asymmetric <u>1-d_1-primary alcohols</u> may be prepared by the reduction of aldehydes with labeled diisopinocampheylborane (Sec. 10.4.3) ($\underline{34}$).

$$(CH_3)_2CHCHO \xrightarrow[\substack{THF \\ 2) \ [O]}]{1) \ (-)-(IPC)_2BD} (-)-(1R)-(CH_3)_2CHCHDOH$$

81% (27% optically pure)

Use of (+)-reagent gives the (1S)-1-d_1-alcohol ($\underline{34}$). However, reduction of benzaldehyde with (-)-diisopinocampheyldeuterioborane is reported to give (1R)-1-d_1-benzyl alcohol ($\underline{35}$) in contrast to

an earlier report which claims that the (1S)-1-d_1-alcohol is
formed (36). 1-d_1-Primary alcohols have also been prepared by
reaction of 1-d_1-alkenes with diisopinocampheylborane (37,38).

$$C_4H_9{\equiv}CH \xrightarrow[\text{2) }CH_3CO_2D]{\text{1) }Sia_2BH} \underset{H}{\overset{H_9C_4}{\diagdown}}C{=}C\underset{D}{\overset{H}{\diagup}} \xrightarrow[\text{[O]}]{(IPC)_2BH} (S)\text{-}C_5H_{11}CHDOH$$

Ref. 37   (86% optically pure)

(S)-1-d_1-Ethanol has been prepared via the reaction of
cis-2-butene with (-)-diisopinocampheyldeuterioborane (39).

$$CH_3CH{=}CHCH_3 \xrightarrow[\text{2) [O]}]{\text{1) (-)-(IPC)}_2BD} \quad \text{(structure)} \xrightarrow[\substack{H_2O_2 \\ \text{3) OH}^{\ominus}}]{\substack{\text{1) }O_2,\ Pt \\ \text{2) }(CF_3CO_2)O}}$$

CH$_3$ structure with HO——H, D——H, CH$_3$

OH structure: D——H, CH$_3$

12%

4-d_1-1-Butanol has been prepared by deuterolysis of the
hydroboration product of 1,4-butadiene (40).

$$3(CH_2{=}CH)_2 \xrightarrow{2BH_3\text{-THF}} \text{B-}C_4H_8\text{-B} \xrightarrow{D_2O,\ THF}$$

$$D(CH_2)_4\text{-B-}C_4H_8\text{-B-}(CH_2)_4D \xrightarrow{\text{[O]}} 2DCH_2CH_2CH_2CH_2OH$$

                 OH   OH

                              40% (95% isotopically pure)

REFERENCES

1.   J. G. Sharefkin and H. D. Banks, J. Org. Chem., 30, 4313 (1965).

2.   A. G. Davies, S. C. W. Hook, and B. P. Roberts, J. Organometal. Chem., 23, C11 (1970).

3.   C. F. Lane, J. Organometal. Chem., 31, 421 (1971).

4.   J. J. Tufariello and M. M. Hovey, J. Am. Chem. Soc., 92, 3221 (1970).

5.   C. F. Lane and H. C. Brown, J. Am. Chem. Soc., 92, 7212 (1970).

6.   C. F. Lane and H. C. Brown, J. Organometal. Chem., 26, C51 (1971).

7.   H. C. Brown and C. F. Lane, J. Am. Chem. Soc., 92, 6660 (1970).

8.   H. C. Brown and C. F. Lane, Chem. Commun., 1971, 521.

9.   H. C. Brown, M. W. Rathke, and M. M. Rogić, J. Am. Chem. Soc., 90, 5038 (1968).

10.   A. Suzuki, S. Nozawa, M. Harada, M. Itoh, H. C. Brown, and M. M. Midland, J. Am. Chem. Soc., 93, 1508 (1971).

11.   H. C. Brown, W. R. Heydkamp, E. Breuer, and W. S. Murphy, J. Am. Chem. Soc., 86, 3565 (1964).

12.   M. W. Rathke, N. Inoue, K. R. Varma, and H. C. Brown, J. Am. Chem. Soc., 88, 2870 (1966).

13.   L. Verbit and P. J. Heffron, J. Org. Chem., 32, 3199 (1967).

14.   L. A. Levy and L. Fishbein, Tetrahedron Letters, 1969, 3773.

15.   (a) A. Suzuki, S. Sono, M. Itoh, H. C. Brown, and M. M. Midland, J. Am. Chem. Soc., 93, 4329 (1971); (b) H. C. Brown, M. M. Midland, and A. B. Levy, ibid, 94, 2114 (1972).

16.   P. Nelson and A. Pelter, J. Chem. Soc., 1965, 5142.

17.   A. Pelter and T. E. Levitt, Tetrahedron, 26, 1899 (1970).

18.   A. Pelter, T. E. Levitt, and P. Nelson, Tetrahedron, 26, 1539 (1970).

19.   H. C. Brown and M. M. Midland, J. Am. Chem. Soc., 93, 3291 (1971).

20.   P. M. Draper, T. H. Chan, and D. N. Harpp, Tetrahedron Letters, 1970, 1687.

21.   D. J. Pasto, J. Am. Chem. Soc., 84, 3777 (1962).

22.   Z. Yoshida, T. Okushi, and O. Manabe, Tetrahedron Letters, 1970, 1641.

23.   R. H. Cragg, J. P. N. Husband, and A. F. Western, Chem. Commun., 1970, 1701.

24.   R. H. Cragg and M. F. Lappert, Organometal. Chem. Rev., 1, 43 (1966); B. M. Mikhailov, Progress in Boron Chemistry, (R. J. Brotherton and H. Steinberg, eds), Vol. 3, Pergamon Press, Oxford, 1970, Chap. 5.

25.   R. H. Cragg and J. P. N. Husband, Inorg. Nucl. Chem. Letters, 7, 221 (1971), and references cited therein.

26.   R. C. Larock and H. C. Brown, J. Am. Chem. Soc., 92, 2467 (1970).

27.   (a) R. C. Larock and H. C. Brown, J. Organometal. Chem., 26, 35 (1971); (b) R. C. Larock and H. C. Brown, ibid, 36, 1 (1972); (c) R. C. Larock, S. K. Gupta, and H. C. Brown, J. Am. Chem. Soc., 94, 4371 (1972); (d) J. D. Buhler and H. C. Brown, J. Organometal. Chem., 40, 265 (1972).

28.   D. S. Matteson and J. G. Shdo, J. Org. Chem., 29, 2742 (1964).

29.   M. F. Lappert, The Chemistry of Boron and Its Compounds (E. L. Muetterties, Ed.), Wiley, New York, 1967, pp. 566-571, Chap. 8.

30.   K. T. Finley and W. H. Saunders, J. Am. Chem. Soc., 89, 898 (1967).

31.   D. S. Bailey and W. H. Saunders, J. Am. Chem. Soc., 92, 6904 (1970); M. P. Cooke and J. L. Coke, ibid., 90, 5556 (1968).

32.   R. W. Murray and G. J. Williams, J. Org. Chem., 34, 1896 (1969).

33.   J. Hooz and D. M. Gunn, J. Am. Chem. Soc., 91, 6195 (1969).

34.   K. R. Varma and E. Caspi, J. Org. Chem., 34, 2489 (1969).

35.   K. R. Varma and E. Caspi, Tetrahedron, 24, 6365 (1968).

36.   S. Wolfe and A. Rauk, Can. J. Chem., 44, 2591 (1966).

37.   H. Weber, P. Loew, and D. Arigoni, Chimia, 19, 595 (1965).

38.   A. Streitwieser, L. Verbit, and R. Bittman, J. Org. Chem., 32, 1530 (1967).

39.   H. Weber, J. Seibl, and D. Arigoni, Helv. Chim. Acta., 49, 741 (1966).

40.   H. C. Brown, E. Negishi, and S. K. Gupta, J. Am. Chem. Soc. 92, 2460 (1970).

Chapter 10

REDUCTION OF FUNCTIONAL GROUPS

A wide variety of reducing agents are available to the organic chemist which permit the ready reduction of most functional groups. The majority of these reagents may be classed as nucleophilic reducing agents which attack groups at positions of low electron density. Among the reagents which have been extensively studied, and which fall into the above class, are sodium borohydride ($\underline{1}$), lithium aluminum hydride ($\underline{3}$), lithium trimethoxyaluminum hydride ($\underline{4}$), lithium tri-tert-butoxyaluminumhydride ($3,5$), and sodium bis(2-methoxyethoxy)aluminum hydride ($\underline{6}$).

In contrast to the above nucleophilic reducing agents, aluminum hydride ($\underline{2}$), diborane and substituted boranes, such as disiamylborane, function as electrophilic reducing agents and attack groups at positions of high electron density (Sec. 10.1). These reagents thus possess some unique reducing characteristics. Carboxylic acids, which are not reduced by sodium borohydride or alkoxyaluminum hydrides, are readily reduced by diborane under mild conditions (Sec. 10.6.1), while acid chlorides are particularly unreactive (Sec. 10.6.3). The degree of stereoselectivity achieved in the reduction of cyclic ketones using various substituted boranes is far superior to that obtained with other reagents (Sec. 10.4.2). In addition, a number of substituted borohydride reagents which exhibit a high degree of stereoselectivity in the asymmetric reduction of ketones have been developed (Sec. 10.4.3). Hydro-

boration-protonolysis provides a useful means of achieving cis-
addition of hydrogen to carbon-carbon double bonds (Sec. 10.11).
Finally the selectivity of these reagents in the reduction of
various functional groups is discussed in Sec. 10.13).

10.1 MECHANISTIC CONSIDERATIONS

Diborane, being a Lewis acid, reacts by attacking groups at
positions of high electron density. Thus, in the reduction of
compounds containing a carbonyl group, the first step has usually
been considered to involve attack by borane at the basic oxygen
to form complex (1) ($\underline{7}$).

$$\sideset{}{}{\mathop{C}}=O + BH_3 \rightleftharpoons \sideset{}{}{\mathop{C}}=O:BH_3$$
$$(1)$$

Originally the second step in the mechanism was proposed to
involve an intramolecular hydride transfer from boron to carbon
to give an alkoxyborane ($\underline{7}$).

$$\sideset{}{}{\mathop{C}}\!=\!\overset{..}{\underset{H\!-\!BH_2}{O}} \longrightarrow \sideset{}{}{\mathop{C}}\!-\!OBH_2$$

However, later studies on the rates of reduction of substituted
cyclohexanones in tetrahydrofuran have shown that the reaction
is first order in ketone and three-halves order in the boron
species in solution ($\underline{8a}$). It was initially assumed that the
reducing species is diborane and, on the basis of the above
results, a step involving the rate-determining attack by diborane
on the complex (1) was proposed ($\underline{8a}$).

$$\sideset{}{}{\mathop{C}}=O:BH_3 + B_2H_6 \xrightarrow[\text{Slow}]{\text{THF}} \quad \text{Products}$$

It has now been firmly established that diborane in tetrahydrofuran exists as a borane-tetrahydrofuran adduct, BH_3:THF (Sec.

2.1). The rate of reduction is thus three-halves order in borane.
The kinetics of the reaction can be explained by the existence of
an equilibrium between 3 moles of borane-tetrahydrofuran and
the species $BH_2^{\oplus}$ and $B_2H_7^{\ominus}$ (8b).

$$3 BH_3:THF \rightleftharpoons (THF)_2BH_2^{\oplus} + B_2H_7^{\ominus} + THF$$

$$B_2H_7^{\ominus} + ketone \xrightarrow{\text{slow}} products$$

Support for this explanation is provided by the fact that the
reduction in the presence of lithium borohydride in first order
in ketone, borane, and borohydride (8b).

$$BH_3:THF + BH_4^{\ominus} \longrightarrow B_2H_7^{\ominus} \xrightarrow[\text{slow}]{\text{Ketone}} products$$

The exact structure of the transition state formed in the
reduction of the carbonyl group by the $B_2H_7^{\ominus}$ species is not known,
but it is possible that boron is coordinated by the carbonyl
oxygen in a transition state in which hydride attacks the carbonyl
carbon atom.

Since interaction between boron of the $B_2H_7^{\ominus}$ species and the
carbonyl oxygen atom is probably involved in the transition state,
any factors which reduce the basicity of the oxygen atom will
reduce the ease of reduction of the carbonyl group. This is
illustrated by the relative inertness of acid chlorides to
reduction by borane (Sec. 10.6.3) and the relatively slow rate
of reduction of esters compared with aldehydes and ketones (Sec.
10.6.4). The low reactivity of acid chlorides and esters can also
be explained in terms of the formation of complex (1) between the
carbonyl oxygen atom and borane as discussed earlier in this
section. However, in order to account for the observed kinetics
in the reduction of ketones (8a) it must be assumed that, in the
case of ketones, the equilibrium involving formation of complex
(1) lies completely to the right. In the case of acid chlorides

and esters the equilibrium will be unfavorable resulting in a low
degree of complex formation.

An alternative explanation of the low reactivity of esters
involves stabilization of complex (1) by the alkoxy group as shown
below (7).

Such stabilization results in a decrease in the electrophilic
character of the carbonyl carbon atom, thereby lowering its
susceptibility to attack by hydride in subsequent steps.

The reduction of esters by borane has also been explained in
terms of the coordination of the borane with the alkoxyl oxygen
atom (8c).

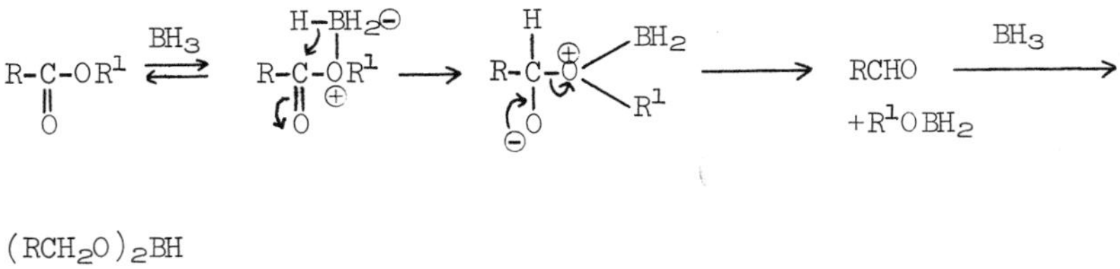

$(RCH_2O)_2BH$

In cases where coordination with the alkoxyl oxygen atom is
hindered the reaction proceeds via coordination with the carbonyl
oxygen atom and results in ether formation (Sec. 10.6.4) (8c).

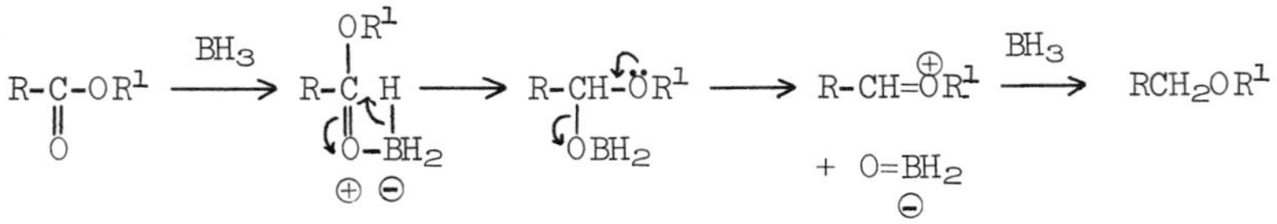

(R or R′=CMe₃)

The above mechanism also accounts for the formation of lactols
and pyran derivatives in the reduction of various lactones
(Sec. 10.6.5) (8c).

The reduction of many of the other functional groups dealt with in this chapter may likewise be envisaged as proceeding via coordination of the group by borane or some other boron species.

10.2 PRACTICAL CONSIDERATIONS

The reductions using borane or a selective reagent, such as disiamylborane, are carried out using the usual hydroboration procedures (Sec. 2.2.1), followed by hydrolysis of the reaction mixture with water and ether extraction of the product.

However, in certain reductions, such as those in which nitrogen derivatives are formed, intermediates which are difficult to hydrolyze are produced. In such cases hydrolysis may be achieved by refluxing with 10% sodium hydroxide or 20% hydrochloric acid (9a), or in certain instances, refluxing with concentrated hydrochloric acid (11). After neutralizing the acid solution with base, the mixture is then extracted with a hydrocarbon solvent (pentane) for prolonged periods (48-90 hr) (9a). A more convenient procedure involves neutralization of the acid solution with a basic ion exchange resin, followed by filtration and removal of the solvent. Extraction of the residue with a hydrocarbon solvent gives the product (9b).

10.3 REACTION WITH ALCOHOLS, PHENOLS, THIOLS, AND AMINES

Alcohols, phenols, and thiols all react with borane-tetrahydrofuran to liberate 1 mole equivalent of hydrogen (10). The rate of evolution at 0° is found to decrease in the order: Primary alcohols > secondary alcohols > tertiary alcohols ~ phenols ~ thiols. These results have been explained in terms of prior coordination of the borane by the oxygen or sulfur followed by loss of hydrogen to form the corresponding oxyborane (10).

$$\text{ROH} \xrightarrow{\text{BH}_3} \underset{\overset{|}{\text{H}}}{\text{R-O}:\overset{\oplus}{\overset{..}{\text{BH}}}_3} \xrightarrow{\hspace{2cm}} \text{ROBH}_2 + \text{H}_2\uparrow$$

Amines readily form addition compounds with borane, but liberate
hydrogen extremely slowly due to the low acidity of the N-H
atoms (10).

Primary and secondary alcohols, and phenols react rapidly
with disiamylborane liberating 1 mole equivalent of hydrogen,
while tertiary alcohols, thiols, and amines fail to react (11).

Carbinols capable of forming relatively stable carbonium
ions, such as benzylic alcohols, tend to undergo hydrogenolysis
on treatment with diborane in the presence of boron trifluoride
(12-14). In the case of certain tertiary benzylic alcohols
containing an available β-hydrogen elimination to form alkenes
occurs in preference to hydrogenolysis (15).

10.4 REDUCTION OF ALDEHYDES AND KETONES

10.4.1 Formation of Alcohols

In general, aldehydes and ketones react rapidly with
borane-tetrahydrofuran at $0°$ to give the corresponding
alcohols (10). Similar results are obtained using disiamylborane
in tetrahydrofuran though these reactions are usually slower (11).

The reduction of benzophenone with either borane or
disiamylborane is considerably slower than that of other aldehydes
and ketones (10,11); this result has been attributed to electronic
and steric factors. In the reduction of alkyl methyl ketones with
disiamylborane, the nature of the alkyl substituent has relatively
little effect on the rate of reduction, an exception being the
tertiary butyl group which exerts a marked retarding influence
(16). Steric factors do, however, play an important role in
determining the stereoselectivity of reduction of cyclic
ketones (Sec. 10.4.2).

In the case of α,β-unsaturated carbonyl compounds, such as 3-phenylpropenal (cinnamalydehyde), a rapid reduction of the carbonyl group and hydroboration of the double bond occur on reaction with borane-tetrahydrofuran (Sec. 5.4) (10). With disiamylborane, however, reaction of the double bond in 3-phenylpropenal is relatively slow suggesting that selective reduction of aldehyde groups in the presence of double bonds should be possible (Sec. 5.4) (11).

In the reaction of certain "electron-rich" aldehydes and ketones, complete reduction to the corresponding hydrocarbons has been observed. Such hydrogenolysis reactions are discussed in Sec. 10.4.4.

Aldehydes and ketones have also been reduced by amine-boranes, such as pyridine- (17), isopropylamine- (18), phenylhydrazine- (19a), and morpholine-borane (19b).

10.4.2 Stereoselective Reduction of Ketones

The reduction of "unhindered" cyclic ketones with borane-tetrahydrofuran generally leads to the predominant formation of the more stable equatorial alcohols (8a,20). Use of sterically more demanding reagents, however, leads to the formation of increasing amounts of the less stable axial alcohols due to attack from the less hindered equatorial direction [steric approach control (21)]. Thus, use of dialkylboranes, such as disiamylborane and diisopinocampheylborane, provides consistent steric control in the reduction of both flexible monocyclic and rigid bicyclic ketones (22). A disadvantage of using such reagents is the slow rate of reduction achieved; thus reduction of camphor using diisopinocampheylborane requires 24 hr for completion of the reaction (23). This problem has been overcome by the use of lithium perhydro-9b-boraphenalylhydride (LiBPH) (2), which is prepared by refluxing cis,cis,trans-perhydro-9b-boraphenalene (Sec. 6.8) with lithium hydride in tetrahydrofuran (23).

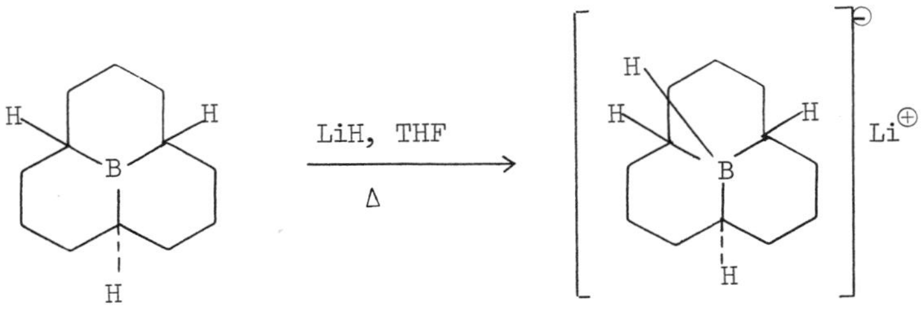

(2) LiBPH

This reagent exhibits high stereoselectivity in the reduction of
cyclic and bicyclic ketones and completely reduces hindered
ketones, such as camphor, within 0.5 hr at 0° (23).

   Some results of the reduction of cyclic and bicyclic ketones
with a variety of reagents are given in Table 10.1
In general, the stereoselectivity achieved using the above
dialkylboranes and LiBPH (2) exceeds that obtained using
substituted lithium aluminum hydride reagents such as lithium
trimethoxy- and tri-tert-butoxyaluminohydride (24a). It is
noteworthy that even in the case of relatively unhindered ketones,
such as 3- or 4-methylcyclohexanones, a significant shift from
the usual predominant attack from the axial direction to attack
from the equatorial direction occurs with LiBPH (2) (23). In
addition, the presence of boron trifluoride etherate in the
reaction mixture increases the percentage of axial alcohol
obtained in the diborane reduction of 3,3,5-trimethylcyclohexanone
from 66% to 85% (8a). LiBPH (2) has also been used in the
stereospecific reduction of E prostaglandins to Fα prostaglandins
(24b).

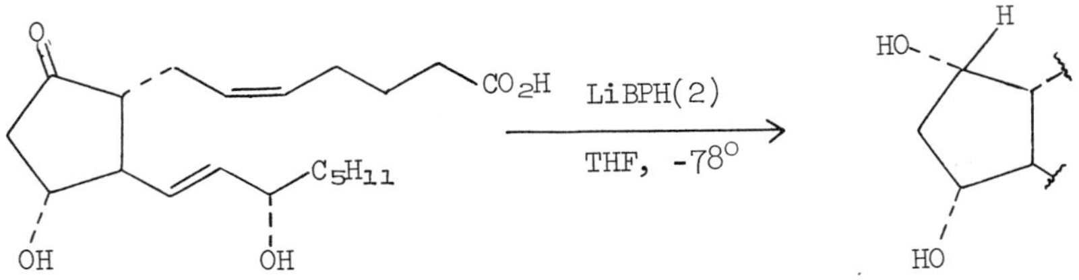

99%

Table 10.1

Reduction of Representative Cyclic and Bicyclic Ketones

Ketone	Axial alcohol: epimer	% axial alcohol formed[a]				
		BH$_3$ (22)	Sia$_2$BH (22)	(C$_6$H$_{11}$)$_2$BH (22)	(IPC)$_2$BH (22)	LiBPH(2) (23)
2-Methylcyclopentanone	Cis	25	78	80	94	94
2-Methylcyclohexanone	Cis	26	79	94	94	97
3-Methylcyclohexanone	Trans	12[b]	--	--	35	59
4-Methylcyclohexanone	Cis	15[b]	--	--	33	52
3,3,5-Trimethyl-cyclohexanone	Trans	66[b,c]	--	--	--	99
2-Methylcycloheptanone	Cis	74	64	97	98	--
Norcamphor	Endo	98	92	94	94	99
Camphor	Exo	52	65[d]	93[d]	100[d]	99

[a]Determined by vpc; total yields usually >85%

[b]Ref. 8a.

[c]Addition of BF$_3$·Et$_2$O raises % to 85%; Ref. 8a

[d]Very slow reduction.

10.4.3 Asymmetric Reduction of Ketones

Though diisopinocampheylborane (Sec. 3.5) has been used in
the stereoselective reduction of cyclic ketones (Sec. 10.4.2),
difficulty in separating the isomeric alcohols formed has
precluded the determination of the optical purities of the
individual compounds (20). The reagent has, however, been used
in the asymmetric reduction of a number of acyclic ketones, while
the deuterioreagent [(IPC)$_2$BD] has been used in the asymmetric
synthesis of 1-d-alcohols from aldehydes (Sec. 9.5). Some
results are summarized in Table 10.2.

It is noteworthy that reduction of the first two ketones in
Table 10.2 leads to alcohols of opposite configurations depending
on whether diglyme or tetrahydrofuran is used as solvent. If it
is assumed that the reduction of ketones with (-)-diisopino-
campheylborane proceeds in a manner analogous to the
hydroboration of terminal alkenes, then application of the Brown
transition-state model (54; Sec. 3.7) should allow predictions to
be made of the configurations of the alcohols formed in these
reductions. This model predicts the formation of (S)-alcohols as
found experimentally using tetrahydrofuran as solvent (25), but
in view of the probable nonconcerted nature of these reductions
(Sec. 10.1) use of such transition-state models is questionable.
The contradictory results might possibly be associated with
excessive dissociation of the reagent to triisopinocampheyldiborane
(Sec. 2.2.1d) or ageing of the reagent before use (see end of
Sec. 3.7). Diisopinocampheylborane has been used in the
resolution of a racemic mixture of ketones by selective reduction
of one enantiomer using a deficiency of the reagent (28).

Reaction of (-)-diisopinocampheylborane with methyl or
butyllithium gives the corresponding optically active
diisopinocampheylalkylborohydrides (3). Reduction of a series
of unsymmetrical ketones using these reagents in diglyme as
solvent gives optically active alcohols in optical yields of

Table 10.2

Asymmetric Reduction of Ketones Using Diisopinocampheylborane

Ketone	$(IPC)_2BH$ used	Solvent	Configuration of Product	Optical Purity (%)	Reference
$CH_3COC_2H_5$	(-)	DG	(R)-(-)	11	20
	(+)	THF	(R)-(-)	7	25
$CH_3COCH(CH_3)_2$	(-)	DG	(R)-(-)	17	20
	(-)	THF	(S)-(+)	12	25
	(+)	THF	(R)-(-)	20	25
$CH_3COC(CH_3)_3$	(-)	DG	(S)-(+)[a]	30	20
	(-)	THF	(S)-(+)	12	25
CH_3COPh	(-)	DG	(R)-(+)	14	20
$C_2H_5COCH(CH_3)_2$	(-)	THF	(S)-(-)	62	25
$(CH_3)_2CHCO(CH_2)_2OTHP$	(-)	THF	(S)-(-)	--	26
	(+)	THF	(R)-(+)	--	26

[a]The (S) configuration of (+)-3,3-dimethyl-2-butanol (pinacolyl alcohol) has been proved (27).

5-45% (29a). However, during studies of the synthesis of
prostaglandins (29b), it has been observed that reduction of
enone (6) with reagent (3) in the presence of hexamethylphosphoramide (HMPA) gives a mixture of the 15α- and 15β-alcohols
[(7) and (8), respectively] with the 15α-alcohol predominating
to the extent of >2 to 1 (29b). An even more selective reagent,
(5), is prepared by converting the trialkylborane (4)(formed
by treatment of either racemic or (+)-limonene with thexylborane,
Sec. 6.2) to the corresponding borohydride by reaction with
tert-butyllithium. Reduction of enone (6) with this reagent in
the presence of HMPA gives a predominance of the 15α-alcohol
over the 15β-alcohol of 4.5 to 1 (29b). Both of the above
reagents, (3) and (5), give only minor quantities (<3%) of the
α,β-reduction product (9), but use of (-)-diisopinocampheylborane
itself results in 90% reduction of the carbon-carbon double bond.

$$(-)-(IPC)_2BH \xrightarrow[\text{THF-Et}_2O]{\text{RLi}} (IPC)_2 \overset{\ominus}{B}\text{-H Li}^{\oplus} \quad (R=CH_3; \text{ n- or t-}C_4H_9)$$
$$\underset{R}{|}$$

(3)

(4)   (5)

PB   (6)   (7)

(PB=p-Phenylbenzoyl)

$$CH=CH-\underset{\underset{HO}{|}}{\overset{}{C}}-C_5H_{11} \quad + \quad CH_2CH_2COC_5H_{11}$$

(8)   (9)

Reagent	Reaction conditions	Products
$ZnBH_4$	$(CH_2OCH_3)_2$, $20°$	(7); 49%; (8), 49%
$(-)-(IPC)_2BH$	THF, $-45°$	(9), 90%
(3; $R=CH_3$)	THF-Et_2O, HMPA	(7), 66%; (8), 31%;
	-97 to $-100°$	(9), 3%
(3; $R=t-C_4H_9$)	THF-Et_2O, HMPA	(7), 68%; (8), 31%;
	-97 to $100°$	(9), 1%
(5)	THF-Et_2O-Pentane	(7) + (8); Ratio
	HMPA, $-120°$	4.5 to 1

Asymmetric reduction of ketones in low optical yields (3-5%) has also been achieved using optically active amine-boranes, such as S-amphetamine-borane or deoxyephedrine-borane (30).

10.4.4 Hydrogenolysis of Aldehydes and Ketones

In situ reduction of aromatic aldehydes and ketones containing electron-releasing substituents generally results in hydrogenolysis to give the corresponding hydrocarbon (12,31). However, varying results have been reported using externally generated diborane. Thus, in one case reduction of 4,4'-dimethoxy-diphenylketone is reported to give 85% of the corresponding hydrocarbon (12), while in another case the exclusive formation of the alcohol is reported (13). Addition of boron trifluoride to the reaction mixture results in exclusive formation of the hydrocarbon (13), indicating the hydrogenolysis might be attributed to the catalytic action of boron trifluoride which is carried over into the reaction mixture from the diborane generator.

Reduction using diborane generated from sodium borohydride and
iodine (Sec. 2.2.1a) gives only the alcohol, thus confirming the
catalytic action of boron trifluoride in the hydrogenolysis (32).
Some results are summarized in Table 10.3.

Table 10.3

Reduction of Substituted
Aromatic Aldehydes and Ketones with Diborane

Compound	Products[a] (% yield)			Reference
	In situ redn.	External redn.	External redn. + BF_3	
$4\text{-}CH_3OC_6H_4CHO$	Complex	A (92)		12
$3,4\text{-}(CH_3O)_2C_6H_3CHO$	H (95)	A (>95)		12, 13
$4\text{-}(CH_3)_2NC_6H_4CHO$	H (72)	H (76)	H (100)	13
$4\text{-}CH_3OC_6H_4COCH_3$	H (88)			31
$4\text{-}CH_3C_6H_4COCH_3$	H (91)			31
PhCOPh	H (90)			31
$(4\text{-}CH_3OC_6H_4)_2CO$	H (65)	A (100)	H (100)	12, 13
$(4\text{-}(CH_3)_2NC_6H_4)_2CO$	H (78)	H (90)		12

[a]A=alcohol; H=hydrocarbon.

Hydrogenolysis probably occurs via formation of an
intermediate alkoxyborane (10) which undergoes slow C-O bond
fission to give a positively charged species (11), which is
then rapidly reduced by hydride transfer (12).

$$X\text{-}C_6H_4COR \longrightarrow X\text{-}C_6H_4CHR \xrightarrow{-BO^{\ominus}} X\text{-}C_6H_4CHR \xrightarrow{H^{\ominus}} X\text{-}C_6H_4CH_2 R$$
$$\underset{(10)}{\overset{OB}{|}} \qquad (11)$$

The extent of hydrogenolysis is dependent on the stability of ion
(11), and is thus increased by strongly electron-donating

substituents, such as a para-dimethylamino group. The catalytic action of boron trifluoride involves coordination of the oxygen atom thereby promoting the C-O bond fission ($\underline{13}$).

Aromatic-cyclopropyl ketones (12) also undergo hydrogenolysis on reaction with borane-tetrahydrofuran in the presence of boron trifluoride ($\underline{13}$).

$$\text{ArCO-}\triangleleft \xrightarrow[\text{BF}_3 \cdot \text{Et}_2\text{O}]{\text{BH}_3\text{-THF}} \text{ArCH}_2\text{-}\triangleleft$$

(12) r.t. >70% (Ar = Ph, $4\text{-CH}_3\text{OC}_6\text{H}_4$,

$4\text{-ClC}_6\text{H}_4$)

Dicyclopropyl ketone is also hydrogenolyzed under these conditions, while cyclopropylmethylketone gives the corresponding alcohol ($\underline{13}$).

3-Acetyl- and benzoylindoles (13; R'=CH$_3$ or Ph) undergo rapid hydrogenolysis using either in situ conditions or externally generated diborane ($\underline{33}$). In the case of 3-glyoxylamides (13; R' = CONR$_2''$) hydrogenolysis of both carbonyl groups occurs ($\underline{33}$).

(13) (R = H, CH$_3$)   54-98% (R' = CH$_3$, Ph)

3-Formylindoles (13; R'=H) generally give low yields of the corresponding 3-methylindoles together with substantial quantities of dimeric products ($\underline{33}$). Reduction of 2-acetylindoles gives mainly the 2-ethyl derivatives, while 2-formylindole gives mainly polymeric material (60%) together with some hydroxymethylindole (32%) and 2-methylindole (8%) ($\underline{33}$).

2-Formylpyrrole and its N-methyl derivative give largely polymeric material on reduction with diborane, but when all the remaining nuclear positions are substituted, reasonable yields

(49-93%) of the corresponding methylpyrroles are isolated (33).
In the presence of another electron-withdrawing substituent on the
ring in direct conjugation with the formyl group the corresponding
hydroxymethylpyrrole is formed, while prolonging the reaction time
gives the corresponding methylpyrrole (33). Likewise, reduction
of pyrroketones (14) gives the corresponding pyrromethane (34).

(14)

This method has been used in the reduction of tetrapyrrolic ketone
intermediates in the synthesis of porphyrins (35). Similar
hydrogenolysis has been reported in the reduction of other
"electron-rich" ketones, such as 10-benzylidene-9-anthrones (36),
1-acylazulenes (37a), and xanthones and related compounds (Sec.
10.7.1).

10.4.5 Reduction of Cycloalkanediones

The hydroboration-oxidation of 1,3-cyclohexanediones gives
mainly trans-1,2-cyclohexanediols, and probably proceeds via the
formation of an allylic borate (Secs. 5.3.2 and 5.4) (37b).

R=H	62%	12%
R=Ph	25%	22%
R=CMe₃	30%	20%

In the hydroboration-oxidation of 1,2-cyclohexanedione the cis-1,2-diol is the major product (37b).

10.5 REDUCTION OF QUINONES

Reduction of p-benzoquinone with borane-tetrahydrofuran at 0° proceeds slowly to give hydroquinone, while the reaction with anthraquinone is extremely slow (10). Reaction with anthraquinone under in situ conditions gives anthracene (10). The reaction of quinones with disiamylborane is extremely slow (11).

10.6 REDUCTION OF CARBOXYLIC ACIDS AND DERIVATIVES

10.6.1 Reduction of Carboxylic Acids

The reaction of aliphatic and aromatic carboxylic acids with borane-tetrahydrofuran results in the uptake of 3 equivalents of hydride to give the corresponding alcohols (10). The reaction with aliphatic acids is extremely fast, being essentially complete in 15 min, while, in the case of aromatic acids, 12-24 hr are required for complete reaction (10).

The reduction of aliphatic carboxylic acids proceeds via the initial formation of triacyloxyboranes (15), which, in the case of lower acids such as ethanoic and propanoic acids, are further reduced (38). With higher aliphatic acids the triacyloxyboranes

rapidly dismute to form anhydrides and oxybisdiacyloxyboranes (15a)
which are then reduced (38).

$$6RCO_2H + B_2H_6 \longrightarrow 2(RCO_2)_3B \xrightarrow{B_2H_6} 2(RCH_2O)_3B$$

$$(15) \qquad\qquad (R = CH_3,\ C_2H_5,\ C_3H_7)$$

$$(RCO_2)O + [(RCO_2)_2B]_2O \xrightarrow{B_2H_6} 2(RCH_2O)_3B$$

$$(15a) \qquad\qquad (R = n\text{-}C_4H_9,\ n\text{-}C_5H_{11})$$

In the case of benzoic acid the course of the reaction is solvent
dependent, with reduction occurring via the dibenzoyloxyborane
in tetrahydrofuran and the tribenzoyloxyborane (15; R=Ph) in
diglyme (38).

The rapid reaction of carboxylic acids compared to esters
(Sec. 10.1) has been associated with interaction of the boron atom
in the triacyloxyborane with the neighboring oxygen atom as shown
in structure (16). Such interaction prevents a reduction in the
electrophilic character of the carbonyl carbon atom as occurs with
esters (Sec. 10.1) (7).

(16)

Reaction of carbohydrate carboxylic acids with diborane,
generated in situ, gives the corresponding hydroxymethyl
derivatives (39). Pyrrole 2-carboxylic acids are reduced to the
corresponding 2-methyl derivatives using externally generated
diborane, while indole 2-carboxylic acid gives mainly polymeric
material (33).

In the reaction with disiamylborane, both aliphatic and
aromatic acids liberate hydrogen quantitatively to form the
anhydride of the acid and hydroxydisiamylborane (disiamylborinic
acid), but no reduction occurs, probably due to steric effects (11).

10.6.2 Reduction of Carboxylic Acid Anhydrides

Acyclic anhydrides derived from aliphatic or aromatic carboxylic acids are readily reduced by borane-tetrahydrofuran to the corresponding alcohols (10,38). The reaction with cyclic anhydrides, such as succinnic and phthalic anhydride is, however, very slow (10). With disiamylborane, acetic anhydride rapidly consumes 2 equivalents of hydride with only slow reduction occurring subsequently, while cyclic anhydrides fail to react (11).

10.6.3 Reduction of Acid Chlorides

In general, the reduction of aliphatic and aromatic acid chlorides with borane-tetrahydrofuran proceeds very slowly, being only 50% complete in 48 hr (10). However, it has been found that acid chlorides containing electronegative substituents on the α-carbon atom are more readily reduced than the parent acid chlorides (40). Thus, the following rates of reduction are observed:

$$Cl_3CCOCl > ClCH_2COCl \ggg ClCH_2CH_2COCl \gg CH_3CH_2COCl, CH_3COCl >$$
$$PhCOCl$$

This unexpected reactivity can possibly be attributed to nucleophilic attack by the $B_2H_7^{\ominus}$ ion which exists in equilibrium with borane in tetrahydrofuran solution (Sec. 10.1) (41).

The reaction of α,β-unsaturated acid chlorides with borane in tetrahydrofuran or with disiamylborane has been reported to result in a limited amount of reduction of the acid chloride group (42). However, aliphatic and aromatic acid chlorides are generally inert to disiamylborane (11).

10.6.4 Reduction of Esters

Reaction of esters of aliphatic acids with borane-tetrahydrofuran at 0° results in the relatively slow (12-24 hr) uptake of 2

equivalents of hydride to give the corresponding alcohols (10).
Esters of aromatic acids, such as ethyl benzoate, react extremely
slowly (ethyl benzoate consumes 0.18 equivalents of hydride in
24 hr)(10). The presence of electron-withdrawing substituents in
the aromatic nucleus enhances the reactivity of aromatic esters,
while electron-releasing substituents retard the reaction (32).
These results are in accordance with the expected influence of
the aryl substituents on the electrophilic character of the
carbonyl carbon atom in the intermediate complex between the
esters and borane (Sec. 10.1). In addition, nuclear ester
substituents in both pyrroles and indoles are inert to borane
reduction (33).

The ester groups in ω-unsaturated esters are unusually
reactive toward reduction by diborane due to the operation of an
intramolecular reduction mechanism (Sec. 5.6). Treatment of
aliphatic esters with diborane in the presence of boron trifluoride
results in reduction to the corresponding ethers, the yields
depending on the nature of the alcohol segment of the ester (43).
In general, the extent of conversion into ether derivatives is
increased substantially as the alcohol segment is varied from
primary to tertiary. Thus, in the reduction of the three
isomeric butyl esters of 5β-cholanic acid the yields of the
corresponding ethers formed are: 7%, R=n-C$_4$H$_9$; 41%, R=s-C$_4$H$_9$;
76%, R=t-C$_4$H$_9$ (43). Esters of trimethylacetic acid (pivalic acid)
are likewise reduced to the corresponding neopentyl ethers using
in situ conditions; prolonged treatment of the ester of
trimethylacetic acid and tetrahydrolanosterol (17) with 3 mole
equivalents of borane-tetrahydrofuran also gives the corresponding
neopentyl ether in 70% yield (8c,44).

$$\text{Me}_3\text{CCO}- \quad \xrightarrow[\text{3 days}]{\text{BH}_3\text{-THF, r.t.}} \quad \text{Me}_3\text{CCH}_2\text{O}-$$

(17)   70%

Benzoates (15) and carbonates (45) are usually largely unaffected by treatment with diborane under in situ conditions.

In general, esters are inert to disiamylborane in tetrahydrofuran at $0°$ (11).

10.6.5 Reduction of Lactones

Treatment of γ-butyrolactone with borane-tetrahydrofuran at $0°$ for 24 hr results in the uptake of 2 equivalents of hydride to give the corresponding diol (10). However, treatment of steroid lactones with 1 equivalent of borane-tetrahydrofuran for 0.5 hr at room temperature gives good yields of the epimeric hemiacetal derivatives (8c,44,46). Prolonging the reaction period for 3 to 4 days give the corresponding vinyl ethers (8c,44), while addition of a further 2 equivalents of borane-tetrahydrofuran gives the corresponding ethers (8c,44).

Ethers are also formed when the reaction is carried out in the presence of boron trifluoride (47,48), while in situ reaction using sodium borohydride-aluminum chloride gives the corresponding diols (18) (47).

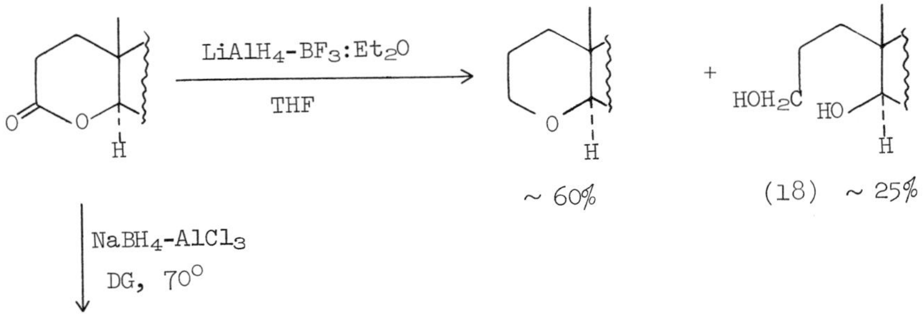

$\sim 60\%$ (18) $\sim 25\%$

(18) $\sim 80\%$

Attempted reduction of 2-oxybicyclo[2.2.2]octan-3-one (19) under similar conditions failed to give any ether derivative (49).

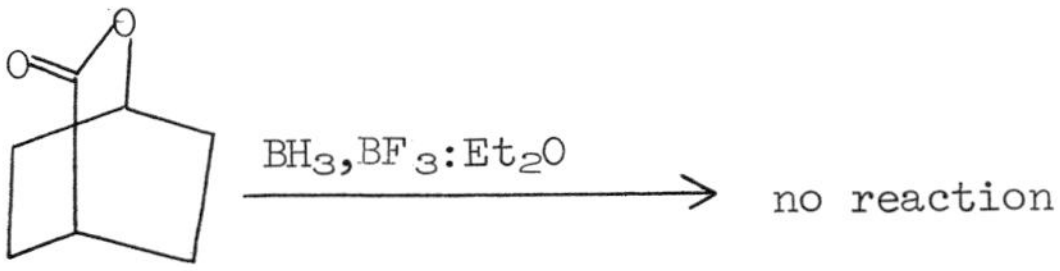

(19)

Reaction of γ-butyrolactone with disiamylborane gives the corresponding hydroxyaldehyde (11), while, in the case of steroidal γ-lactones high yields of the corresponding lactols are obtained (50,51).

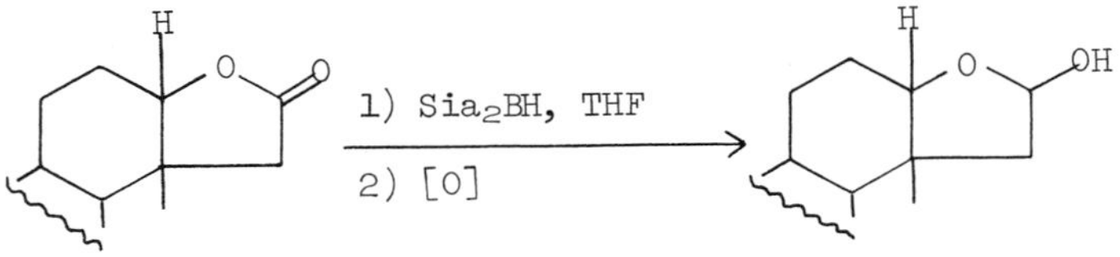

> 90%

Disiamylborane has also been extensively used in the reduction of free (52) and fully acylated (53) aldono-γ-lactones. In the latter cases high yields (80-100%) of the corresponding tetraacylfuranoses are obtained.

97%

The use of other dialkylboranes in the reduction of aldono-γ-
lactones has been investigated, but disiamylborane is generally
found to give the most satisfactory results (54).

The formation of a variety of products has been reported in
the hydroboration of coumarins. Thus, reaction of coumarin (20)
with diborane (externally generated) has been reported to give
products (21) to (23).

(20)   (21) .50%   (22) 57%   (23) 90%
            Ref. 55   Ref. 56   Ref. 32

Similar products have been obtained in the reactions of substituted
coumarin derivatives (32,55,56). The nature of the products
obtained appears to depend on the work-up procedure subsequent
to hydroboration, and can also probably be associated with the
catalytic action of boron trifluoride carried over into the
reaction mixture from the diborane generator.

10.6.6   Reduction of Amides

Reaction of primary amides with borane-tetrahydrofuran at
0° proceeds with evolution of hydrogen and very slow reduction
(50% reduction after 48 hr) (10). Reaction of N,N-disubstituted
amides is considerably faster, being complete in 24 hr (10).
However, refluxing the reaction mixture for 1 to 2 hr results in

the reduction of primary, secondary, and tertiary amides to the
corresponding <u>amines</u> in good yields (57).

$$RCONR'R'' \xrightarrow[\substack{1\text{-}2 \text{ hr} \\ 2)HC\ell}]{1)BH_3\text{-}THF, \Delta} \begin{array}{l} RCH_2NR'R'' \\ >80\% \end{array}$$

$$(R = tert\text{-}C_4H_9, \ n\text{-}C_5H_{11}, \ Ar)$$

$$(R', R'' = H, \ CH_3, \ C_2H_5, \ i\text{-}C_3H_7)$$

The reaction has also been successfully applied to the reduction
of N-aryl (58), and N-cyclopropyl (59) amides to the corresponding
amines. Reaction of the N,N-dimethylamide of the substituted
pyrrole 2-carboxylic acid (24) gives a mixture of the corresponding
dimethylaminomethyl, hydroxymethyl, and methylpyrroles (33,34).

$$R = \overset{\oplus}{N}Me_2\overset{\ominus}{B}H_3; \quad 40\%$$
$$R = OH; \quad 20\%$$
$$R = H; \quad 23\%$$

The hydroxymethyl and methyl derivatives probably arise via
elimination of the dimethylamino group as a complex with borane
(34).

Treatment of primary amides with disiamylborane in
tetrahydrofuran at 0° results in the rapid evolution of 2
equivalents of hydrogen, but no reduction is observed (11).
Tertiary amides consume 1 equivalent of hydride to form an
intermediate borane (25) which, on hydrolysis, gives the
corresponding <u>aldehydes</u> (11).

$$RCON(CH_3)_2 \xrightarrow[\text{THF, }0^\circ]{\text{Sia}_2\text{BH}} \underset{(25)}{\overset{\overset{\displaystyle OBSia_2}{\displaystyle |}}{RCH-N(CH_3)_2}} \xrightarrow{H_2O} \underset{72-89\%}{RCHO}$$

$(R = n\text{-}C_5H_{11}, Ph)$

10.6.7 Reduction of Lactams, Imides, and Hydrazides

2-Pyrrolidone is readily reduced to pyrrolidine in 80% yield using borane-tetrahydrofuran (60), and the pyrrolidone carbonyl group is selectively reduced in the presence of peptide bonds in polypeptides and proteins (60).

$$30\text{-}45\%$$

Six- (61) and seven-membered (62) lactams are also reduced to the corresponding cyclic amines. Attempted reduction of a nine-membered lactam in the synthesis of the indole alkaloid, quebrachamine, failed (63).

N-aryl succinimides (26) are reduced by borane-tetrahydrofuran at 0° to the corresponding pyrrolidines in 70-80% yields (64), while piperazine-2,6-diones (27) are reduced to the corresponding piperazines in about 60% yields on refluxing with borane in tetrahydrofuran (65). Likewise, 1,2-disubstituted perhydropyri-dazine-3,6-diones (28) give the corresponding perhydropyridazines in 70-85% yields (66); use of a large excess of borane (10 equivalents) results in cleavage of the N-N bond to give N,N'-disubstituted 1,4-diaminobutanes (66).

(26)   (27)   (28)

1,2-Diacylhydrazines are converted to 1,2-dialkylhydrazines in
50-65% yields on treatment with borane in diglyme at 130° (66),
while 1,2-diacyl-1,2-dialkylhydrazines form the corresponding
tetraalkylhydrazines in 60-80% yields on refluxing with borane in
tetrahydrofuran (66).

10.7   REDUCTION OF OTHER OXYGEN COMPOUNDS

10.7.1   Reduction of Pyrones

Hydroboration-oxidation of chromone (29; R=H) gives 79% of
3-chromanol (30) (55,56), while 2-methylchromone (29; R=CH₃) gives
45% of the diol (31) (55).   In the reaction of flavone (29; R=Ph)
ring cleavage occurs to give (32) in 73% yield (56).

(29)   (30)   (31)   (32)

The formation of flav-2-en-4-ol (67), as well as a number of
ring-cleaved products differing from (32) (32,55), has also been
reported in the reaction of flavone. The reaction of isoflavone
is reported to give isoflav-2-en-4-ol in 70% yield (67).

Xanthone (33; X=O), thioxanthone (33; X=S), and 9(10H)-
acridone (33; X=NH) all undergo hydrogenolysis (Sec. 10.4.4) to
give the corresponding methylene derivatives (34) (<u>68</u>).

(33)   (34)

10.7.2 Reduction of Epoxides

The rate of reduction of epoxides with borane-tetrahydrofuran
at 0° is dependent on the structure of the epoxide. Thus,
1,2-butene oxide and cyclohexene oxide consume 1 equivalent of
hydride in 2 to 3 days, while styrene oxide consumes 2.13
equivalents of hydride in 6 hr at 25° (<u>10</u>). Only 18% of
2-phenylethanol is formed from styrene oxide (<u>10</u>), and the
reaction has been shown to proceed with extensive rearrangement
(<u>69</u>).

Reaction of 1-methyl-1,2-cyclohexene oxide with borane-
tetrahydrofuran at 25° results in the uptake of 1.92 equivalents
of hydride in 24 hr and the evolution of 1 mole equivalent of
hydrogen (<u>10</u>); oxidation gives a mixture of 2-hydroxymethylcyclo-
hexanols in 60-70% yield (<u>10</u>).

60-70%

Similar results have been obtained in the reaction of epoxides
derived from alkenes related to α- and β-pinene (<u>70</u>). In these

cases the formation of intermediate allylic alcohols has been observed.

The formation of allylic alcohols has been shown to require the presence of neighboring hydrogen atom cis to the epoxide oxygen atom (70a). Thus, epoxide (35) reacts via the formation of the allylic alcohol (36), while the epimeric epoxide (37) fails to react.

(35)   (36)   (37)

The rearrangement possibly proceeds via initial coordination of the borane by the epoxide oxygen, followed by evolution of hydrogen and epoxide ring cleavage (38) (70a).

(38)

Prolonged treatment with an excess of borane can lead to hydroboration of the intermediate allylic species followed by elimination-rehydroboration (70b).

The presence of small amounts of sodium or lithium borohydride greatly increases the rate of reaction of epoxides with borane-tetrahydrofuran and results in the formation of mono-alcohols in yields of 41-100% (71). The alcohol formed by anti-Markownikoff opening of the epoxide ring predominates.

This method thus provides a convenient route to cis-2-substituted cycloalkanols and supplements the synthesis of trans-2-substituted

cycloalkanols by hydroboration-oxidation of 1-substituted
cycloalkenes (Sec. 3.1.5).

Reaction of epoxides derived from arylethenes with borane-
tetrahydrofuran in the presence of boron trifluoride also results
in anti-Markownikoff reductive opening of the epoxide ring to give
high yields of the corresponding mono-alcohols (72).

The reaction probably proceeds via boron trifluoride catalyzed
rearrangement of the epoxide to the carbonyl compound, followed
by reduction of the carbonyl group (72).

In general the reaction of epoxides with disiamylborane is
slow (11).

10.7.3   Reduction of Ethers and Acetals

Under the usual hydroboration conditions ethers are generally
stable. However, at high temperatures significant reaction with
diborane can occur. Thus, heating of tetrahydrofuran with
diborane at $60°$ for 64 hr gives about 20% of tributoxyborane (73);
similar results have been reported for various other ethers (74).
In situ generation of borane in diglyme at room temperature,
followed by oxidation, gives some diethylene glycol monoether (75),
while, in the presence of a thiol, formation of the corresponding
methyl sulfide has been observed (76).

$$PhSH + NaBH_4 + BF_3 \cdot Et_2O \xrightarrow[\text{20 hrs}]{\text{DG, r.t.}} PhSCH_3 \; 89\%$$

The cleavage of acetals to hydroxy-ethers using in situ
conditions and externally generated diborane has been reported (77).

Similar cleavage of steroidal ethylene acetals has also been reported, but, if precautions are taken to exclude boron trifluoride from the reaction mixture, no cleavage occurs (78).

10.8 REDUCTION OF NITROGEN COMPOUNDS

10.8.1 Reduction of Nitro Compounds

Aliphatic and aromatic nitro compounds are relatively inert to borane reduction (10), an exception being 2-nitrophenylethene (β-nitrostyrene) which is reduced to 2-phenylethylhydroxylamine in good yield by borane-tetrahydrofuran (79a). Aliphatic nitro compounds fail to react with disiamylborane, but slow reaction occurs with nitrobenzene (11).

Salts of primary and secondary nitro compounds are, however, reduced to the corresponding N-monosubstituted hydroxylamines in 30-60% yields (79b).

$$R_2CHNO_2 \xrightarrow[\text{r.t., 24 hr}]{\text{KOH, EtOH}} [R_2C=NO_2]^{\ominus} K^{\oplus} \xrightarrow[\text{2)KOH}]{\text{1)BH}_3\text{-THF}} R_2CHNHOH$$

$$[R_2 = -(CH_2)_5-]\ 50\%$$

In the case of potassium 9-fluorenenitronate, 39% of the corresponding oxime is obtained indicating that the reduction probably proceeds via the oxime (79b).

10.8.2 Reduction of Nitroso Compounds

Aromatic nitroso compounds are reduced by borane in tetrahydrofuran at room temperature to the corresponding amines in 60-90% yields (80).

$$ArN=O \xrightarrow[\text{2)H}^{\oplus}\text{ or OH}^{\ominus}]{\text{1)BH}_3\text{-THF}} ArNH_2$$
$$(60-90\%)$$

Gem-nitronitroso and chloronitroso compounds are reduced to the corresponding hydroxylamines in 60-70% yields (80).

$$R_2\overset{\overset{\displaystyle X}{\displaystyle |}}{C}-N=O \xrightarrow[\text{THF}]{2BH_3} \underset{60\text{-}70\%}{R_2CHNHOH}$$

$(X=NO_2, \; C\ell)$

$(R=CH_3, \; -(CH_2)_5-)$

10.8.3 Reduction of Oximes and Derivatives

Treatment of aldoximes and ketoximes with borane-
tetrahydrofuran at $<5°$, followed by acid or base hydrolysis,
gives the corresponding N-monosubstituted hydroxylamines in 50-90%
yields (9a,81,82). Similar results have been obtained using
sodium borohydride on silica gel and a benzene solution of the
oxime (83).

$$RR'C=NOH \xrightarrow[\substack{<5° \\ 2)H^{\oplus} \text{ or } OH^{\ominus}}]{1)BH_3\text{-}THF, \; 4 \text{ hr}} \underset{50\text{-}90\%}{RR'CHNHOH}$$

$(R,R'=H, \text{ Alkyl})$

Use of the above reaction conditions fails to reduce diaryl-
ketoximes, such as the oximes of benzophenone and fluorenone (9a).
However, reaction in diglyme-tetrahydrofuran at $110°$ reduces
benzophenone oxime to the hydroxylamine (84). Likewise,
1,3-diphenylpropanone oxime is not reduced below $65°$; at $85\text{-}90°$
it is converted to the corresponding hydroxylamine while at
$105\text{-}110°$ the only product is the amine (84).

$$\underset{77\%}{(PhCH_2)_2CHNHOH} \xleftarrow[85\text{-}90°]{BH_3,THF\text{-}DG} (PhCH_2)_2C=NOH \xrightarrow[BH_3,THF\text{-}DG]{105°\text{-}110°} \underset{74\%}{(PhCH_2)_2CHNH_2}$$

Reduction of dialkylketoximes or their hydroxylamine derivatives
at $105\text{-}110°$ also gives the corresponding amines (84). In general,
treatment of oximes with disiamylborane results in hydrogen
evolution, but no reduction occurs (11).

Reaction of _oxime ethers_ with diborane in refluxing tetrahydrofuran, followed by basic hydrolysis, gives the corresponding _amines_ in 50-90% yields (84). Since O-methyl oximes are readily prepared in high yield by treatment of aldehydes and ketones with methoxyamine hydrochloride (85), the above reductions are usually carried out using the methyl derivatives.

$$RR'C=O \xrightarrow[\text{py., r.t.}]{CH_3ONH_2 \cdot HC\ell} RR'C=NOCH_3 \xrightarrow[\substack{\Delta,\ 2\ hr \\ 2)\ OH^{\ominus}}]{1)BH_3-THF} RR'CHNH_2 \quad 50\text{-}90\%$$

Similar results are obtained on treating _oxime-esters_ with borane-tetrahydrofuran at room temperature (84,86). Thus, ketoxime acetates or tosylates give the corresponding amines in 60-75% yields.

$$R_2C=NOCOR' \xrightarrow[2)\ H^{\oplus}\ or\ OH^{\ominus}]{1)\ BH_3-THF} R_2CHNH_2 \quad 60\text{-}75\%$$

(R = Alkyl or aryl)
(R' = CH_3)

10.8.4 Reduction of Imines and Schiff Bases

Reaction of 3,4-dihydroisoquinolines (39) with borane-

(39) (R', R'' = H, OCH_3)   (40) 69-82% (R = H, CH_3)

(41) 78-87%

tetrahydrofuran gives the corresponding amine-boranes (40) (87a).
Treatment of the adducts (40; R=H) with acid gives the
corresponding saturated compounds (41; R=H), while the l-methyl
derivatives (40; R=CH$_3$) give only the original dihydroisoquinolines
(39; R=CH$_3$).

Similar treatment of the N-methyl derivatives gives the
corresponding saturated compounds (87a).

Reaction of 2-alkyl-1-piperideines (42) with either
(-)-diisopinocampheylborane or (-)-triisopinocampheyldiborane
results in asymmetric reduction to give the corresponding
piperidines of optical purity 2-11% (88). Use of lithium
diisopinocampheylbutylborohydride (Li$^\oplus$ (IPC)$_2$BuBH$^\ominus$), formed by
reaction of (-)-diisopinocampheylborane and butyllithium (Sec.
10.4.3), gives improved optical yields, but the piperidines are
of opposite configuration to those obtained using the borane
reagents mentioned above (88).

R=CH$_3$; 100% (~20% opt. pure)

R=C$_3$H$_7$; 30-50% (~4% opt. pure)

Schiff bases are readily reduced by either borane-tetrahydrofuran (87b) or dimethylamine-borane in glacial acetic acid (89a)
to give high yields (>80%) of the corresponding secondary
amines.

$$\text{ArCH=NAr'} \quad \xrightarrow[\substack{(CH_3)_2NH\text{-}BH_3,\ HOAc \\ \Delta,\ 15\ min}]{BH_3,\ THF,\ 3\text{-}5^\circ\ \underline{or}} \quad \begin{array}{c} \text{ArCH}_2\text{NHAr'} \\ >80\% \end{array}$$

Prolonged refluxing of Schiff bases with dimethylamine-borane in glacial acetic acid results in reductive acetylation to give the corresponding N-acetyl amines (89b). Treatment of the tosylhydrazone of 5α-cholestan-3-one with diborane, followed by refluxing with water, gives cholestane in high yields (89c).

10.8.5 Reduction of Nitriles and Isonitriles

Reaction of nitriles with borane-tetrahydrofuran gives the corresponding N,N,N-trialkylborazoles which, on acid hydrolysis, give the corresponding amines (10).

$$3RCN \xrightarrow[0^{\circ}]{3 BH_3\text{-}THF} \quad \text{(borazole)} \quad \xrightarrow{H^{\oplus}} 3RCH_2NH_2$$

The reaction with disiamylborane is very slow, eventually giving the corresponding amines (11).

Isonitriles react with diborane or disiamylborane to give the corresponding 2,5-diboradihydropyrazines (90).

$$2R\text{-}N\equiv C \xrightarrow[Et_2O,\,-80^{\circ}]{BH_3} \quad \text{(diboradihydropyrazine)} \quad (R=CH_3,\ C_6H_{11},\ Ph)$$

10.8.6 Reduction of Azo-Compounds, Azides, and N-Oxides

Azobenzene reacts with borane-tetrahydrofuran at 0° with the uptake of 2 equivalents of hydride (10), but is inert to disiamylborane (11). Only 50% aniline is isolated in the former reduction, possibly due to the formation of stable nitrogen intermediates (10).

Treatment of β-iodo azides with diborane in tetrahydrofuran
at 50° gives the corresponding amines which form aziridines on
treatment with base (91).

87%

Azoxybenzene fails to react with borane-tetrahydrofuran (10)
but forms azobenzene on treatment with disiamylborane (11).
Pyridine-N-oxide reacts slowly with borane-tetrahydrofuran to
give products in which attack on the aromatic ring is evident (10).
Reaction with disiamylborane, however, gives only pyridine (11).

10.9 REDUCTION OF SULFUR COMPOUNDS

Disulfides, sulfides, sulfones, and tosylates are inert to
both diborane and disiamylborane in tetrahydrofuran at 0° (10,11).
However, bis(triphenylmethyl)disulfide and triphenylmethylthiol
both undergo C-S bond cleavage on treatment with borane in
benzene (92). Thioketals derived from saturated steroid ketones
are inert to diborane in ether at 20°, but those derived from
α,β-unsaturated steroid ketones react to give complex mixtures
of alcohols (93). Sulfonic acids evolve hydrogen on treatment
with borane-tetrahydrofuran but are not reduced, while dimethyl
sulfoxide is reduced to dimethyl sulfide (10). Thioxanthone
(33; X=S) undergoes hydrogenolysis to the corresponding methylene
derivative (34; X=S) (Sec. 10.7.1) (68).

10.10 REDUCTION OF HALOGEN COMPOUNDS

Alkyl chlorides and bromides are generally inert to borane-
tetrahydrofuran, but the N(2-iodoethyl)amide (43) reacts with
borane-tetrahydrofuran at 60° to give the secondary amine (44) (94).

$$3,4\text{-}Cl_2C_6H_3CONHCH_2CH_2I \xrightarrow[\text{2) KOH \quad 3) HBr}]{\text{1) BH}_3\text{-THF, }60^\circ} 3,4\text{-}Cl_2C_6H_3CH_2\overset{\oplus}{\underset{H}{N}}HCH_2CH_3 \ Br^{\ominus}$$

$$(43) \hspace{6cm} (44)$$

Treatment of aralkyl chlorides and bromides with diborane in nitromethane (95) or triethylamine-borane in nitromethane (96) gives the corresponding hydrocarbons in yields of 50-98%. The rates of reduction and yields of hydrocarbons are dependent on the stability of the carbonium ions formed from the halides, being greater for more stable ions.

$$(CH_3)_2\overset{Ph}{\underset{|}{C}}\text{-}Cl \xrightarrow[15^\circ, \ 16 \ hr]{B_2H_6, \ CH_3NO_2} (CH_3)_2CHPh \hspace{1cm} 82\%$$

$$Ph_2CHBr \xrightarrow{\hspace{3cm}} Ph_2CH_2 \hspace{1cm} 92\%$$

10.11 REDUCTION OF CARBON-CARBON DOUBLE BONDS: PROTONOLYSIS OF ORGANOBORANES

The scope of the hydroboration reaction as applied to alkenes is discussed in Sec. 3.1.

Organoboranes, on refluxing with propanoic acid in diglyme, give the corresponding hydrocarbons in high yields (97). Faster reaction and higher yields can often be achieved using octanoic acid and triglyme (97). The reaction thus provides a useful method for the reduction of alkenes to alkanes via hydroboration.

$$3RCH=CH_2 \xrightarrow{B_2H_6, \ DG} (RCH_2CH_2)_3B \xrightarrow[\Delta, \ 2 \ hr]{C_2H_5CO_2H, \ DG} 3RCH_2CH_3$$
$$70\text{-}95\%$$

Secondary alkyl groups undergo protonolysis less readily than primary groups (97). However, in the reaction of unsymmetrical trialkylboranes little selectivity is exhibited in the protonolysis of primary and secondary alkyl groups, but tertiary alkyl groups react far more slowly because of steric hindrance to proton transfer in the transition state (45) (see below) (98).

Protonolysis proceeds with retention of configuration (99,100). The mechanism of the reaction probably involves prior coordination of the boron atom by the acid carbonyl group which simultaneously weakens the B-C bond of the organoborane and probably increases the electrophilic character of the acid hydrogen (101). Rate-determining proton transfer occurs via transition state (45).

$$R_3B + C_2H_5CO_2H \longrightarrow R \overset{B \leftarrow O}{\underset{H-O}{\diagup}} C-C_2H_5 \rightleftharpoons R \overset{B\cdots\cdots O}{\underset{H\cdots\cdots O}{\diagup}} C-C_2H_5$$

$$\downarrow (45)$$

$$RH + R_2BO_2CC_2H_5$$

The removal of the first alkyl group is faster than the remaining two, presumably as a result of the lower coordinating ability of the di- and monoalkylboron propanoate species.

Since protonolysis of organoboranes generally proceeds with retention of configuration, the reduction of double bonds by hydroboration-protonolysis results in cis addition of hydrogen to the double bond. Thus, (+)-3-carene [(8); Sec. 3.1.6a] gives (-)-cis-carane (102), but attempted protonolysis of the organoborane derived from (+)-2-carene [(7); Sec. 3.1.6 a] failed (103). However, hydroboration-protonolysis of 1,2,4-trimethyl-3-phenyl-3-pyrazoline (46) unexpectedly gives a mixture of the cis- and trans-pyrazolidines (47 and 48) (104).

| 1)BH$_3$,Et$_2$O |
| 2)C$_2$H$_5$OH |
| 3)CH$_3$CO$_2$H |

(46)   (47) 30%   (48) 20%

Hydroboration of acenaphthylene (49), followed by treatment with acetic acid, gives very little acenaphthene but substantial amounts of 3-acetylacenaphthene (50) (105a). The formation of (50) probably arises from intramolecular acylation involving the intermediate formed from coordination between the organoborane and acetic acid.

(49)   (50) 25-45%

The protonolysis of certain allylic organoboranes has been observed to proceed with allylic rearrangement. Thus, protonolysis of the 6α-boryl derivative of 5α-cholest-7-en-3β-ol gives 5α-cholest-6-en-3β-ol (Sec. 6.4) (105b), while allylic organoboranes derived from the monohydroboration of monosubstituted and 1,1-disubstituted acyclic allenes likewise undergo allylic rearrangement (105c).

The protonolysis of vinylboranes is usually achieved using refluxing acetic acid (Chap. 7). Vinylboranes containing functional groups which are destroyed by this treatment are, however, readily cleaved by treatment with aqueous 2 M silver ammonium nitrate complex at 75-80° (Sec. 7.1.1) (106a).

In cases where a relatively stable carbanion can be formed, the carbon-boron bond of organoboranes is cleaved by dilute alkali or even water. Thus, phenyl-substituted fulvene

derivatives (51; R=H or Ph), on hydroboration followed by
treatment with water, give the corresponding benzyl or benzhydryl
derivatives (52; R=H or Ph) (106b).

(51) (R=H, Ph; X=H, Br, OCH₃)   (52) 50-75%

Hydroboration of cis- and trans-1,2-dimethyl-1,2-diphenyl-
ethene, followed by treatment with dilute sodium hydroxide, gives
meso- and d,ℓ-2-3-diphenylbutane, respectively, with a high degree
of stereoselectivity (107a). However, treatment of dibutoxy-1-
phenylethylborane with sodium deuteroxide in deuterium oxide
proceeds by an S_E1 mechanism to give racemic deuterated
product (100).

Treatment of indoles with borane-tetrahydrofuran, followed
by sodium methoxide in methanol, gives the corresponding
indolines (107b); however, treatment with acetone or methanol in
place of sodium methoxide in methanol, gives the unchanged
indoles (107b).

(R, R′=H, CH₃, C₂H₅)

In the case of N-methylindoles no reduction to indolines occurs.
These reactions have been explained in terms of formation of an
intermediate aminoborane species (107b).

The protonolysis of certain unhindered trialkylboranes, such
as triethylborane, by water or alcohols is greatly accelerated by
catalytic amounts of diethylboryl pivalate (107c).

$$ROH + (C_2H_5)_3B \xrightarrow{Et_2BOCOCMe_3} ROB(C_2H_5)_2 + C_2H_6$$

The reaction proceeds smoothly for a variety of alcohols and
polyhydroxy compounds and can be used for the determination of
the hydroxyl content of such molecules (107c). The reaction also
proceeds with amino compounds (107c), and gives diethylboryl
enolates with enolizable ketones (107d).

$$RCH_2COR^1 + (C_2H_5)_3B \xrightarrow[50\text{-}70^\circ]{Et_2BOCOCMe_3} RCH=COB(C_2H_5)_2 + C_2H_6$$
$$\overset{}{\underset{R^1}{|}}$$

(R=CH_3, C_2H_5; R^1=C_2H_5, C_3H_7, Ph)   60-82%

With five- and six-membered cycloalkanones, aldol condensations
occur during the reaction (107d).

10.12 REDUCTIVE CLEAVAGE OF CYCLOPROPANES

The vapor phase reaction of cyclopropane with diborane at
95°, followed by treatment with water, gives propane in 52%
yield (108). In the case of bicyclo[4.1.0]heptane, reaction at
100° in the absence of solvent, followed by oxidation, gives a
good yield of cyclohexylmethanol (109); no reaction occurs in the
presence of tetrahydrofuran. When 1-methylbicyclo[4.1.0]heptane
(51) is treated in a similar manner, 2 methylcyclohexylmethanol
is the main product with none of the 1-methyl derivative being
formed (109). This indicates a preference for fission of the

primary-tertiary rather than the primary-secondary carbon-carbon
bond, but the cleavage is not stereospecific (109).

Some cleavage of the cyclopropyl ring occurs in the
hydroboration-protonolysis of the vinylcyclopropane (52) (110).

Cleavage under far milder hydroboration conditions occurs in the
case of α-gurjunene [(17); Sec. 3.1.6a] though similar
hydroboration of a number of vinylcyclopropanes proceeds without
ring cleavage (Secs. 3.1.1 and 3.1.3).

10.13 SELECTIVE REDUCTIONS OF FUNCTIONAL GROUPS

     Although diborane reduces a wide variety of functional
groups, the reaction conditions required vary considerably from
group to group. The rates of reaction of a number of organic
compounds containing representative functional groups with
diborane (10) and disiamylborane (11) in tetrahydrofuran at 0°
have been studied. The approximate relative reactivities of some
representative compounds with borane-tetrahydrofuran and
disiamylborane in tetrahydrofuran at 0° are given in Tables 10.4
and 10.5, respectively. The compounds are listed in approximate
order of decreasing reactivity.

TABLE 10.4

Reactivity of Some Representative Organic

Compounds With Borane-Tetrahydrofuran at 0° (10)

Compound	Relative reactivity[a]	Compound	Relative reactivity[a]
$C_5H_{11}CO_2H$	<0.25	Ph_2CO	24
PhCHO	0.25	$PhCON(CH_3)_2$	
Alkynes	b	$(CH_3CO)_2O$	~24
Alkenes	c	CH_3CO_2Ph	>24
$CH_3COC_5H_{11}$, Norcamphor	1	Cyclic anhydrides	
$C_5H_{11}CHO$, PhCH=CHCHO		Acid chlorides	
$PhCOCH_3$	2	$PhCO_2C_2H_5$	>>>24
$C_5H_{11}CON(CH_3)_2$		Primary amides	
$C_5H_{11}CO_2C_2H_5$	12	Nitriles	
$PhCO_2H$		Epoxides	
γ-Butyrolactone	~12	Ethers, ketals, halides	
		Nitro-compounds, disulfides	Inert
		Sulfides, sulfones	

[a]Time in hours required for complete reduction.

[b]Ref. 111. Triple bonds are selectively reduced in enyne systems; Sec. 7.5.

[c]Reactivity determined by competition experiments (112). However, it must be noted that reduction of α,β-unsaturated aldehydes and ketones proceeds with prior reduction of the carbonyl group (Sec. 5.4).

It must be emphasized that the reactivities quoted in Tables 10.4 and 10.5 are very approximate. In many cases initial reaction is rapid but completion of reduction is slow (10, 11). In addition, the reactivity of a group can be greatly affected by

TABLE 10.5

Reactivity of Some Representative Organic Compounds
With Disiamylborane in Tetrahydrofuran at $0°$ (11)

Compound	Relative reactivity[a]	Compound	Relative reactivity
PhCH=CHCHO[b]	0.5	Ph$_2$CO	24
PhCHO, C$_5$H$_{11}$CHO		(CH$_3$CO)$_2$O	
CH$_3$COC$_5$H$_{11}$	1	Nitriles	>>>24
Norcamphor		PhNO$_2$	
γ-Butyrolactone		Carboxylic acids[e]	
Alkynes	c	Acid chlorides	
Alkenes	d	Cyclic anhydrides	
PhCON(CH$_3$)$_2$	3	Esters, primary amides	Inert
C$_5$H$_{11}$CON(CH$_3$)$_2$		Epoxides[f]	
PhCOCH$_3$	6	Oximes	

[a]See (a) in Table 10.4.

[b]Selective reduction of the aldehyde group occurs (See 10.4.1).

[c]Ref. 111; Triple bonds are selectively reduced in enyne
systems; Sec. 7.5.

[d]The reactivity is strongly dependent on alkene structure
(Sec. 3.3). Reduction of α,β-unsaturated aldehydes and
ketones proceeds with prior reduction of the carbonyl group
(Sec. 5.4).

[e]Rapid liberation of hydrogen is observed (Sec. 10.6.1).

[f]1-Methyl-1,2-cyclohexene oxide and styrene oxide do
react (11).

its molecular environment, and such factors must be considered in
attempting selective reductions of one group in the presence of
others.

Use of controlled reaction conditions and limited quantities of diborane do permit the selective reduction of various functional groups. Thus, selective reduction of carboxyl groups in peptides and proteins (113), and in the presence of esters (114) and keto groups (115) has been reported. The selective reduction of N-monosubstituted and N,N-disubstituted amides in the presence of ester groups has been achieved (116), while side-chain esters are reduced in the presence of nuclear benzyl ester groups in substituted pyrroles (117). The reduction of aliphatic esters in the presence of more reactive groups can be suppressed by carrying out the reductions in the presence of ethyl acetate (33,35).

Use of disiamylborane permits a marked degree of selectivity to be achieved in the reduction of functional groups. In addition, amine-boranes, such as phenylhydrazine-borane, exhibit a high degree of selectivity; thus, aldehydes, ketones, and acid chlorides are readily reduced, while esters, carboxylic acids, amides, nitro-compounds, and halides are not affected (19a).

10.14 PROTECTION OF GROUPS DURING HYDROBORATION

Keto groups may be protected by the formation of ethylene acetal groups though precautions must be taken to exclude traces of boron trifluoride from the reaction mixture (Sec. 10.7.3). 2,4-Dinitrophenylhydrazones are also effective protecting groups; the group is removed by ozonolysis in ethyl acetate at -78° (118).

Carboxylic acids may be protected by formation of the sodium salt (7) or the acid chloride. Alternatively, use of disiamyl-borane avoids reduction of the carboxyl group (11).

Triple bonds may be protected by formation of the corresponding alkyne-dicobalt hexacarbonyl complexes (119). Reaction of enynes with dicobalt octacarbonyl in a hydrocarbon solvent at room temperature gives the enyne complexes in yields of 70-90%. Hydroboration of the complexes results in exclusive reaction at the double bond. The triple bond may be regenerated

by oxidative degradation of the complex with ferric nitrate in
95% ethanol (119).

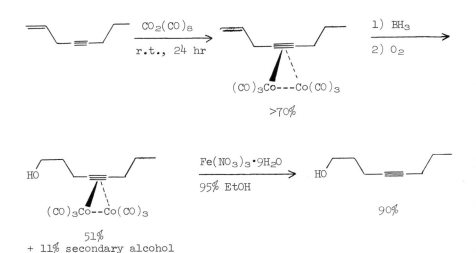

(CO)$_3$Co---Co(CO)$_3$

>70%

HO

Fe(NO$_3$)$_3$·9H$_2$O
————————————→
95% EtOH

HO

(CO)$_3$Co--Co(CO)$_3$

51%
+ 11% secondary alcohol

90%

REFERENCES

1.   H. C. Brown, Hydroboration, Benjamin, New York, 1962, p. 242.

2.   H. C. Brown and N. M. Yoon, J. Am. Chem. Soc., 88, 1464 (1966);
N. M. Yoon and H. C. Brown, ibid., 90, 2927 (1968).

3.   H. C. Brown, P. M. Weissman, and N. M. Yoon, J. Am. Chem.
Soc., 88, 1458 (1966).

4.   H. C. Brown and P. M. Weissman, J. Am. Chem. Soc., 87, 5614
(1965).

5.   H. C. Brown and P. M. Weissman, Israel J. Chem., 1, 430 (1963).

6.   J. Málek and M. Cérny, Synthesis, 1972, 217.

7.   H. C. Brown and B. C. Subba Rao, J. Am. Chem. Soc., 82, 681
(1960).

8.   (a) J. Klein and E. Dunkelblum, Tetrahedron, 23, 205 (1967);
(b) J. Klein, personal communication;   (c) J. R. Dias and
G. R. Pettit, J. Org. Chem., 36, 3485 (1971).

9. (a) H. Feuer, B. Vincent, and R. S. Bartlett, J. Org. Chem., 30, 2877 (1965); (b) A. S. Howard, personal communication.

10. H. C. Brown, P. Heim, and N. M. Yoon, J. Am. Chem. Soc., 92, 1637 (1970).

11. H. C. Brown, D. B. Bigley, and N. M. Yoon, J. Am. Chem. Soc., 92, 7161 (1970).

12. K. M. Biswas, L. E. Houghton, and A. H. Jackson, Tetrahedron, Suppl. 7, 261 (1966).

13. E. Breuer, Tetrahedron Letters, 1967, 1849.

14. G. R. Pettit, B. Green, P. Hofer, D. C. Ayres, and P. J. S. Pauwels, Proc. Chem. Soc., 1962, 357.

15. G. R. Pettit, B. Green, G. L. Dunn, P. Hofer, and W. J. Evers, Can. J. Chem., 44, 1283 (1966).

16. R. Fellous, R. Luft, and A. Puill, Tetrahedron Letters, 1970, 1509.

17. R. P. Barner, J. H. Graham, and M. D. Taylor, J. Org. Chem., 23, 1561 (1958); E. M. Fedneva, Zhur. Obshchei Khim., 30, 2818 (1960); Chem. Abstr., 55, 16461 (e) (1961).

18. H. Nöth and H. Beyer, Chem. Ber., 93, 1078 (1960).

19. (a) R. J. Baumgarten and M. C. Henry, J. Org. Chem., 29, 3400 (1964); (b) S. S. White and H. C. Kelly, J. Am. Chem. Soc., 92, 4203 (1970).

20. H. C. Brown and D. B. Bigley, J. Am. Chem. Soc., 83, 3166 (1961).

21. E. L. Eliel and Y. Senda, Tetrahedron, 26, 2411 (1970).

22. H. C. Brown and V. Varma, J. Am. Chem. Soc., 88, 2871 (1966).

23. H. C. Brown and W. C. Dickason, J. Am. Chem. Soc., 92, 709 (1970).

24. (a) H. C. Brown and H. R. Deck, J. Am. Chem. Soc., 87, 5620 (1965); (b) E. J. Corey and R. K. Varma, ibid., 93, 7319 (1971).

25. K. R. Varma and E. Caspi, Tetrahedron, 24, 6365 (1968).

26. E. Caspi and K. R. Varma, J. Org. Chem., 33, 2181 (1968).

27. J. Jacobus, Z. Majerski, K. Mislow, and P. R. v. R. Schleyer, J. Am. Chem. Soc., 91, 1998 (1969).

28. A. G. Brook, H. W. Kucera, and D. M. MacRae, Can. J. Chem.,
48, 818 (1970).

29. (a) M. F. Grundon, W. A. Khan, D. R. Boyd, and W. R. Jackson,
J. Chem. Soc., C1971, 2557; (b) E. J. Corey, S. M. Albonico,
U. Koelliker, T. K. Schaaf, and R. K. Varma, J. Am. Chem. Soc., 93,
1491 (1971).

30. J. C. Fiaud and H. B. Kagan, Bull. Soc. Chim. France, 1969,
2742.

31. G. P. Thakar and B. C. Subba Rao, J. Sci. Ind. Res. (India),
21B, 583 (1962); Chem. Abstr., 59, 5117(g) (1963).

32. K. M. Biswas and A. H. Jackson, J. Chem. Soc., C1970, 1667.

33. K. M. Biswas and A. H. Jackson, Tetrahedron, 24, 1145 (1968).

34. J. A. Ballantine, A. H. Jackson, G. W. Kenner, and
G. McGillivray, Tetrahedron, Suppl. 7, 241 (1966).

35. A. H. Jackson, G. W. Kenner, and G. S. Sach, J. Chem. Soc.,
C1967, 2045.

36. M. Rabinovitz and G. Salemnik, J. Org. Chem., 33, 3935 (1968).

37. (a) A. G. Anderson and R. Breazeale, J. Org. Chem., 34,
2375 (1969); (b) E. Dunkelblum, R. Levene, and J. Klein,
Tetrahedron, 28, 1009 (1972).

38. A. Pelter, M. G. Hutchings, T. E. Levitt, and K. Smith,
Chem. Commun., 1970, 347.

39. F. Smith and A. M. Stephen, Tetrahedron Letters, 1960, 17.

40. S. L. Ioffe, V. A. Tartakovskii, and S. S. Novikov, Izv.
Akad. Nauk. SSSR, Ser. Khim., 1964, 622; Chem. Abstr., 61, 8154(a)
(1964).

41. O. P. Shitov, S. L. Ioffe, V. A. Tartakovskii, and S. S.
Novikov, Russ. Chem. Rev., 1970, 905.

42. K. Kratzl and P. Claus, Monats. Chem., 94, 1140 (1963).

43. G. R. Pettit and D. M. Piatak, J. Org. Chem., 27, 2127 (1962).

44. G. R. Pettit and J. R. Dias, Chem. Commun., 1970, 901.

45. G. R. Pettit and W. J. Evers, Can. J. Chem., 44, 1293 (1966).

46. G. R. Pettit, J. C. Knight, and W. J. Evers, Can. J. Chem.,
44, 807 (1966).

47.   G. R. Pettit and T. R. Kasturi, J. Org. Chem., 26, 4557 (1961).

48.   G. R. Pettit, U. R. Ghatak, B. Green, T. R. Kasturi, and D. M. Piatak, J. Org. Chem., 26, 1685 (1961).

49.   T. A. Giudici and T. C. Bruice, J. Org. Chem., 35, 2386 (1970).

50.   R. E. Ireland, D. A. Evans, D. Glover, G. M. Rubottom, and H. Young, J. Org. Chem., 34, 3717 (1969).

51.   R. W. Kierstead and A. Faraone, J. Org. Chem., 32, 704 (1967).

52.   T. A. Giudici and A. L. Fluharty, J. Org. Chem., 32, 2043 (1967).

53.   P. Kohn, R. H. Samaritano, and L. M. Lerner, J. Am. Chem. Soc., 87, 5475 (1965).

54.   P. Kohn, L. M. Lerner, A. Chan. S. D. Ginocchio, and C. A. Zitrin, Carbohyd. Res., 7, 21 (1968).

55.   B. S. Kirkiacharian and D. Raulais, Bull. Soc. Chim. France, 1970, 1139.

56.   W. Clark Still and D. J. Goldsmith, J. Org. Chem., 35, 2282 (1970).

57.   H. C. Brown and P. Heim, J. Am. Chem. Soc., 86, 3566 (1964).

58.   P. L. Warner and T. J. Bardos, J. Med. Chem., 13, 407 (1970).

59.   H. J. Brabander and W. A. Wright, J. Org. Chem., 32, 4053 (1967).

60.   S. Takahashi and L. A. Cohen, Biochem., 8, 864 (1969).

61.   W. F. Gannon, J. D. Benigni, and J. Suzuki, Tetrahedron Letters, 1967, 1531; K. H. Shin, L. Fonzes, and L. Marion, Can. J. Chem., 43, 2012 (1965).

62.   O. Yonemitsu, T. Tokuyama, M. Chaykovsky, and B. Witkop, J. Am. Chem. Soc., 90, 776 (1968).

63.   F. E. Ziegler, J. A. Kloek, and P. A. Zoretic, J. Am. Chem. Soc., 91, 2342 (1969).

64.   W. G. Duncan and D. W. Henry, J. Med. Chem., 12, 25 (1969).

65.   D. W. Henry, J. Hetero. Chem., 3, 503 (1966).

66.   H. Feuer and F. Brown, J. Org. Chem., 35, 1468 (1970).

67.   G. P. Thakar, N. Janaki, and B. C. Subba Rao, Indian J. Chem., 3, 74 (1965).

68.   W. J. Wechter, J. Org. Chem., 28, 2935 (1963).

69.   D. J. Pasto, C. C. Cumbo, and J. Hickman, J. Am. Chem. Soc., 88, 2201 (1966).

70.   (a) Y. Bessière-Chretién and B. Meklati, Tetrahedron Letters, 1971, 621; (b) Y. Bessière-Chretién, M. M. El Gaied, and B. Meklati, Bull. Soc. Chim. France, 1972, 1000.

71.   H. C. Brown and N. M. Yoon, J. Am. Chem. Soc., 90, 2686 (1968).

72.   H. C. Brown and N. M. Yoon, Chem. Commun., 1968, 1549.

73.   J. Kollonitsch, J. Am. Chem. Soc., 83, 1515 (1961).

74.   J. Kollonitsch, U.S. Patent 3,112,336, Jan. 3, 1961; Chem. Abstr., 60, 2766(c) (1964); B. Stibr, S. Hermanek, J. Plesek, and J. Stuchlik, Coll. Czech. Chem. Commun., 33, 976 (1968).

75.   R. E. Lyle and C. K. Spicer, Chem. and Ind. (London), 1963, 739.

76.   D. J. Pasto, J. Am. Chem. Soc., 84, 3777 (1962).

77.   N. Janaki, K. D. Pathak, and B. C. Subba Rao, Indian J. Chem., 3, 123 (1965).

78.   A. D. Cross, E. Denot, H. Carpio, R. Acevedo, and P. Crabbé, Steroids, 5, 557 (1965); M. Nussim, Y. Mazur, and F. Sondheimer, J. Org. Chem., 29, 1120 (1964).

79.   (a) J. Klein, personal communication; (b) H. Feuer, R. S. Bartlett, B. F. Vincent, and R. S. Anderson, J. Org. Chem., 30, 2880 (1965).

80.   H. Feuer and D. M. Braunstein, J. Org. Chem., 34, 2024 (1969).

81.   S. L. Ioffe, V. A. Tartakovskii, A. A. Medvedeva, and S. S. Novikov, Izv. Akad. Nauk. SSSR, Ser. Khim., 1964, 1537; Chem. Abstr., 64, 14114(e) (1966).

82.   H. K. Kim, H. H. Yaktin, and R. E. Bamburg, J. Med. Chem., 13, 238 (1970).

83.   F. Hodosan and V. Ciurdaru, Tetrahedron Letters, 1971, 1997.

84.   H. Feuer and D. M. Braunstein, _J. Org. Chem._, $\underline{34}$, 1817 (1969).

85.   H. M. Fales and T. Luukainen, _Anal. Chem._, $\underline{37}$, 955 (1965).

86.   A. Hassner and P. Catsoulacos, _Chem. Commun._, $\underline{1967}$, 590.

87.   (a) S. Yamada and S. Ikegami, _Chem. Pharm. Bull._, $\underline{14}$, 1382 (1966); (b) S. Ikegami and S. Yamada, _ibid._, $\underline{14}$, 1389 (1966); B. M. Mikhailov and L. S. Povarov, _Zh. Obshch. Khim._, $\underline{41}$, 1540 (1971); _Chem. Abstr._, $\underline{75}$, 129425v (1971).

88.   J. F. Archer, D. R. Boyd, W. R. Jackson, M. F. Grundon, and W. A. Khan, _J. Chem. Soc._, C1971, 2560.

89.   (a) J. H. Billman and J. W. McDowell, _J. Org. Chem._, $\underline{26}$, 1437 (1961); (b) J. H. Billman and J. W. McDowell, _ibid._, $\underline{27}$, 2640 (1962); (c) L. Caglioti, _Tetrahedron_, $\underline{22}$, 487 (1966).

90.   A. Meller and H. Batka, _Monats. Chem._, $\underline{101}$, 648 (1970); and references cited therein.

91.   A. Hassner, G. J. Matthews, and F. W. Fowler, _J. Am. Chem. Soc._, $\underline{91}$, 5046 (1969).

92.   J. Tanaka and A. Risch, _J. Org. Chem._, $\underline{35}$, 1015 (1970).

93.   D. W. Theobald, _J. Org. Chem._, $\underline{31}$, 3929 (1965).

94.   G. R. Pettit, S. K. Gupta, and P. A. Whitehouse, _J. Med. Chem._, $\underline{10}$, 692 (1967).

95.   S. Matsumura and N. Tokura, _Tetrahedron Letters_, $\underline{1969}$, 363.

96.   S. Matsumura and N. Tokura, _Tetrahedron Letters_, $\underline{1968}$, 4703.

97.   H. C. Brown and K. J. Murray, _J. Am. Chem. Soc._, $\underline{81}$, 4108 (1959).

98.   D. B. Bigley and D. W. Payling, _J. Inorg. Nucl. Chem._, $\underline{33}$, 1157 (1971).

99.   H. C. Brown and K. J. Murray, _J. Org. Chem._, $\underline{26}$, 631 (1961).

100. A. G. Davies and B. P. Roberts, _J. Chem. Soc._, C1968, 1474.

101. L. H. Toporcer, R. E. Dessy, and S. I. E. Green, _J. Am. Chem. Soc._, $\underline{87}$, 1236 (1965).

102. H. C. Brown and A. Suzuki, _J. Am. Chem. Soc._, $\underline{89}$, 1933 (1967).

103. S. P. Acharya and H. C. Brown, _J. Am. Chem. Soc._, $\underline{89}$, 1925 (1967).

104. J. L. Aubagnac, J. Elguero, and R. Jacquier, Bull. Soc. Chim.,
France, 1969, 3316.

105. (a) H. W. Whitlock, C. Y. Hsu, and K. Sundaresan,
Tetrahedron Letters, 1965, 4821; (b) L. Caglioti, G. Cainelli,
and G. Maina, Tetrahedron, 19, 1057 (1963); (c) I. Mehrotra
and D. Devaprabhakara, J. Organometal. Chem., 33, 287 (1971).

106. (a) E. J. Corey and T. Ravindranathan, J. Am. Chem. Soc.,
94, 4013 (1972); (b) M. Rabinovitz, G. Salemnik, and E. D.
Bergmann, Tetrahedron Letters, 1967, 3271.

107. (a) A. J. Weinheimer and W. E. Marsico, J. Org. Chem., 27,
1926 (1962); (b) S. A. Monti and R. R. Schmidt, Tetrahedron, 27,
3331 (1971); (c) R. Köster, K. Amen, H. Bellut, and W. Fenz,
Angew. Chem. Intern. Ed., 10, 748 (1971); (d) W. Fenzl and
R. Köster, ibid., 10, 750 (1971).

108. W. A. G. Graham and F. G. A. Stone, Chem. and Ind. (London),
1957, 1096.

109. B. Rickborn and S. E. Wood, J. Am. Chem. Soc., 93, 3940
(1971).

110. J. B. Pierce and H. M. Walborsky, J. Org. Chem., 33, 1962
(1968).

111. H. C. Brown and A. W. Moerikofer, J. Am. Chem. Soc., 85,
2063 (1963).

112. H. C. Brown and W. Kortnyk, J. Am. Chem. Soc., 82, 3866
(1960).

113. O. Yonemitsu, T. Hamada, and Y. Kanaoka, Chem. Pharm. Bull.,
17, 2075 (1969); M. Z. Atassi and A. F. Rosenthal, Biochem. J.,
111, 593 (1969).

114. N. L. Allinger and L. A. Tushaus, J. Org. Chem., 30, 1945
(1965): M. Fetizon and N. Moreau, Bull. Soc. Chim. France, 1969
4385.

115. B. C. Subba Rao and G. P. Thakar, Current Sci. (India), 32
404 (1963); Chem. Abstr., 60, 438(g) (1966).

116. M. J. Kornet, P. A. Thio, and S. I. Tan, J. Org. Chem., 33,
3637 (1968).

117.   R. P. Carr, A. H. Jackson, G. W. Kenner, and G. S. Sach, J. Chem. Soc., C1971, 487.

118.   J. E. McMurray, Chem. Commun., 1968, 433; J. Am. Chem. Soc., 90, 6821 (1968); K. H. Baggaley, S. G. Brooks, J. Green, and B. T. Redman, J. Chem. Soc., C1971, 2671.

119.   K. M. Nicholas and R. Pettit, Tetrahedron Letters, 1971, 3475.

AUTHOR INDEX

Numbers in parentheses are reference numbers
and indicate that an author's work is referred
to although his name is not cited in the text.
Underlined numbers give the page on which the
complete reference is listed.

A

Abraham, M. H., 122(6), 124(6),
 134
Acevedo, R., 349(78), 368
Acharya, S. P., 51(56), 61,
 66(11), 71(11), 74(25),
 76(48), 78(54), 114, 115,
 116, 356(103), 369
Acton, E. M., 145(18), 167(62),
 192, 195
Adams, R. M., 33(14b), 58,
 170(66b), 195
Albonico, S. M., 330(29b), 366
Allies, P. G., 122(7,9,11), 134
Allinger, N. L., 363(114), 370
Allred, E. L., 190(112,113),
 191(112), 197
Amen, K., 359(107c), 370
Anderson, A. G., 334(37a), 366
Anderson, C. L., 190(113), 197
Anderson, R. S., 349(79b), 368
ApSimon, J. W., 79(59), 117
Arase, A., 253(8), 254(10a),
 271(10a), 278(8), 279(10a),
 295
Archer, J. F., 352(88), 369
Arigoni, D., 315(37,39), 318
Arkell, A., 64(5), 114
Armour, A.G., 128(27), 135
Aronovich, P. M., 155(37,40),
 193
Arora, S. K., 34(21), 46(48),
 58, 61, 69(21), 86(21),
 115, 129(28), 130(33),
 135, 149(23), 155(38),
 192, 193

Arzoumanian, H., 145(18),
 167(62), 192,195, 231(4a),
 232(4a), 233(6a), 234(6a),
 237(6a,10), 238(10), 239(11),
 241(10), 246, 247
Ashby, E. C., 32(14a), 58
Atassi, M. Z., 363(113), 370
Aubagnac, J. L., 356(104), 370
Avrahami, D., 69(20), 74(20),
 92(20), 115, 173(84),
 177(84), 178(84), 196
Ayres, D. C., 324(14), 365
Ayyangar, N. R., 33(17),
 35(22b,23a), 58, 59, 66(12),
 88(12), 97(91), 98(95),
 99(91), 100(95), 101(95),
 102(95), 103(98), 105(91),
 (106(98), 107(95,98),
 108(95), 111(91), 112(91),
 113(91), 114, 118, 119

B

Babler, J. H., 183(98), 184(101),
 196, 197
Baggaley, K. H., 363(118), 371
Bagli, J. F., 131(38), 136
Baig, M. I., 186(106b), 197
Bailey, D. S., 313(31), 317
Bailey, K., 173(83a), 196
Ballantine, J. A., 334(34),
 342(34), 366
Bamburg, R. E., 350(82), 368
Banks, H. O., 302(1), 316
Bardos, T. J., 342(58), 367
Barieux, J.J., 150(25a,25b),
 153(27a,27b), 174(25a), 193

C

103(98), 105(91), 106(98),
107(95,98), 108(95),
111(91), 112(91), 113(91),
114, 116, 118, 119, 127(22),
128(22,24), 130(34), 135,
185-188(102), 197, 200(1),
203(1), 207(1,15), 208(16),
209(1,16), 211(1,15,16),
212(1,15), 213(1,15),

214(1,15), 215(30,32), 223,
224, 225, 228(1a,1b),
229(1a,1b), 230(1a), 231(4a),
232(4a,4b,5), 233(5,6a),
234(4b), 235(6b,7), 236(1a,
7), 237(6a,10), 238(10),
239(11), 241(10), 242(22),
243(22), 244(22), 245(1b),
246, 247, 292(77), 300

A

synthesis of, 33-34
see also, Dithexylborane
Thioacetals, 310
Thioboranes, 44, 309-310
Thioenol ethers, see Enethiol
    ethers
Thioesters, 310
Thioethers, allylic, see Hydro-
    boration of
Thioketals, see Reduction of
Thiols
    reaction with
        borane, 309, 323
        disiamylborane, 324
Thioxanthone, see Reduction of
Thujene,
    hydroboration of, 71-72
    hydroboration-isomerization of,
        51
Thujopsene, hydroboration of, 78
Tosylhydrazones, see Reduction of
1-Tosyloxy-3-butyne, conversion
    to cyclobutanol, 6, 25, 240
Transfer reactions
    α-
        application to synthesis of
            carbon chains and
            rings, 249-287
        of α-substituted organobor-
            anes, 138-140, 142,
            148
    β-
        of β-substituted organobor-
            anes, 141, 151, 187
Transposition of keto groups, 15,
    150-151
Trialkylaluminums, 47
Trialkylboranes
    mixed, synthesis of, 40-43
    reactions of, see Organoboranes,
        reactions of; reactions
        with
    structure of, 29
    synthesis of, 29-36, 84-85, 86
    see also, Organoboranes
Trialkylcarbinols, see Alcohols,
    tertiary
Trialkylcyanoborates
    application to the synthesis
        of

alcohols, tertiary, 5, 265-
    266
ketones
    cyclic, 14, 275
    symmetrical, 11, 274-275
    unsymmetrical, 12, 275
Triallylborane, see Tris(al-
    lylic)boranes
Tricyclic alkenes, see Hydro-
    boration of alkenes, tri-
    cyclic
Tricyclic organoboranes, 40,
    267
Tricycloekasantalol, 236
Trienes, hydroboration of, 221-
    223
Triethylamine-borane, 170, 181,
    222-223
Triethylborane
    application to synthesis of
        alcohols, 5, 263
        amines, 17-18, 307
        coupled products, 294-295
        diethylboryl derivatives of
            alcohols, 359
            amines, 359
            ketones, 359
Trifluoroacetic anhydride, 5,
    11-12, 266, 274-275
Triisobutylborane, reaction
    with alkynes, 230, 237
Triisopinocampheyldiborane
    formation of, 35, 98
    hydroboration of alkenes, 98,
        100-102, 103
    hydroboration procedure, 35
Trimesitylborane, stability of,
    122
Trimethoxyborane, see Alkoxy-
    boranes, trialkoxy
2,4,5-Trimethoxyaniline, 306
2,4,5-Trimethoxypropenylbenzene,
    306
Trimethylamine-borane, 170
Trimethyamine tert-butylborane,
    43
Trimethylamine N-oxide, see Oxi-
    dation of organoboranes,
    amine N-oxides
Trimethyleneborate, 45